RING THEORY

This is Volume 44 in
PURE AND APPLIED MATHEMATICS
A Series of Monographs and Textbooks
Editors: PAUL A. SMITH AND SAMUEL EILENBERG
A complete list of titles in this series appears at the end of this volume

RING THEORY

ERNST-AUGUST BEHRENS
Department of Mathematics
McMaster University
Hamilton, Ontario, Canada

TRANSLATED BY

CLIVE REIS
Department of Mathematics
University of Western Ontario
London, Ontario, Canada

1972

ACADEMIC PRESS New York and London

THIS IS THE ONLY AUTHORIZED ENGLISH EDITION OF *Ring Theory,* ORIGINALLY PUBLISHED AND COPYRIGHTED IN THE GERMAN LANGUAGE BY BIBLIOGRAPHISCHES INSTITUT, MANNHEIM, 1972.

ACADEMIC PRESS, INC.
111 Fifth Avenue, New York, New York 10003

United Kingdom Edition published by
ACADEMIC PRESS, INC. (LONDON) LTD.
24/28 Oval Road, London NW1 7DD

LIBRARY OF CONGRESS CATALOG CARD NUMBER: 76-154366

AMS(MOS) 1970 Subject Classifications: 16-02, 06-A50, 20-M25, 17-A99

PRINTED IN THE UNITED STATES OF AMERICA

Contents

Preface

The first five chapters of this book give an introduction to the theory of rings (the Jacobson density theorem, semisimple artinian rings, completely reducible modules, and representation theory). Chapters VI–XI are devoted to various aspects of the theory of associative rings with radical different from zero, and the last chapter (Chapter XII) deals with rings of continuous functions, culminating with the theorems of Arens–Kaplansky and of Stone.

Except for some minor changes and corrections, the first part of the book is a direct translation of the author's "Algebren," Mannheim, 1965. The author is indebted to Dr. E. Hotzel for pointing out some errors in the German edition. In the second half of the book, semiperfect rings play the most important role, particularly in the chapter dealing with distributively representable rings. The section on arithmetic rings which forms part of this chapter contains a number of hitherto unpublished results.

In order to give the broadest possible view of ring theory compatible with keeping the book to a reasonable size, I have had to omit a number of important topics, among them division rings—a treatment of which will be found in Jacobson's "Structure of Rings"—and the application of homological methods to ring theory. However, projective and injective modules are introduced in Chapter VII and the results obtained are applied to the theory of Asano orders and self-injective rings.

Once again, I would like to express my appreciation to the "Bibliographisches Institut" in Mannheim. I would also like to thank Academic Press for undertaking to publish the book and for their patience in the face of many delays.

Finally, my sincere thanks go to Professor Clive Reis of the University of Western Ontario for translating the German manuscript.

RING THEORY

Basic Concepts

CHAPTER I

1. Embedding a ring R in the ring of endomorphisms of an abelian group

A *mapping* of a set M into itself is a correspondence which assigns to each element m in M a uniquely determined element m' also in M. The image of m under the mapping a is denoted by am. If a and b are mappings of M into itself, the *product* (or composite) mapping ab is obtained by the successive application of the mappings b and a in that order:

$$m \rightarrow bm \rightarrow a(bm) \qquad \text{for all} \quad m \in M \tag{1}$$

The associative law,

$$a(bc) = (ab)c \tag{2}$$

holds for composition of mappings, since by the definition of the product,

$$(a(bc))\, m = a((bc)\, m) = a(b(cm))$$

and

$$((ab)\, c)\, m = (ab)(cm) = a(b(cm)) \text{ for all } m \in M$$

Definition 1(a). A *binary operation* defined on a set S is a correspondence which assigns to each ordered pair of elements a, b from M a uniquely determined element denoted ab in M.

Definition 1(b). A *semigroup* is a system consisting of a nonempty set S and an associative binary operation defined on S.

The associative law (2) holds for composition of mappings of a set M into itself. Hence:

Theorem 1. The mappings $a, b, \ldots$ of a set M into itself form a semigroup with composition of mappings as binary operation. ▮

We may also define a sum of mappings which will have the usual properties of ordinary addition, provided the set M itself possesses a suitable algebraic structure. To this end, let us recall the following:

Definition 2. A *group* G is a semigroup with the additional property that for any arbitrary pair of elements $a, b \in G$, both the equations

$$a \circ x = b \qquad \text{and} \qquad y \circ a = b$$

have solutions in G. Here $\circ$ denotes the operation of the semigroup.

Let e be an element of a semigroup S such that

$$e \circ a = a \qquad \text{and} \qquad a \circ e = a \qquad \text{for all} \quad a \in S$$

Then e is called an *identity* of the semigroup. *We may show that a group possesses a unique identity and that the solutions of $a \circ x = b$ and $y \circ a = b$ are unique.* This may be seen as follows:

Let c and a be arbitrary elements of G. Let e_c be a solution of $e_c \circ c = c$ and x a solution of $c \circ x = a$. Then

$$e_c \circ a = e_c \circ c \circ x = c \circ x = a$$

and hence

$$e_c \circ a = a \qquad \text{for all} \quad a \in G$$

Let a^{-1} be a solution of $x \circ a = e_c$. It follows that $a^{-1} \circ a \circ a^{-1} = a^{-1}$. Therefore if y is a solution of $x \circ a^{-1} = e_c$,

$$e_c \circ a \circ a^{-1} = y \circ a^{-1} \circ a \circ a^{-1} = y \circ a^{-1} = e_c$$

This implies

$$a \circ a^{-1} = e_c$$

which in turn gives

$$a \circ e_c = a \circ a^{-1} \circ a = e_c \circ a = a \qquad \text{for all} \quad a \in G$$

The uniqueness of the identity $e = e_c$ in G follows from the fact that if e and e' are identity elements in a semigroup S, then

$$e = e \circ e' = e' \tag{3}$$

Given $a \in G$, a^{-1} is unique: for suppose a' is also an inverse of a, i.e., $a' \circ a = e$. Then

$$a^{-1} = ea^{-1}(a'a)\, a^{-1} = a'(aa^{-1}) = a'e = a'$$

The uniqueness of the solutions of the equations $a \circ x = b$ and $y \circ a = b$ follows from

$$x = a^{-1} \circ a \circ x = a^{-1} \circ b \qquad \text{and} \qquad y = y \circ a \circ a^{-1} = b \circ a^{-1}$$

where a^{-1} is as above.

It is clear that a semigroup with identity e is a group if each element a has an inverse a^{-1}.

An *abelian group* is a group in which the operation $\circ$ is commutative:

$$a \circ b = b \circ a \qquad \text{for all} \quad a, b$$

We often call the operation of an abelian group addition and denote it by $+$. The identity element is then generally denoted by 0.

We may define the sum of two mappings of an additively written abelian group into itself as in the case of real-valued functions:

Definition 3. Let M be an additively written abelian group and let a and b be mappings of M into itself. Their *sum* $a + b$ is defined to be the mapping

$$a + b \colon m \to am + bm \qquad \text{for all} \quad m \in M \tag{4}$$

This addition satisfies the usual rules, i.e. we have

Theorem 2. The mappings $a, b, \ldots$ of an abelian group M into itself, under the operation of addition defined above, form an abelian group $\mathfrak{A}$.

Proof. The associativity of the addition in $\mathfrak{A}$ follows from the associativity of the operation in M:

$$((a + b) + c)\, m = (a + b)\, m + cm = am + bm + cm$$

and

$$(a + (b + c))\, m = am + (b + c)\, m = am + bm + cm$$

As a consequence of the commutativity in M we have

$$(a + b)\, m = am + bm = bm + am = (b + a)m$$

Hence the addition in $\mathfrak{A}$ is commutative. If 0 denotes the identity of M, the zero mapping defined by

$$m \to 0 \qquad \text{for all} \quad m \in M$$

is the identity for $+$ in $\mathfrak{A}$ and the mapping.

$$-a\colon m \to -(am) \qquad \text{for all} \quad m \in M$$

is the additive inverse of a. We have therefore shown that $\mathfrak{A}$ under $+$ is an abelian group. ∎

Theorems 1 and 2 show that the mappings of an abelian group into itself almost form a ring in the sense of the following:

Definition 4. A *ring* R is a system consisting of a set R and two binary operations in R called addition $(+)$ and multiplication $(\cdot)$ such that

1. R together with addition is an abelian group
2. R together with multiplication is a semigroup
3. $a \cdot (b + c) = a \cdot b + a \cdot c$ and $(b + c) \cdot a = b \cdot a + c \cdot a$ (distributivity of $\cdot$ over $+$)

The two distributive laws in 3 link together the two operations $+$ and $\cdot$. We often omit the dot in $a \cdot b$ and write simply ab.

We have already proved Properties 1 and 2 of Definition 4 for the system consisting of the set of mappings of an abelian group into itself under addition and multiplication. We may further show right distributivity

$$(b + c)\, a = ba + ca \tag{5}$$

For,

$$((b + c)\, a)\, m = (b + c)(am) = bam + cam \qquad \text{for all} \quad m \in M$$

On the other hand, if we try to prove left distributivity

$$a(b + c) = ab + ac \tag{6}$$

we find that the proof breaks down since, in general,

$$a(b + c)\, m = a((b + c)\, m) = a(bm + cm) \tag{7}$$

is not the same as $abm + acm$ because the equality

$$a(m_1 + m_2) = am_1 + am_2 \qquad \text{for all} \quad m_1, m_2 \in M$$

need not hold. This suggests that we should single out those mappings $a, b, \ldots$ of M into itself which are additive in the sense of the following:

Definition 5. A mapping of the abelian group M into itself is said to be *additive* if

$$a(m_1 + m_2) = am_1 + am_2 \qquad \text{for all} \quad m_1, m_2 \in M \tag{8}$$

Such mappings are called *endomorphisms* of M.

The left distributivity of multiplication over addition in the set of endomorphisms of M now follows from (7). In order to show that this subset of $\mathfrak{A}$ together with addition and multiplication is a ring, we still have to show closure of the two operations; i.e. we must show that if a and b are arbitrary endomorphisms, then both $a - b$ and ab are also endomorphisms. In the proof that $a - b$ is an endomorphism, the commutativity of the group M is essential.

1. $$\begin{aligned}(a - b)(m_1 + m_2) &= a(m_1 + m_2) - b(m_1 + m_2)\\ &= am_1 + am_2 - bm_1 - bm_2\\ &= (a - b)\, m_1 + (a - b)\, m_2\end{aligned}$$

2. $$ab(m_1 + m_2) = a(b(m_1 + m_2)) = a(bm_1 + bm_2) = abm_1 + abm_2$$

We have therefore proved the following:

Theorem 3. The endomorphisms of an abelian group M form a ring, denoted Hom(M, M). The endomorphism

$$e\colon m \to m \qquad \text{for all} \quad m \in M \tag{9}$$

is an identity in Hom(M, M), i.e. $ea = ae = a$ for all $a \in \text{Hom}(M, M)$. ∎

Multiplication in the ring of endomorphisms of an abelian group M is not in general commutative as the following example shows: Let M be the set of ordered pairs (m_1, m_2) of integers. Define addition by

$$(m_1, m_2) + (m_1', m_2') = (m_1 + m_1', m_2 + m_2')$$

The mappings a and b defined by

$$\begin{aligned} a&\colon (m_1, m_2) \to (m_2, m_1)\\ b&\colon (m_1, m_2) \to (m_1, 0)\end{aligned}$$

are both endomorphisms of M. Now

$$(ab)(m_1, m_2) = a(m_1, 0) = (0, m_1)$$

but

$$(ba)(m_1, m_2) = b(m_2, m_1) = (m_2, 0)$$

Rings of endomorphisms of abelian groups are not the best known examples of rings. The integers $\mathbb{Z}$, for example, form a ring. However, we shall soon show that, from an algebraic point of view, an arbitrary abstract ring may be considered as a subring of the ring of endomorphisms of a suitable abelian group. A *subring* of a ring R is, of course, a subset of R which is itself a ring under the operations inherited from R.

Two rings R and R' are indistinguishable "from an algebraic point of view" if they have the same algebraic structure, i.e. if there is a one-to-one correspondence between their elements which preserves the operations. This concept, fundamental to the whole of algebra, is precisely formulated in the following:

Definition 6. A *homomorphism* from a semigroup $\mathfrak{A}$ to a semigroup $\mathfrak{A}'$ is a correspondence ϕ which assigns to each element $a \in \mathfrak{A}$ a uniquely determined element $a\phi$ in $\mathfrak{A}'$ such that

$$(a \cdot b)\,\phi = (a\phi) \cdot (b\phi) \qquad \text{for all} \quad a, b \in \mathfrak{A} \tag{10}$$

$a\phi$ is called the image of a under ϕ and we say that a is mapped to $a\phi$. In (10), the dot on the left-hand side denotes the operation in $\mathfrak{A}$ and on the right-hand side the operation in $\mathfrak{A}'$. If the only element mapped to $a\phi$ is a, i.e. if for all $a, b \in \mathfrak{A}$,

$$a\phi = b\phi \qquad \text{implies} \qquad a = b$$

then we call ϕ a *monomorphism*. Under a homomorphism not every element of $\mathfrak{A}'$ need occur as the image of some element in $\mathfrak{A}$. If, however, this does occur, then we call ϕ an *epimorphism* of $\mathfrak{A}$ onto $\mathfrak{A}'$. A homomorphism which is both a monomorphism and an epimorphism is called an *isomorphism* of $\mathfrak{A}$ onto $\mathfrak{A}'$.

Since a group is a semigroup, the above definitions apply to groups as well. As an example, consider the isomorphism between the additive group of real numbers and the multiplicative group of positive real numbers given by

$$\phi\colon a \to r^a \qquad \text{where} \qquad 0 < r \neq 1, \quad r \text{ fixed}$$

Every endomorphism of an abelian group M is a homomorphism of M to M. The same concepts as those introduced above are defined for rings R except that we require (10) to hold for both operations. Thus we require

$$(a + b)\,\phi = a\phi + b\phi \qquad \text{and} \qquad (a \cdot b)\,\phi = (a\phi) \cdot (b\phi) \qquad \text{for all} \quad a, b \in R$$

Example α. The additive group $\mathbb{Z}$ of integers may be mapped isomorphically onto the additive group $\mathbb{Z}'$ of even integers by means of

$$\phi\colon \zeta \to 2\zeta \qquad \text{for all} \quad \zeta \in \mathbb{Z}$$

The mapping ϕ does not preserve multiplication. For example, $2 \cdot (2 \cdot 3) \neq (2 \cdot 2) \cdot (2 \cdot 3)$.

Example β. Let n be a fixed natural number. Consider the residue classes mod n of the integers $\mathbb{Z}$, where two integers ζ_1 and ζ_2 belong to the same residue class if and only if $\zeta_1 - \zeta_2$ is divisible by n. Let $\bar{\zeta}$ denote the residue class to which ζ belongs. We may define two binary operations, $+$ and $\cdot$ on the set of residue classes as follows:

$$\bar{\zeta}_1 + \bar{\zeta}_2 = \overline{\zeta_1 + \zeta_2} \qquad \text{and} \qquad \bar{\zeta}_1\bar{\zeta}_2 = \overline{\zeta_1\zeta_2}$$

These operations are well defined (i.e. the sums and products are independent of the choice of representatives ζ from the residue classes) and the system consisting of the set of residue classes together with these operations is a ring denoted by $\mathbb{Z}/n$. The mapping

$$\phi\colon \zeta \to \bar{\zeta} \qquad \text{for all} \quad \zeta \in \mathbb{Z}$$

is an epimorphism of the ring $\mathbb{Z}$ onto the ring $\mathbb{Z}/n$ which is not a monomorphism because $\zeta\phi = (\zeta + n)\phi$.

Example γ. Not every ring possesses an identity, i.e. an element e such that

$$ae = ea = a \qquad \text{for all} \quad a \in R$$

For example, $2\mathbb{Z}$ has no identity. The identity endomorphism

$$\text{Id}\colon m \to m \qquad \text{for all} \quad m \in M$$

is the identity for the ring $\text{Hom}(M, M)$. Let R be an arbitrary ring. *We may "adjoin" an identity to R as follows*: Define two binary operations on the set

$$R^* = \{(\zeta, a) \mid \zeta \in \mathbb{Z}, \quad a \in R\}$$

by

$$(\zeta_1, a_1) + (\zeta_2, a_2) = (\zeta_1 + \zeta_2, a_1 + a_2)$$

and

$$(\zeta_1, a_1)(\zeta_2, a_2) = (\zeta_1\zeta_2, \zeta_1 a_2 + \zeta_2 a_1 + a_1 a_2)$$

where $\zeta a = 0$, the zero element of R if $\zeta = 0$, $\zeta a = a + \cdots + a$ to ζ terms if $\zeta \geqslant 1$ and $\zeta a = |\zeta|(-a)$ if $\zeta \leqslant -1$. Under these operations we may easily verify that R^* is a ring with $(1, 0)$ as identity. The mapping

$$\phi: a \to (0, a) \qquad \text{for all} \quad a \in R$$

is a monomorphism of R into R^* and an isomorphism of R onto the subring of elements of R^* of the form $(0, a)$. *We say that R is isomorphically embedded in R^*.*

If ϕ is an isomorphism of R onto R', we may define a unique structure-preserving inverse mapping from R' to R by

$$\phi^{-1}: a' \to a, \qquad \text{where} \quad \phi a = a' \tag{11}$$

Here ϕ^{-1} satisfies

$$\phi^{-1}(a' + b') = \phi^{-1}a' + \phi^{-1}b', \qquad \phi^{-1}(a'b') = (\phi^{-1}a')(\phi^{-1}b') \tag{12}$$

Proof of (12). The image of $\phi^{-1}(a' + b')$ under ϕ is $\phi\phi^{-1}(a' + b') = a' + b'$. On the other hand, since ϕ is an isomorphism

$$\phi(\phi^{-1}a' + \phi^{-1}b') = \phi\phi^{-1}a' + \phi\phi^{-1}b' = a' + b'$$

Hence $\phi^{-1}(a' + b')$ and $\phi^{-1}a' + \phi^{-1}b'$ have the same image under ϕ. Since ϕ is a monomorphism, the first equality in (12) is proved. The second follows similarly.

If, from an algebraic point of view, we do not distinguish between isomorphic rings, the remarks preceding Definition 6 may be more precisely stated in the following:

Theorem 4. 1. Any ring R is isomorphic to a subring of the ring of endomorphisms Hom(M, M) of a suitable abelian group M.

2. If R possesses an identity 1, then R may be embedded in the ring of endomorphisms of the group $(R, +)$. The embedding mapping ϕ is defined by

$$\phi: a \to \mathfrak{L}_a$$

where $\mathfrak{L}_a$ is the endomorphism of $(R, +)$ given by

$$\mathfrak{L}_a: x \to ax \qquad \text{for all} \quad x \in R$$

$\mathfrak{L}_a$ is called the left multiplication of R by a.

3. If R has no identity we may first embed R in R^* (cf. Example γ) and then apply Statement 2 to R^*.

Proof. α. The left distributivity in R gives

$$\mathfrak{L}_a(x_1 + x_2) = a(x_1 + x_2) = ax_1 + ax_2 = \mathfrak{L}_a x_1 + \mathfrak{L}_a x_2$$

which proves $\mathfrak{L}_a \in \mathrm{Hom}((R, +), (R, +))$

β. That ϕ is a homomorphism follows from

$$\begin{aligned}(\phi(a + b))\, x &= \mathfrak{L}_{a+b} x = (a + b)\, x = ax + bx = \mathfrak{L}_a x + \mathfrak{L}_b x \\ &= (\phi a)\, x + (\phi b)\, x = (\phi a + \phi b) x\end{aligned}$$

$$\begin{aligned}(\phi(a \cdot b))\, x &= \mathfrak{L}_{ab} x = (ab)\, x = a(bx) = \mathfrak{L}_a(\mathfrak{L}_b x) \\ &= ((\phi a)(\phi b)) x\end{aligned}$$

γ. If R has an identity, ϕ is a monomorphism. For if $\mathfrak{L}_a = \mathfrak{L}_b$, then $ax = bx$ for all $x \in R$. In particular, $a \cdot 1 = b \cdot 1$, i.e. $a = b$. ∎

The image of R under ϕ is a subring of $\mathrm{Hom}(M, M)$. In general, however, the image is not the whole of $\mathrm{Hom}(M, M)$. In other words, ϕ is not in general an epimorphism. In fact, let R be any noncommutative ring with 1 and let f, g be elements with $fg \neq gf$. Because of the right distributivity in R, the mapping

$$\mathfrak{R}_g \colon x \to xg \qquad \text{for} \quad x \in (R, +)$$

is an endomorphism of $(R, +)$. Suppose a is an element in R with $\phi a = \mathfrak{R}_g$, i.e. $\mathfrak{L}_a = \mathfrak{R}_g$. Then

$$a = \mathfrak{L}_a \cdot 1 = \mathfrak{R}_g \cdot 1 = g$$

and hence $\mathfrak{L}_g = \mathfrak{L}_a = \mathfrak{R}_g$. But then

$$\mathfrak{L}_g f = \mathfrak{R}_g f$$

i.e. $gf = fg$, a contradiction. Thus $\mathfrak{R}_g$ is not the image of any element in R under ϕ. The existence of noncommutative rings was established after the end of Theorem 3.

Definition 7. a. A *division ring* is a ring in which the set of nonzero elements form a group under multiplication.

b. A field is a division ring with commutative multiplication.

The rationals $\mathbb{Q}$, the real numbers $\mathbb{R}$ and the complex numbers $\mathbb{C}$ are examples of fields. In Section VI.3 we shall give an example of a division ring which is not a field, namely the quaternions.

2. *R*-linear mappings of an *R*-module into itself

According to Theorem I.1.4, an arbitrary ring is isomorphic to a subring of the ring of endomorphisms Hom(M, M) of a suitable abelian group M. We must therefore find means of describing such subrings. Particularly suited to this purpose are the R-linear mappings of a module. Both the concept of a module and that of an R-linear map have arisen as direct generalizations of a vector space and its usual linear transformations.

Definition 1. Let M be an abelian group with operation $+$ and let R be a ring.

α. M is an *R-module*, or, more precisely, a *left R-module*, if to each ordered pair (a, m), $a \in R$ and $m \in M$, there corresponds a uniquely determined element am in M such that

1. $a(m_1 + m_2) = am_1 + am_2$ for all $m_1, m_2 \in M$
2. $(a + b)m = am + bm$ for all $m \in M$
3. $(a \cdot b)m = a(bm)$ for all $m \in M$

If R has an identity 1, and if $1m = m$ for all $m \in M$, we call M unital.

β. We define a *right R-module* similarly. In this case, the elements a, b of R act on M from the right and hence we must have

1′. $(m_1 + m_2)a = m_1a + m_2a$ for all $m_1, m_2 \in M$
2′. $m(a + b) = ma + mb$ for all $m \in M$
3′. $m(a \cdot b) = (ma)b$ for all $m \in M$

γ. If R is a division ring, we call an R-module M a *left* (resp. *right*) *vector space* over R.

δ. The R-module M is *faithful* if no nonzero element of R annihilates the whole of M, i.e. $am = 0$ for all $m \in M$ implies $a = 0$.

Because of Statement 1 (resp. 1′), each element of R acts as an operator on M. More precisely, each element of R gives rise to an endomorphism of the abelian group $(M, +)$ of the R-module M. An abelian group M is a Hom(M, M)-module and also a $\mathbb{Z}$-module where $\mathbb{Z}$ is the ring of integers. We make $(R, +)$ into a left or right R-module, defining the action of R on $(R, +)$ by

$$ax = a \cdot x \quad \text{and} \quad xa = x \cdot a$$

respectively, where $\cdot$ is the multiplication in R. It is convenient to denote these left and right R-modules by ${}_RR$ and R_R, respectively.

We have already made use of ${}_RR$ in Theorem I.1.4 without explicitly mentioning it. If M is a K-module, K a ring, then it is natural to consider those endomorphisms of the group $(M, +)$ which preserve the K-module structure of M in the sense of the following:

Definition 2. Let K be a ring and M a right K-module. A mapping $a \in \text{Hom}(M, M)$ is said to be *K-linear* if it satisfies the further condition

1. $a(mk) = (am)k \quad \text{for all} \quad m \in M \quad \text{and} \quad k \in K$

In a similar fashion, if M is a left K-module, it is convenient to write the mappings of $\text{Hom}(M, M)$ on the right of M.

Theorem 1. Let K be a ring. The K-linear mappings of the right K-module M into itself form a subring $\text{Hom}_K(M, M)$ of $\text{Hom}(M, M)$. We call $\text{Hom}_K(M, M)$ the *centralizer* of the K-module M. A similar theorem may be proved for left modules.

Proof. We have only to show that the difference and product of two K-linear maps are also K-linear. Let $a, b \in \text{Hom}_K(M, M)$. Then

$$(a - b)(mk) = a(mk) - b(mk) = (am)\,k - (bm)\,k = ((a - b)\,m)k$$

and

$$(a \cdot b)(mk) = a(b(mk)) = a((bm)\,k) = (a(bm))\,k = ((a \cdot b)\,m)k$$

for all $m \in M$ and $k \in K$. ∎

The reason for calling $\text{Hom}_K(M, M)$ the "centralizer" of the K-module M becomes apparent if we write the elements $a \in \text{Hom}_K(M, M)$ on the right of M. Then, instead of

$$a(mk) = (am)k \tag{1}$$

we get

$$mka = mak \quad \text{for all} \quad m \in M \quad \text{and} \quad k \in K$$

The elements of $\text{Hom}_K(M, M)$ thus centralize K, i.e.

$$ka = ak \quad \text{for all} \quad a \in \text{Hom}_K(M, M) \quad \text{and} \quad k \in K$$

We now show that an arbitrary ring R is isomorphic to a subring of the ring of all K-linear mappings of a suitable right K-module M. This result is a corollary to the following:

Theorem 2. Let M be a left R-module and K its centralizer. Then the elements of R induce K-linear mappings on M, i.e. we have

$$a(m_1k_1 + m_2k_2) = (am_1)\,k_1 + (am_2)\,k_2 \tag{2}$$

for all $a \in R$, $m_i \in M$, and $k_i \in K$.

Proof. By Theorem 1, the centralizer of the left R-module M forms a ring K. Definition 2, applied to the left R-module M, gives

$$(am)\,k = a(mk) \qquad \text{for all} \quad a \in R, \quad m \in M, \quad \text{and} \quad k \in K \tag{3}$$

a is clearly additive since M is an R-module and Eq. (3), read from right to left, completes the proof of the K-linearity of a. ▮

Corollary An arbitrary ring R is isomorphic to a subring of the ring $\mathrm{Hom}_K(M, M)$ of K-linear mappings of a suitable right K-module M.

Proof. See Theorem I.1.4. ▮

A closely related theorem of interest is the following:

Theorem 3. Let R be a ring with 1. Then
1. The centralizer $K = \mathrm{Hom}_R({}_RR, {}_RR)$ of the left R-module ${}_RR$ is the ring of right multiplications $\mathfrak{R}_s$ of R and
2. R is isomorphic to the full ring of K-linear mappings of ${}_RR$ into itself under the isomorphism $a \to \mathfrak{L}_a$.

Proof. 1. Let k be an element of the centralizer K and let $1k = s \in R$. Then

$$ak = (a \cdot 1)\,k = a \cdot (1k) = as \qquad \text{for all} \quad a \in R$$

Hence $k = \mathfrak{R}_s$. Conversely, for each $s \in R$, the right multiplication $\mathfrak{R}_s$ is an element of $\mathrm{Hom}_R({}_RR, {}_RR)$ as may be seen from

$$(ax)\,\mathfrak{R}_s = (ax)\,s = a(xs) \qquad \text{for all} \quad a, x \in R$$

The second equality follows from the associativity of R.

2. By Theorem I.1.4 (2), the mapping $a \to \mathfrak{L}_a$ is a monomorphism of R into $\mathrm{Hom}_K({}_RR, {}_RR)$. Let A be a K-linear map of ${}_RR$ into itself and let $A1 = a$. Then

$$Ax = A(1 \cdot x) = A(1 \cdot \mathfrak{R}_x) = (A1)\,\mathfrak{R}_x = ax = \mathfrak{L}_a x$$

for all $x \in R$. Therefore $A = \mathfrak{L}_a$. ▮

In actual fact, Theorem 2 gives little information about the structure of R unless we know more about K and the subring of $\mathrm{Hom}_K(M, M)$ which

R induces. In Chapter II we shall characterize those rings R with the following two properties:

1. R possesses a faithful module, the centralizer of which is a division ring;
2. the subring of $\mathrm{Hom}_K(M, M)$ which R induces is (in a sense to be precisely defined) "dense" in $\mathrm{Hom}_K(M, M)$.

To arrive at this characterization we must first extend the well-known theorems about finite-dimensional vector spaces over division rings to vector spaces of arbitrary dimension.

3. Vector spaces

Definition 1. Let K be a division ring and M a right vector space over K. The set $\{m_i \mid i \in I\}$ of elements of M is *linearly dependent* over K if there is a finite subset I_0 of the index set I together with a corresponding set $\{k_i \mid i \in I_0\}$ of elements of K, not all zero, such that

$$\sum_{i \in I_0} m_i k_i = 0$$

Otherwise, the set $\{m_i \mid i \in I\}$ is *linearly independent.*

In the next theorem we shall prove that every vector space M possesses a maximal linearly independent set $\{m_i \mid i \in I\}$, i.e. a linearly independent set with the property that any set properly containing it is linearly dependent. *Each element m of the vector space is then uniquely expressible as a finite linear combination of elements of the K-basis* $\{m_i \mid i \in I\}$, i.e.

$$m = \sum_{\lambda \in \Lambda} m_\lambda k_\lambda\,, \qquad \Lambda \subseteq I, \qquad |\Lambda| < \infty \tag{1}$$

Proof. Since $m \cup \{m_i \mid i \in I\}$ is linearly dependent, there exist elements ζ, ζ_λ, $\lambda \in \Lambda \subseteq I$, $|\Lambda| < \infty$, not all zero, such that

$$m\zeta + \sum_{\lambda \in \Lambda} m_\lambda \zeta_\lambda = 0$$

$\{m_i \mid i \in I\}$ is linearly independent and therefore $\zeta \neq 0$. Since K is a division ring,

$$m = \sum_{\lambda \in \Lambda} m_\lambda k_\lambda\,, \qquad \text{where} \quad k_\lambda = -\zeta_\lambda \zeta^{-1}$$

To show the uniqueness of (1) we may assume without loss of generality that

$$m = \sum_{\lambda \in \Lambda} m_\lambda k_\lambda = \sum_{\lambda \in \Lambda} m_\lambda k_\lambda{}'$$

where each summation is taken over the same index set. This implies

$$\sum_{\lambda \in \Lambda} m_\lambda(k_\lambda - k_\lambda') = 0$$

and hence $k_\lambda = k_\lambda'$ for all $\lambda \in \Lambda$. ▮

Theorem 1. Every vector space $M \neq (0)$ over a division ring K possesses a K-basis, i.e. a maximal linearly independent set of vectors. The cardinality of this set is called the *dimension* of M over K and is denoted by $\dim_K M$.

Proof. We first show the existence of a basis. M contains at least one element $m_1 \neq 0$. Clearly $\{m_1\}$ is a linearly independent set. Let

$$\cdots \subseteq \mathcal{M}_\mu \subseteq \mathcal{M}_\nu \subseteq \cdots \tag{2}$$

be a *chain* of K-linearly independent subsets $\mathcal{M}_i$ of M. That is to say, for each pair $\mathcal{M}_\alpha$ and $\mathcal{M}_\beta$ of members of (2), either $\mathcal{M}_\alpha \subseteq \mathcal{M}_\beta$ or $\mathcal{M}_\beta \subseteq \mathcal{M}_\alpha$. Let V denote the set-theoretic union of the sets occurring in this chain. V is also a K-linearly independent set since for each finite set of elements $u_1, u_2, \ldots, u_r \in V$, there is a member $\mathcal{M}_{i(r)}$ of the chain (2) which contains each u_t, $t = 1, \ldots, r$. The existence of a maximal linearly independent set $\{m_i \mid i \in I\}$ in M now follows. ▮ This last step in the argument is a consequence of Zorn's lemma. Zorn's lemma, an equivalent form of the axiom of choice, is of great importance in modern algebra and we shall often make use of it in this book. We pave the way for a statement of the lemma with the following:

Definition 2. 1. A *relation* defined on a set H is a subset ρ of the set $H \times H$ of ordered pairs (h_1, h_2), $h_1, h_2 \in H$. If (h_1, h_2) is in ρ we say h_1 is in the relation ρ to h_2 or $h_1 \rho h_2$.

2. A *partial order* $(H, \subseteq)$, is a set H together with a relation $\subseteq$ satisfying the following:

α. $h \subseteq h$ for all $h \in H$ (reflexivity)
β. if $h_1 \subseteq h_2$ and $h_2 \subseteq h_1$, then $h_1 = h_2$ (antisymmetry)
γ. if $h_1 \subseteq h_2$ and $h_2 \subseteq h_3$, then $h_1 \subseteq h_3$ (transitivity).

3. Let H be a partial order and U a subset of H. An element $s \in H$ is an *upper bound* of U if $u \subseteq s$ for all $u \in U$.

4. An element m in a partial order H is *maximal* if $m \subseteq h$ for some $h \in H$ implies $m = h$.

The set $P(\mathcal{M})$ of subsets of a set $\mathcal{M}$, together with set theoretic inclusion as the relation, is an example of a partial order. The set of natural numbers with $h_1 \subseteq h_2$ if h_1 is not greater than h_2 is another example. In this case the partial order has the additional property that any two elements are related. Accordingly, we have:

Definition 3. A partial order is a *chain* (also an *order*) if any two of its elements are comparable, i.e. either $h_2 \subseteq h_1$ or $h_1 \subseteq h_2$ for any $h_1, h_2 \in H$.

Using the terminology introduced, we may now state:

Zorn's lemma If every nonempty chain of elements of a nonempty partial order H has an upper bound *in* H, then H contains at least one maximal element. ∎

Example. In the proof of Theorem 1, the elements of H are the linearly independent subsets of M partially ordered by set-theoretic inclusion. An arbitrary chain (2) has an upper bound, namely V, which is in H.

The equivalence of Zorn's lemma and the axiom of choice will not be proved here. The reader is referred to B. H. Hermes (1955) for a proof.

Many authors [e.g. N. Bourbaki (1947), Chapter III] call a partial order an order and a chain a total order.

We shall now use Zorn's lemma to complete the proof of Theorem 1 by showing that the dimension of M is independent of the choice of basis for M. If $\{b_\lambda \mid \lambda \in \Lambda\}$ is a set of elements of M, $\langle b_\lambda \mid \lambda \in \Lambda \rangle$ will denote the subspace generated by $\{b_\lambda \mid \lambda \in \Lambda\}$ over K. This subspace consists of all finite linear combinations of the b_λ with coefficients from K. Two subsets $\{b_\lambda \mid \lambda \in \Lambda\}$, $\{c_\mu \mid \mu \in M\}$ are said to be linearly equivalent if $\langle b_\lambda \mid \lambda \in \Lambda \rangle = \langle c_\mu \mid \mu \in M \rangle$. Let $\mathcal{M}$ and $\mathcal{N}$ be two bases for M and let $\mathcal{B}$ and $\mathcal{C}$ be subsets of $\mathcal{M}$ and $\mathcal{N}$, respectively. We shall write $(\mathcal{B}; \mathcal{C})_f$ if

i. $\mathcal{B}$ and $\mathcal{C}$ are linearly equivalent and
ii. there exists a fixed one-one mapping f from $\mathcal{B}$ onto $\mathcal{C}$.

We partially order the set $\{(\mathcal{B}; \mathcal{C}_f \mid) \; \mathcal{B} \subseteq \mathcal{M}, \mathcal{C} \subseteq \mathcal{N}\}$ by the relation $\leqslant$ defined as follows:

$$(\mathcal{B}; \mathcal{C})_f \leqslant (\mathcal{B}'; \mathcal{C}')_{f'},$$

if

i. $\mathcal{B} \subseteq \mathcal{B}'$, $\mathcal{C} \subseteq \mathcal{C}'$ and
ii. f' restricted to $\mathcal{B}$ is precisely f.

A chain **C**

$$\cdots \leqslant (\mathcal{B}_\mu\,;\, \mathcal{C}_\mu)_{f_\mu} \leqslant (\mathcal{B}_\nu\,;\, \mathcal{C}_\nu)_{f_\nu} \leqslant \cdots$$

in this partial order H, possesses an upper bound $(\mathcal{B}; \mathcal{C})_f$ in H, where

$$\mathcal{B} = \bigcup \{\mathcal{B}_\mu \mid (\mathcal{B}_\mu\,;\, \mathcal{C}_\mu) \in \mathbf{C}\}, \qquad \mathcal{C} = \bigcup \{\mathcal{C}_\nu \mid (\mathcal{B}_\nu\,;\, \mathcal{C}_\nu) \in \mathbf{C}\}$$

and f is the mapping defined by

$$f(b) = f_\mu(b) \qquad \text{if} \quad b \in \mathcal{B}_\mu$$

Clearly f is well defined, since for any pair of mappings occurring in the chain **C**, one is the restriction of the other. By Zorn's lemma there exists a maximal element $(\mathcal{B}^*; \mathcal{C}^*)_{f^*}$ in H. We must now show that $\mathcal{B}^* = \mathcal{M}$ and $\mathcal{C}^* = \mathcal{N}$. Assume there is an element $b_1 \in \mathcal{M} - \mathcal{B}^*$. Since $\langle \mathcal{B}^* \rangle = \langle \mathcal{C}^* \rangle$ and $\mathcal{N}$ is a basis for $\mathcal{M}$, there are elements $c_1, ..., c_i$ in $\mathcal{N}$ such that $b_1 \in \langle \mathcal{C}^*, c_1, ..., c_i \rangle$. Let i be minimal for b_1. We see that $i > 1$, since otherwise $(\mathcal{B}^* \cup b_1; \mathcal{C}^* \cup c_1)_{f_1} \in H$, where f_1 restricted to $\mathcal{B}^*$ is f^* and $f_1(b_1) = c_1$. This contradicts the maximality of $(\mathcal{B}^*; \mathcal{C}^*)_{f^*}$ in H. Now let j be minimal subject to

$$\langle c_1, ..., c_i \rangle \subseteq \langle \mathcal{B}^*, b_1, ..., b_j \rangle, \qquad b_i \in \mathcal{M}$$

If $j > 1$, let $\langle b_1, ..., b_j \rangle \subseteq \langle \mathcal{C}^*, c_1, ..., c_i, c_{i+1}, ..., c_k \rangle$, $c_{i+1}, ..., c_k \in \mathcal{N}$, again choosing k minimal. Continue in this fashion. The process either continues without stopping or it stops after finitely many steps. In the former case we obtain two infinite sequences $b_1, b_2, ...$ and $c_1, c_2, ...$ in $\mathcal{M}$ and $\mathcal{N}$, respectively. Clearly

$$(\mathcal{B}^* \cup b_1 \cup \cdots; \mathcal{C}^* \cup c_1 \cup \cdots)_{f_0} \in H$$

where f_0 restricted to $\mathcal{B}^*$ is f^* and $f_0(b_i) = c_i$ and hence we arrive at a contradiction. In the latter case, we obtain the following equality after a finite number of steps

$$\langle \mathcal{B}^*, b_1, ..., b_m \rangle = \langle \mathcal{C}^*, c_1, ..., c_n \rangle$$

for some pair of natural numbers m and n. To complete the proof we need only show that $m = n$. It then follows that

$$(\mathcal{B}^* \cup b_1 \cup \cdots \cup b_n; \mathcal{C}^* \cup c_1 \cup \cdots \cup c_n)_{\hat{f}} \in H$$

where $\hat{f}$ is defined in the obvious way, contradicting the maximality of $(\mathcal{B}^*, \mathcal{C}^*)_{f^*}$. To show that $m = n$, we shall use induction on n. For each $i = 1, ..., n$,

$$c_i = b_i^* + \sum_{j=1}^{m} b_j k_{ij}, \qquad b_i^* \in \langle \mathcal{B}^* \rangle, \quad k_{ij} \in K$$

for uniquely determined k_{ij} and b_i^*. Define a mapping ϕ from $\langle c_1, ..., c_n \rangle$ to $\langle b_1, ..., b_m \rangle$ by

$$\phi(c_i) = \sum_{j=1}^{m} b_j k_{ij} = b_i', \qquad i = 1, ..., n$$

and extend linearly. The linear mapping ϕ is well defined and $\{b_1', ..., b_n'\}$ is a linearly independent set since $\mathcal{M}$ and $\mathcal{N}$ are bases. Furthermore, ϕ is onto $\langle b_1, ..., b_m\rangle$, since each element of $\langle b_1, ..., b_m\rangle$ is expressible as a sum of an element from $\langle \mathcal{C}^*\rangle$ and an element from $\langle c_1, ..., c_n\rangle$. Hence $\{b_1', ..., b_n'\}$ is a basis for $\langle b_1, ..., b_m\rangle$. Therefore b_m is a linear combination of $b_1', ..., b_n'$, where we may assume, without loss of generality, that the coefficient of b_n' is not zero. We may then solve for b_n' in terms of $b_1', ..., b_{n-1}', b_m$. Thus $\{b_1', ..., b_{n-1}', b_m\}$ is a basis for $\langle b_1, ..., b_m\rangle$. Now construct a linear mapping ψ from $\langle b_1', ..., b_{n-1}'\rangle$ to $\langle b_1, ..., b_{m-1}\rangle$ in the same way ϕ was constructed above. Let $\psi(b_i') = b_i''$, $i = 1, ..., n-1$. Then as above, $\{b_1'', ..., b_{n-1}''\}$ is a linearly independent set and $\langle b_1'', ..., b_{n-1}''\rangle = \langle b_1, ..., b_{m-1}\rangle$. Assume inductively that if a vector space has a basis of $n-1$ vectors, then every basis has $n-1$ vectors. It follows from the induction hypothesis that $n-1 = m-1$ and hence $m = n$. As we remarked above, this leads to a contradiction and therefore the assumption that $\mathcal{B}^*$ is not the whole of $\mathcal{M}$ is false. Similarly, we may show that $\mathcal{C}^* = \mathcal{N}$ and the theorem is proved. ∎

The existence of a basis in an arbitrary vector space not only allows us to express an arbitrary element of the space as a finite linear combination of the basis elements, but also to describe K-linear mappings by means of matrices with entries from K. By the K-linearity of $A \in \mathrm{Hom}_K(M, M)$ and on account of (1), the action of A is completely determined once we know the images Am_λ of a fixed K-basis $\{m_\lambda \mid \lambda \in I\}$. Since $Am_\lambda \in M$

$$Am_\lambda = \sum_\iota m_\iota \alpha_{\iota\lambda}, \qquad \alpha_{\iota\lambda} \in K, \quad \lambda \in I \tag{3}$$

where, for a fixed λ, only finitely many $\alpha_{\iota\lambda}$ are different from zero. With respect to a fixed basis $\{m_\iota \mid \iota \in I\}$, each K-linear mapping of M into itself gives rise to a matrix $(\alpha_{\iota\lambda})$, $\iota, \lambda \in I$, $\alpha_{\iota\lambda} \in K$ under the correspondence

$$A \to (\alpha_{\iota\lambda}) \tag{4}$$

$(\alpha_{\iota\lambda})$ is an $|I| \times |I|$ matrix ($|I|$ possibly infinite) with only finitely many nonzero entries in each column. Conversely, it is easily verified that each such matrix represents (again, relative to a fixed basis) a K-linear mapping A of M into itself, where A is defined as in (3). The sum and the product in $\mathrm{Hom}_K(M, M)$ correspond, respectively, to the sum and the product of the corresponding matrices:

$$A + B \to (\alpha_{\iota\lambda} + \beta_{\iota\lambda})$$

$$AB \to \Big(\sum_\kappa \alpha_{\iota\kappa}\beta_{\kappa\lambda}\Big)$$

We note that the sum $\sum_\kappa \alpha_{\iota\kappa}\beta_{\kappa\lambda}$ makes sense since only finitely many terms are nonzero. Of course we define the sum of infinitely many zeros to be zero. We have therefore proved

Theorem 2. Let M be a right K-module over the division ring K. Let $|I|$ be the cardinality of a K-basis of M. Then the mapping

$$\phi: A \to (\alpha_{\iota\lambda}), \qquad A \in \mathrm{Hom}_K(M, M) \tag{5}$$

is an isomorphism of the ring $\mathrm{Hom}_K(M, M)$ onto the ring of $|I| \times |I|$ matrices having only finitely many nonzero entries in each column. If $\dim_K M = n$ is finite, $\mathrm{Hom}_K(M, M)$ is isomorphic to the full ring of $n \times n$ matrices over K. ▮

4. Algebras

In Theorem I.1.3 we used the additive structure of an abelian group M to show that the endomorphisms of M form a ring $\mathrm{Hom}(M, M)$. If M is also a right K-module, K a ring, it is natural to ask whether the K-linear mappings $\mathrm{Hom}_K(M, M)$ of M into itself may be given the structure of a right K-module by defining the operation of $k \in K$ on $A \in \mathrm{Hom}_K(M, M)$ by

$$Ak: m \to Amk \qquad \text{for all} \quad m \in M \tag{1}$$

Clearly Ak is an additive mapping of M into itself and hence an element of $\mathrm{Hom}(M, M)$. In general, however, it is not K-linear. In fact, if $k' \in K$,

$$(Ak)(mk') = Amk'k$$

but

$$((Ak)\, m)\, k' = Amkk'$$

In particular, if Id is the identity mapping of M,

$$(\mathrm{Id} \cdot k)(mk') = ((\mathrm{Id} \cdot k)\, m)\, k' \qquad \text{for all} \quad k' \in K, \quad m \in M$$

if and only if

$$m(k'k - kk') = 0 \qquad \text{for all} \quad k' \in K \quad \text{and} \quad m \in M$$

If M is a faithful right K-module, this condition implies

$$k'k = kk' \qquad \text{for all} \quad k' \in K \tag{2}$$

Conversely, if k satisfies (2), Ak is K-linear and hence an element of $\mathrm{Hom}_K(M, M)$.

As a consequence of our discussion, we are led to the following:

Definition 1. The *center* Z of a ring R is the subring

$$\{z \mid z \in R,\ za = az \quad \text{for all } a \in R\}$$

Z is indeed a subring, for if z_1, $z_2 \in Z$ and $a \in R$,

$$a(z_1 z_2) = (az_1)\, z_2 = (z_1 a)\, z_2 = z_1(az_2) = z_1(z_2 a) = (z_1 z_2)a$$

Using the concept of the center of K we are led by our discussion above to the following:

Theorem 1. Let M be a right K-module and let Z be the center of the ring K. Then the ring $\mathrm{Hom}_K(M, M)$ may be given the structure of a right Z-module by defining

$$A\zeta\colon m \to Am\zeta \qquad \text{for all} \quad A \in \mathrm{Hom}_K(M, M), \quad \zeta \in Z, \quad \text{and} \quad m \in M \tag{3}$$

The following also holds:

$$(AB)\zeta = A(B\zeta) = (A\zeta)B \qquad \text{for all} \quad A, B \in \mathrm{Hom}_K(M, M) \ \text{ and } \ \zeta \in Z \tag{4}$$

Proof. The module properties

$$(A + B)\,\zeta = A\zeta + B\zeta$$
$$A(\zeta_1 + \zeta_2) = A\zeta_1 + A\zeta_2 \qquad \text{and} \qquad A(\zeta_1\zeta_2) = (A\zeta_1)\,\zeta_2$$

follow immediately from Definition (3); (4) follows from

$$((AB)\,\zeta)\, m = (AB)\, m\zeta = A(Bm\zeta) = A(B\zeta)m$$

and

$$((A\zeta)\, B)\, m = (A\zeta)(Bm) = ABm\zeta \quad ∎$$

If the center Z of K possesses an identity, we have shown that $\mathrm{Hom}_K(M, M)$ *is an algebra over Z in the sense of the following*:

Definition 2. Let R be a ring and Φ a commutative ring with 1. R is an algebra over Φ if R is a unital right Φ-module satisfying

$$(ab)\,\alpha = a(b\alpha) = (a\alpha)b \qquad \text{for all} \quad a, b \in R \quad \text{and} \quad \alpha \in \Phi \tag{5}$$

In Chapter XII we shall consider functions defined on a topological space with values in a noncommutative ring R. We shall then generalize the concept of an algebra to that of a right algebra over R (cf. Definition XII.1.1).

The fact that $\text{Hom}_K(M, M)$ may be considered as an algebra over the center of K has application in the theory of linear transformations of a vector space over a field. For example, the minimal polynomial of a linear transformation is best introduced in this context.

According to the corollary to Theorem I.2.2, every ring R may be isomorphically embedded in the ring $\text{Hom}_K(M, M)$ of a suitable right K-module M. A similar result holds for algebras. To prove this we introduce the concept of an algebra-homomorphism (-isomorphism, -epimorphism, etc.):

Definition 3. A *homomorphism of the R-module M* to the R-module M' is a homomorphism ϕ of the additive group $(M, +)$ into $(M', +)$ satisfying

$$(am)\,\phi = a(m\phi) \qquad \text{for all} \quad a \in R, \quad m \in M$$

Monomorphisms, *epimorphisms*, and *isomorphisms* are defined similarly.

The R-linear mappings of M into itself are therefore R-homomorphisms of M into itself. We denote the abelian group of R-homomorphisms of M into M' by

$$\text{Hom}_R(M, M')$$

Of course, the sum $\phi_1 + \phi_2$ of two elements of $\text{Hom}_R(M, M')$ is defined by

$$m(\phi_1 + \phi_2) = m\phi_1 + m\phi_2 \qquad \text{for all} \quad m \in M$$

Definition 4. Let R and R' be algebras over Φ. An (algebra-) homomorphism of R into R' is a ring-homomorphism ϕ which is also a Φ-module homomorphism satisfying

$$(a\alpha)\,\phi = (a\phi)\alpha \qquad \text{for all} \quad a \in R, \quad \alpha \in \Phi \tag{6}$$

Algebra-monomorphisms, etc. are defined similarly. We may now state and prove:

Theorem 2. 1. A Φ-algebra R with identity is isomorphic to the Φ-algebra $\text{Hom}_K(M, M)$ of a suitable right K-module M.

2. If R has no identity we may isomorphically embed R in a Φ-algebra $R^{(1)}$ with identity.

Proof. If R possesses an identity, we may take the right Φ-module ${}_R R$ for M and the centralizer of ${}_R R$ for K. Then by Theorem I.2.3, R is ring-isomorphic to the full ring $\text{Hom}_K(M, M)$ under the mapping $a \to \mathfrak{L}_a$.

Therefore

$$\operatorname{Hom}_K(M, M) = \{\mathfrak{L}_a \mid a \in R\} \tag{7}$$

If we define

$$\mathfrak{L}_a\alpha = \mathfrak{L}_{a\alpha} \qquad \text{for} \quad a \in R, \quad \alpha \in \Phi$$

condition (5) may be verified thus showing that $\operatorname{Hom}_K(M, M)$ is an algebra over Φ:

$$(\mathfrak{L}_a\mathfrak{L}_b)\,\alpha = \mathfrak{L}_a(\mathfrak{L}_b\alpha) = (\mathfrak{L}_a\alpha)\,\mathfrak{L}_b$$

Furthermore, the mapping $a \to \mathfrak{L}_a$ is an algebra isomorphism by (7).

2. If R has no identity we may embed it as a subalgebra of an algebra $R^{(1)}$ with identity. We proceed as in Example γ of Section I.1: Let

$$R^{(1)} = \{(\alpha, a) \mid \alpha \in \Phi, a \in R\}$$

and define the operations by

$$\begin{aligned} (\alpha_1, a_1) + (\alpha_2, a_2) &= (\alpha_1 + \alpha_2, a_1 + a_2) \\ (\alpha_1, a_1)(\alpha_2, a_2) &= (\alpha_1\alpha_2, a_1\alpha_2 + a_2\alpha_1 + a_1a_2) \\ (\alpha_1, a_1)\,\alpha_2 &= (\alpha_1, a_1)(\alpha_2, 0) \end{aligned}$$

The identity of the algebra $R^{(1)}$ over Φ is $(1, 0)$ and we embed R in $R^{(1)}$ by means of the mapping $a \to (0, a)$. ∎

We have defined an algebra in such a way that every ring may be considered as an algebra over the ring of integers $\mathbb{Z}$. The additive group of R is a right $\mathbb{Z}$-module and (5) is easily verified for all $\alpha \in \mathbb{Z}$ and $a, b \in R$.

The definition we have given of an algebra is more general than that found in the classical theory. There, an algebra or hypercomplex system means an algebra (usually finite-dimensional) over a field Φ. With this restriction we have at our disposal, by Theorem I.3.1, a Φ-basis $\{b_\iota \mid \iota \in I\}$. (5) implies

$$a(b_1\alpha_1 + b_2\alpha_2) = (ab_1)\,\alpha_1 + (ab_2)\,\alpha_2 \qquad \text{for all} \quad a, b_\iota \in R, \quad \alpha \in \Phi$$

showing that the multiplication in R is completely determined once we know the product of every pair of basis elements. With respect to a fixed Φ-basis of R we have

$$b_\iota b_\kappa = \sum_{\lambda \in I} b_\lambda \gamma_{\iota\kappa\lambda} \qquad \text{for} \quad \iota, \kappa \in I \tag{8}$$

where $\gamma_{\iota\kappa\lambda} \in \Phi$ and for fixed (ι, κ), at most finitely many $\gamma_{\iota\kappa\lambda}$ are different from zero. The associativity of multiplication in R implies

$$(b_{\iota\kappa})\,b_\sigma = b_\iota(b_\kappa b_\sigma) \qquad \text{for all} \quad \iota, \kappa, \sigma \in I$$

and this in turn gives

$$\sum_{\lambda} \gamma_{\iota\kappa\lambda}\gamma_{\lambda\sigma\tau} = \sum_{\lambda\in I} \gamma_{\iota\lambda\tau}\gamma_{\kappa\sigma\lambda} \qquad \text{for all} \quad \iota, \kappa, \sigma, \tau \in I \tag{9}$$

The sums occurring on the left and right of (9) make sense since for fixed (ι, κ) (resp. (κ, σ)) only finitely many $\gamma_{\iota\kappa\lambda}$ (resp. $\gamma_{\kappa\sigma\lambda}$) are different from zero.

Conversely, if we are given a field Φ, a set I and a mapping

$$(\iota, \kappa, \lambda) \to \gamma_{\iota\kappa\lambda}, \; \gamma_{\iota\kappa\lambda} \in \Phi, \qquad \iota, \kappa, \lambda \in I$$

such that the following conditions hold:

α. for each fixed pair (ι, κ) only finitely many $\gamma_{\iota\kappa\lambda}$ are different from zero and

β. (9) is satisfied,

we may use (8) to define an algebra over Φ with basis $\{b_\iota \mid \iota \in I\}$, where $\{b_\iota \mid \iota \in I\}$ is a set of symbols in one-one correspondence with I.

In Theorem I.4.2.1, we used the mapping $a \to \mathfrak{L}_a$ to prove that a Φ-algebra R is isomorphic to $\mathrm{Hom}_K({}_RR, {}_RR)$. Relative to the basis $\{b_\iota \mid \iota \in I\}$, $\mathfrak{L}_b$ may be described by

$$\mathfrak{L}_{b_\iota} b_\kappa = \sum_{\lambda\in I} b_\lambda \gamma_{\iota\kappa\lambda}, \qquad \iota, \kappa \in I \tag{10}$$

and it follows from (9) that the mapping

$$\sum_{\iota\in I} b_\iota \zeta_\iota \to \left(\sum_{\iota} \gamma_{\iota\kappa\lambda}\zeta_\iota\right), \qquad \zeta_\iota \in \Phi \tag{11}$$

is a homomorphism of the Φ-algebra R into the algebra of $|I| \times |I|$ matrices over Φ. The image of R under this mapping is the set of matrices with all but finitely many entries zero in each column. For the definitions of sum and product of these matrices see Section I.3.

We call the mapping (11) the *first regular representation* of the algebra R by matrices over the ground field Φ relative to the Φ-basis $\{b_\iota \mid \iota \in I\}$. If R has an identity the representation is *faithful*, i.e. no nonzero element of R is mapped to zero. The general theory of representations of rings by matrices over a field Δ will be presented in Chapters V and VI.

Primitive Rings

CHAPTER II

1. Dense rings of linear transformations of a vector space into itself

We showed in Theorem I.2.2 and its corollary that an arbitrary ring R is isomorphic to a subring of the full ring of K-homomorphisms of a suitable R-module M over its centralizer K. This result, however, gives us little information about the structure of R if we know neither the extent to which the isomorphic image of R "fills" $\mathrm{Hom}_K(M, M)$ nor the structure of K. By assuming that R possesses an *irreducible module* M we shall be able to obtain more precise information about K.

Definition 1. Let M be an R-module.

1. An *R-submodule* U of M is a subgroup of the additive group of M which is *closed* under multiplication by elements of R, i.e. for all $a \in R$ and $u \in U$, $au \in U$.

2. An R-module M is *irreducible* if

α. M and the zero submodule (0) are the only submodules of M and

β. $RM = \{\sum_{\iota=1}^{n} a_\iota m_\iota \mid a_\iota \in R, m_\iota \in M\} \neq (0)$.

If we take $M = {}_RR$ we have the following:

Definition 2. A *left ideal* L of a ring R is an R-submodule of the left R-module ${}_RR$, i.e. L is a subgroup of the additive group of R for which $al \in L$ for all $a \in R$ and $l \in L$. Similarly, a *right ideal* is defined to be a submodule of R_R. A left ideal which is also a right ideal is called an (two-sided) *ideal*.

Theorem 1. Let R be a ring with identity. Then the left R-module ${}_RR$ is irreducible if and only if R is a division ring.

Proof. 1. Let $a \neq 0$ be an element of the division ring R. Then $Ra = \{x \cdot a \mid x \in R\} = R$ is the smallest R-submodule of ${}_RR$ containing a. Furthermore, $R_RR = R^2 \neq (0)$ since $1 \in R$.

2. Let ${}_RR$ be irreducible and let $0 \neq a \in R$. Since $0 \neq a = 1 \cdot a \in Ra$, the R-submodule Ra of ${}_RR$ is the whole of R. Hence the equation $xa = 1$ has a solution x in R; x is clearly different from 0 and as above, the equation $yx = 1$ also has a solution y. Since $xa = 1$, it follows that $xax = x$ and therefore

$$a \cdot x = 1ax = yxax = yx = 1$$

Hence x is an inverse of a. ∎

Theorem 2. (*Schur's lemma*). The centralizer $K = \mathrm{Hom}_R(M, M)$ of an irreducible R-module M is a division ring.

Proof. Let 0 and 1 be the zero and identity, respectively, of K. Let $k \neq 0$ be an element of K. The subset

$$(0: k) = \{m \mid m \in M, \quad mk = 0\}$$

is an R-submodule since $m_1k = m_2k = 0$ imply $(m_1 - m_2)\,k = 0$ and $(am_1)\,k = a(m_1k) = 0$ for all $a \in R$ and $m_1 \in (0: k)$. Hence $(0: k)$ is either the whole of M or (0). The first possibility is untenable since $k \neq 0$. Therefore $(0: k) = 0$ and k is a monomorphism. The image Mk of M under k is an R-submodule of M since $R(Mk) = (RM)\,k \subseteq Mk$. Hence, either $Mk = (0)$ or $Mk = M$. Since $k \neq 0$, $mk \neq (0)$ and therefore $Mk = M$, proving that k is an isomorphism of M onto itself. Thus the inverse mapping k^{-1} of k exists. It belongs to K; for

$$((am)\,k^{-1})\,k = am \qquad \text{and} \qquad (a(mk^{-1}))\,k = a(mk^{-1}k) = am$$

proving $(am)\,k^{-1} = a(mk^{-1})$ since k is one-one. Clearly $kk^{-1} = 1 = k^{-1}k$ and K is a division ring. ∎

According to Schur's lemma, an irreducible left R-module M is a vector space over $K = \mathrm{Hom}_R(M, M)$. M therefore possesses a K-basis. If $\dim_K M$ is finite and $\{u_1, \ldots, u_n\}$ is a K-basis of M, then R will induce the whole ring $\mathrm{Hom}_K(M, M)$ of K-linear transformations if, given arbitrary vectors $v_1, \ldots, v_n$ of u, we can find an element $a \in R$ such that

$$au_1 = v_1, \; au_2 = v_2, \; \ldots, \; au_n = v_n$$

If $\dim_K M$ is not finite we would like to have a measure of how well a subring "fills" $\mathrm{Hom}_K(M, M)$. To this end we introduce the concept of *density*.

Definition 3. Let K be a division ring and M a right vector space over K. Let $\mathfrak{A}$ be a subring of $\mathrm{Hom}_K(M, M)$. If for each finite set $\{u_1, ..., u_n\}$ of linearly independent vectors and each arbitrary finite set $\{v_1, ..., v_n\}$ of vectors there exists $a \in \mathrm{Hom}_K(M, M)$ such that

$$au_1 = v_1, \; au_2 = v_2, \; ..., \; au_n = v_n \tag{1}$$

then we say $\mathfrak{A}$ is *dense* in $\mathrm{Hom}_K(M, M)$.

Clearly, if $\dim_K M$ is finite, $\mathfrak{A}$ is dense in $\mathrm{Hom}_K(M, M)$ if and only if $\mathfrak{A} = \mathrm{Hom}_K(M, M)$. The connection between the concept of density as defined above and the topological concept of density will be established in the next section.

The following Theorem 3, due to Jacobson, implies that when M is a faithful, irreducible R-module, R is dense in $\mathrm{Hom}_K(M, M)$. We may prove a similar result for left ideals B in R. In this case, however, it is clear that B is not necessarily dense in $\mathrm{Hom}_K(M, M)$, since an element $0 \neq u_1 \in M$ may be annihilated by the whole of B, i.e. u_1 may be in the set

$$(0: B) = \{m \mid bm = 0 \quad \text{for all } b \in B\} \tag{2}$$

Clearly, since $b(mk) = (bm)k$ for all $k \in K$, $(0 : B)$ is a K-subspace of M. We therefore replace the linearly independent set $\{u_1, ..., u_n\}$ of Definition 3 by one which is *linearly independent modulo* $(0 : B)$ in the sense of the following:

Definition 4. Let V be a subspace of the vector space M over the division ring K. A set $\{u_1, ..., u_n\}$ of elements of M is *linearly independent modulo* V if

$$u_1\lambda_1 + \cdots + u_n\lambda_n \in V, \qquad \lambda_i \in K$$

implies

$$\lambda_1 = \lambda_2 = \cdots = \lambda_n = 0$$

Theorem 3. (*Jacobson*). Let M be an irreducible R-module, K its centralizer and B a left ideal in R. Let $\{u_1, ..., u_n\}$ be an arbitrary set which is linearly independent modulo $(0 : B)$ and let $\{v_1, ..., v_n\}$ be an arbitrary set of vectors. Then there exists $b \in B$ such that

$$bu_i = v_i, \qquad i = 1, ..., n \tag{3}$$

Proof. 1. If $(0 : B) = M$ there is no set of elements which is linearly independent modulo $(0 : B)$ and the theorem is trivially true. If $(0 : B) \subset M$, it is sufficient to show that, for each given set $\{u_1, ..., u_n\}$ which is linearly independent modulo $(0 : B)$, there exist n elements $b_1, ..., b_n$ in B such that

$$b_\iota u_\kappa = 0 \quad \text{if} \quad \iota \neq \kappa \quad \text{and} \quad b_\iota u_\iota \neq 0 \tag{4}$$

Suppose we have shown the existence of the b_ι. The irreducibility of the R-module M implies that the submodule $Rb_\iota u_\iota$ is either (0) or M. Now

$$(0 : R) = \{m \mid am = 0 \quad \text{for all } a \in R\}$$

is a submodule of M since m_1 and $m_2 \in (0 : R)$ imply $m_1 - m_2 \in (0 : R)$ and $Rxm_1 \subseteq Rm_1 = (0)$ for arbitrary $x \in R$. If $Rb_\iota u_\iota = (0)$ then $0 \neq b_\iota u_\iota \in (0 : R)$ and $(0 : R) = M$, contradicting the irreducibility of M. Hence $Rb_\iota u_\iota = M$ and there exists an element $a_\iota \in R$ such that

$$a_\iota b_\iota u_\iota = v_\iota \quad \text{and} \quad a_\iota b_\iota u_\kappa = 0 \quad \text{if } \iota \neq \kappa \tag{5}$$

The element $b = a_1 b_1 + \cdots + a_n b_n \in B$ satisfies (3).

2. We shall prove the existence of the elements $b_1, ..., b_n$ satisfying (4) by induction on n. When $n = 1$ the existence of b_1 such that $b_1 u_1 \neq 0$ follows since $u_1 \notin (0 : B)$. Assume inductively that for every modulo $(0 : B)$ linearly independent set $\{w_1, ..., w_{n-1}\}$ there exists $b \in B$ such that

$$bw_1 = \cdots = bw_{n-2} = 0, \qquad bw_{n-1} \neq 0 \tag{6}$$

We must show the existence of an element $b_n \in B$ such that

$$b_n u_1 = \cdots = b_n u_{n-2} = b_n u_{n-1} = 0, \qquad b_n u_n \neq 0 \tag{7}$$

The subset $L = \{l \mid l \in B, lu_1 = \cdots = lu_{n-2} = 0\}$ is clearly a left ideal of R. The subset

$$Lu_{n-1} = \{lu_{n-1} \mid l \in L\} \tag{8}$$

of M is an R-submodule of M since $RLu_{n-1} \subseteq Lu_{n-1}$. $\{u_1, ..., u_{n-1}\}$ is linearly independent modulo $(0 : B)$ and by the induction hypothesis there exists $b \in B$ such that

$$bu_1 = \cdots = bu_{n-2} = 0, \qquad bu_{n-1} \neq 0$$

i.e. there exists an element $b \in L$ with $bu_{n-1} \neq 0$. Thus the submodule Lu_{n-1} is not zero and hence

$$Lu_{n-1} = M \tag{9}$$

Assume by way of contradiction that there is no element $b \in B$ satisfying (7), i.e. assume

$$lu_{n-1} = 0 \quad \text{implies} \quad lu_n = 0 \quad \text{for all} \quad l \in L \tag{10}$$

Then

$$\phi\colon lu_{n-1} \to lu_n \quad \text{for all} \quad l \in L \tag{11}$$

is a well-defined mapping of $M = Lu_{n-1}$ into itself since $l_1u_{n-1} = l_2u_{n-1}$ implies $(l_1 - l_2)\, u_{n-1} = 0$, giving $l_1u_n = l_2u_n$ by our assumption. The mapping ϕ is R-linear:

$$(l_1u_{n-1} + l_2u_{n-1})\, \phi = (l_1 + l_2)\, u_{n-1}\phi = l_1u_n + l_2u_n$$

and

$$(a(l_1u_{n-1}))\, \phi = ((al_1)\, u_{n-1})\, \phi = (al_1)\, u_n = a(l_1u_n)$$

for all $l_1, l_2 \in L$. Therefore $\phi \in K = \mathrm{Hom}_R(M, M)$. We may write (11) in the form,

$$lu_{n-1}\phi = lu_n \quad \text{for all} \quad l \in L$$

or as

$$l(u_{n-1}\phi - u_n) = 0 \quad \text{for all} \quad l \in L \tag{12}$$

On the other hand, any nontrivial linear combination of $\{u_1, ..., u_{n-2}, u_{n-1}\phi - u_n\}$ may be considered as a nontrivial linear combination of $\{u_1, ..., u_{n-1}, u_n\}$ since $\phi \in \mathrm{Hom}_R(M, M)$. Hence $\{u_1, ..., u_{n-2}, u_{n-1}\phi - u_n\}$ is linearly independent modulo $(0 : B)$ and we may apply the induction hypothesis to obtain an element $b \in B$ such that

$$bu_1 = \cdots = bu_{n-2} = 0, \qquad b(u_{n-1}\phi - u_n) \neq 0$$

Such an element b lies in L and fails to satisfy (12). Hence our assumption (10) is false and an element $b_n \in B$ satisfying (7) exists. ▌

Put $R = B$ in Theorem 3 and assume that M is a faithful module. Then R is a subring of $\mathrm{Hom}_K(M, M)$ and we have the following corollary to Theorem II.1.3:

Theorem 4 (*Jacobson density theorem*). Let K be the centralizer of the faithful, irreducible R-module M. Then R is dense in $\mathrm{Hom}_K(M, M)$. ▌

This theorem, of fundamental importance in the theory of rings, suggests that we should give a special name to the class of rings possessing a faithful irreducible module.

Definition 5. A ring R is *primitive* if it possesses a faithful, irreducible R-module.

In Theorem 4 we have proven the harder half of Theorem 5.

Theorem 5. A ring R is primitive if and only if it is isomorphic to a dense ring of linear transformations of a vector space $M \neq (0)$ into itself. ▮

To prove the implication the other way we need only observe that by the definition of equality of mappings, M is a faithful module and the density implies $Ru = M$ for any element $u \neq 0$.

2. The finite topology

We shall now justify the use of the term "dense." The toplogical concepts introduced in this section are not essential to the development of the theory in the next seven chapters. In Chapter XII, however, these concepts will play an essential role. Until then the reader is only required to know the following:

1. By a *neighborhood* $U(a; u_1, \ldots, u_n)$ of a linear transformation a of a vector space M into itself we mean the set of all linear transformations c such that

$$cu_i = au_i, \qquad i = 1, \ldots, n$$

2. A subset R of the set $R^0(M)$ of all linear transformations of M into itself is dense in $R^0(M)$ if for each element x of $R^0(M)$ and for each neighborhood $U(x; u_1, \ldots, u_n)$ of x, $U(x; u_1, \ldots, u_n) \cap R \neq \varnothing$. Although this section may be omitted, the connection between Section II.1 and the topological ideas introduced here will shed light on the concept of a subdirect sum in the next two chapters.

We may endow a set T with a topological structure as follows (cf. van der Waerden [1959] Vol. II, Chap. XII, or Franz [1960]): To each point $p \in T$ assign a family $\mathfrak{F}(p)$ of subsets $U(p)$ such that

U_1. $\mathfrak{F}(p) \neq \varnothing$ and $p \in U(p)$ for all $U(p) \in \mathfrak{F}(p)$.
U_2. Given $U(p)$ and $V(p)$ in $\mathfrak{F}(p)$, there exists $W(p) \in \mathfrak{F}(p)$ such that $W(p) \subseteq U(p) \cap V(p)$.
U_3. For each $q \in U(p)$ there is $V(q) \in \mathfrak{F}(q)$ such that $V(q) \subseteq U(p)$.

A set T with a topological structure is called a *topological space*. Each family $\mathfrak{F}(p)$ is called a *neighborhood basis* of p and $U(p) \in \mathfrak{F}(p)$ is called a *basic neighborhood*. A *neighborhood* of p is any set containing at least one basic neighborhood of p. A subset $\mathcal{O}$ of T is said to be *open* if it is the empty set or if for each point $p \in \mathcal{O}$ there exists a neighborhood of p contained in $\mathcal{O}$. A subset $\mathcal{S}$ of T is *closed* if its complement $\mathcal{CS}$ is open.

In the Euclidean plane we may take the set of open disks with center p as a neighborhood basis of p.

Definition 1. A *topological ring* R is a ring such that
α. the set R is a topological space
β. the functions ϕ and ψ from $R \times R$ to R defined by

$$\phi(x, y) = x - y, \qquad \psi(x, y) = xy$$

are *continuous*, i.e. for each pair of elements $a, b \in R$ and each neighborhood $U(a - b)$ (resp. $U(ab)$) there exist neighborhoods $V(a)$ and $W(b)$ such that $x \in V(a)$ and $y \in W(b)$ imply $x - y \in U(a - b)$ [resp. $xy \in U(ab)$].

We may endow the ring $R = \mathrm{Hom}_K(M, M)$ of K-linear mappings of a right K-module M with a topological structure by taking for the family $\mathcal{F}(a)$ of basis neighborhoods all sets of the form

$$U(a; u_1, ..., u_n) = \{c \mid c \in R, cu_1 = au_1, ..., cu_n = au_n\} \tag{1}$$

where $\{u_1, ..., u_n\}$ is an arbitrary finite set of elements of M. The basic neighborhood (1) of a consists therefore of those elements $c \in \mathrm{Hom}_K(M, M)$ whose restriction to $\{u_1, ..., u_n\}$ coincides with that of a. Condition U_1 holds trivially for these families of subsets. For U_2, let $U = U(p; u_1, ..., u_n)$ and $V = U(p; v_1, ..., v_n)$ be two basic neighborhoods of p. Then $W = U(p; u_1, ..., u_n, v_1, ..., v_n) \subseteq U \cap V$. In order to show U_3 let $q \in U(q; u_1, ..., u_n)$ and take $V = U(q; u_1, ..., u_n) \in \mathcal{F}(q)$. If $x \in V(q)$,

$$xu_i = qu_i = pu_i \qquad \text{for} \quad i = 1, ..., n$$

and hence $x \in U(p)$.

The continuity of subtraction relative to this topological structure follows immediately. For let $U(a - b; u_1, ..., u_n)$ be a given neighborhood of $a - b$ and let $x \in U(a; u_1, ..., u_n)$ and $y \in U(b; u_1, ..., u_n)$. Then

$$(x - y)\, u_i = xu_i - yu_i = au_i - bu_i = (a - b)\, u_i, \qquad i = 1, ..., n$$

showing that $x - y \in U(a - b; u_1, ..., u_n)$.

To prove the continuity of multiplication we note first that the product ab of two elements in $\mathrm{Hom}_K(M, M)$ is defined by $m \to a(bm)$. Therefore, if $U(ab; u_1, ..., u_n)$ is given, let $W(b) = U(b; u_1, ..., u_n)$ and $V(a) = U(a; bu_1, ..., bu_n)$. Then if $x \in V(a)$ and $y \in W(b)$

$$x(yu_i) = x(bu_i) = abu_i \qquad \text{for} \quad i = 1, ..., n$$

and hence $xy \in U(ab)$.

We have proved the following:

Theorem 1. $R = \mathrm{Hom}_K(M, M)$ is a topological ring if, for each $a \in R$ we take the family of sets

$$U(a; u_1, ..., u_n) = \{c \mid c \in R, \quad cu_i = au_i, \quad i = 1, ..., n\} \tag{2}$$

as a neighborhood basis of a. This topology is called the *finite topology* on $\mathrm{Hom}_K(M, M)$. ▮

Corollary 2. $\mathrm{Hom}(M, M)$ is a topological ring relative to the topology defined by (2).

To prove this we merely observe that $\mathrm{Hom}(M, M) = \mathrm{Hom}_{\mathbb{Z}}(M, M)$, $\mathbb{Z}$ the ring of integers. ▮

Let $\mathfrak{A}$ be a subset of the topological space T. An element $t \in T$ is *adherent* to $\mathfrak{A}$ if each neighborhood of t contains at least one element of $\mathfrak{A}$ (which may be t itself). The set $\bar{\mathfrak{A}}$ of all elements $t \in T$ adherent to $\mathfrak{A}$ is called the *closure* of $\mathfrak{A}$. A subset $\mathfrak{D}$ of T is *dense* in T if $\bar{\mathfrak{D}} = T$.

In the terminology introduced in the preceding paragraphs, the density theorem (Theorem II.1.3) may be stated as follows: If M is a faithful, irreducible R-module, K its centralizer and a an arbitrary element of $\mathrm{Hom}_K(M, M)$, then each neighborhood $U(a; u_1, ..., u_n)$ of a intersects R nontrivially. In other words, R is *dense in* $\mathrm{Hom}_K(M, M)$ in the usual topological sense. This may be seen by observing that from each finite set of elements $\{u_1, ..., u_n\}$ we may extract a maximal linearly independent subset $\{u_{i_i}, ..., u_{i_s}\}$. By the density theorem there exists $b \in R$ such that $bu_{i_1} = au_{i_1}, ..., bu_{i_s} = au_{i_s}$. It follows that $bu_i = au_i$, $i = 1, ..., n$ since each u_i is a K-linear combination of $u_{i_1}, ..., u_{i_s}$.

Theorem II.1.4 may be restated as follows:

Theorem 3. A primitive ring R is a dense ring of K-linear transformations of a vector space M relative to the finite topology on $\mathrm{Hom}_K(M, M)$. ▮

Theorem II.1.3 gives us some information about the left ideals of R.

This information may be used to characterize the closure in $\mathrm{Hom}_K(M, M)$ of an arbitrary left ideal. To this end we define [complementary to Definition II.1. (2) of $(0 : B)$]

$$(0 : U) = \{A \mid A \in \mathrm{Hom}_K(M, M), \quad Au = 0 \text{ for all } u \in U\} \tag{3}$$

where U is a subspace of the right K-vector space M. Clearly $(0 : U)$ is a subgroup of the additive group of the ring $\mathrm{Hom}_K(M, M)$ and furthermore, since $(CA)\,u = C(Au) = C0 = 0$ for all $C \in \mathrm{Hom}_K(M, M)$ and $A \in (0 : U)$, it is a left ideal in $\mathrm{Hom}_K(M, M)$. Using this notation, $(0 : (0 : B))$ consists of all those elements of $\mathrm{Hom}_K(M, M)$ annihilating all those elements of M annihilated by B. Hence

$$B \subseteq (0 : (0 : B)) \tag{4}$$

Theorem 4. Let K be the centralizer of the faithful irreducible R-module M. Then the closure (relative to the finite topology) of a left ideal B is the left ideal $(0 : (0 : B))$ in $\mathrm{Hom}_K(M, M)$.

Proof. 1. Let $A \in \bar{B}$. If $(0 : B) = 0$, $(0 : (0 : B)) = \mathrm{Hom}_K(M, M)$ contains A. Hence assume $0 \neq u_1 \in (0 : B)$. Then the neighborhood $U(A; u_1)$ contains an element $b \in B$. But $bu_1 = Au_1$, by the definition of the finite topology and $bu_1 = 0$. Therefore $Au_1 = 0$. u_1 is arbitrary in $(0 : B)$ and hence $A \in (0 : (0 : B))$.

2. Conversely, let $A \in (0 : (0 : B))$. We must show that every neighborhood $U(A; u_1, ..., u_n)$ of A intersects B nontrivially. Let $\{u_{i_1}, ..., u_{i_1}\}$ be a maximal modulo $(0 : B)$ linearly independent subset of $\{u_1, ..., u_n\}$. By Theorem II.1.3, there exists $b \in B$ such that $bu_{i_j} = Au_{i_j}$, $j = 1, ..., s$. But

$$u_t = u_{i_1}k_1t + \cdots + u_{i_s}k_st + v_t, \qquad t = 1, ..., n$$

for suitable $k_{jt} \in K$ and $v_t \in (0 : B)$. Therefore $bu_t = Au_t$ for $t = 1, ..., n$ and hence $b \in U(A; u_1, ..., u_n)$. ∎

3. The lattice of left ideals in a primitive Artinian ring

We may simplify the theorems of the last two sections considerably if we assume that the K-vector space M is finite dimensional. In fact, if $\dim_K M = n < \infty$ and $\{u_1, ..., u_n\}$ is a basis of M, an element $A \in \mathrm{Hom}_K(M, M)$ is in the neighborhood $U(B; u_1, ..., u_n)$ if and only if $A = B$. In this case the finite topology is the *discrete* topology, i.e. each singleton $\{A\}$ is a neighborhood. Therefore, if M is a faithful, irreducible

R-module of finite dimension over its centralizer, the density theorem (Theorem II.1.4) implies that $R = \mathrm{Hom}_K(M, M)$ and $\bar{B} = B$, B a left ideal. Hence

$$(0 : (0 : B)) = B$$

by Theorem II.2.4. The mapping

$$B \to (0 : B) \tag{1}$$

of the set of left ideals into the set of subspaces of M is onto, i.e. every subspace U is of the form $(0 : B)$ for a suitable left ideal B. This follows from

$$(0 : (0 : U)) = U \tag{2}$$

which is proved as follows: $(0 : (0 : U))$ consists of those elements $m \in M$ annihilated by $(0 : U)$. Therefore

$$U \subseteq (0 : (0 : U))$$

Assume that $w \in (0 : (0 : U)) - U$. Since $\dim_K M < \infty$, let $\{u_1, ..., u_r\}$ be a basis of U. Then $\{u_1, ..., u_r, w\}$ is a linearly independent set since $w \notin U$. By Theorem II.1.4 there exists $a \in R = \mathrm{Hom}_K(M, M)$ such that

$$au_1 = \cdots = au_r, \qquad aw \neq 0$$

This implies $a \in (0 : U)$ and $w \notin (0 : (0 : U))$, a contradiction.

Let U_1 and U_2 be two subspaces of M. Then

$$U_1 \subseteq U_2 \Rightarrow (0 : U_1) \supseteq (0 : U_2) \tag{3}$$

and similarly for left ideals

$$B_1 \subseteq B_2 \Rightarrow (0 : B_1) \supseteq (0 : B_2) \tag{4}$$

The set of subspaces of M and the set of left ideals of R form partial orders under set theoretic inclusion. The implications (3) and (4) together with $(0 : (0 : U)) = U$ and $(0 : (0 : B)) = B$ prove the following:

Theorem 1. Let M be a faithful, irreducible R-module of finite dimension over its centralizer K. Then the mappings

$$\phi : B \to (0 : B), \qquad \psi : U \to (0 : U) \tag{5}$$

are inverses of each other and reverse inclusion, i.e.

$$B_1 \subseteq B_2 \Leftrightarrow (0 : B_1) \supseteq (0 : B_2). \quad \blacksquare$$

The mapping ϕ is called an *antiisomorphism* of the partial order of left ideals in R onto the partial order of subspaces of M.

Definition 1. Let M be an R-module. The *meet* $U_1 \cap U_2$ of two submodules U_1 and U_2 is the set theoretic intersection

$$U_1 \cap U_2 = \{m \mid m \in U_i, \quad i = 1, 2\} \tag{6}$$

Their *join* $U_1 \cup U_2$ is the smallest R-submodule of M containing U_1 and U_2, namely

$$U_1 \cup U_2 = \{u_1 + u_2 \mid u_i \in U_i, \quad i = 1, 2\} \tag{7}$$

The join is also called the sum because of (7) and is denoted by $U_1 + U_2$.

We may easily verify that $U_1 \cup U_2$ is the smallest submodule containing U_1 and U_2 and that dually, $U_1 \cap U_2$ is the largest submodule contained in both U_1 and U_2. It follows that the mapping ϕ of Theorem 1 has the following two properties:

α. $\phi(B_1 \cap B_2) = \phi(B_1) \cup \phi(B_2)$
β. $\phi(B_1 \cup B_2) = \phi(B_1) \cap \phi(B_2)$

We are of course applying Definition 1 to the submodules B of the left R-module ${}_RR$ and the K-subspaces of the right vector space M. We may further show without much difficulty that the submodules of an R-module M under the operations of join and meet defined above form a lattice in the sense of the following:

Definition 2. A *lattice* $\mathfrak{L}$ is a set of elements $v_1, v_2, \ldots$ with two operations $\cup$ and $\cap$ satisfying

$$\begin{array}{lll} v_1 \cup v_1 = v_1, & v_1 \cup v_2 = v_2 \cup v_1, & v_1 \cup (v_2 \cup v_3) = (v_1 \cup v_2) \cup v_3 \\ v_1 \cap v_1 = v_1, & v_1 \cap v_2 = v_2 \cap v_1, & v_1 \cap (v_2 \cap v_3) = (v_1 \cap v_2) \cap v_3 \\ & v_1 \cap (v_1 \cup v_2) = v_1, & v_1 \cup (v_1 \cap v_2) = v_1 \end{array} \tag{8}$$

In the lattice $V(M)$ of R-submodules of an R-module M, we may define the partial ordering $\subseteq$, i.e. set theoretic inclusion, in terms of the operations $\cap$ and $\cup$ as follows:

$$U_1 \subseteq U_2 \Leftrightarrow U_1 = U_1 \cap U_2 \Leftrightarrow U_2 = U_1 \cup U_2 \tag{9}$$

It has been shown that $V(M)$ is a partial order $(\mathfrak{L}, \subseteq)$ with the following properties:

α. if v_1 and $v_2 \in \mathfrak{L}$, there exists an element denoted $v_1 \cup v_2 \in \mathfrak{L}$ such that (i) $v_1 \subseteq v_1 \cup v_2$ and $v_2 \subseteq v_1 \cup v_2$ and (ii) $v_1 \subseteq v$ and $v_2 \subseteq v$, implies $v_1 \cup v_2 \subseteq v$

β. the dual of Property α, i.e. the statement obtained from α by replacing $\cup$ by $\cap$ and $\subseteq$ by $\supseteq$. $\mathfrak{L}$ is a lattice under $\cap$ and $\cup$.

The set of subsets of an arbitrary set M under set theoretic union and intersection forms a lattice $P(M)$. The corresponding partial ordering is set theoretic inclusion. In $P(M)$ the *distributive laws*

$$V_1 \cup (V_2 \cap V_3) = (V_1 \cup V_2) \cap (V_1 \cup V_3) \tag{10_1}$$

and the corresponding dual

$$V_1 \cap (V_2 \cup V_3) = (V_1 \cap V_2) \cup (V_1 \cap V_3) \tag{10_2}$$

hold for arbitrary subsets V_1, V_2, V_3.

In the lattice of R-submodules these laws do not in general hold. However, (10_1) is valid in the particular case that $V_1 \subseteq V_3$.

Theorem 2. The lattice $V(M)$ of submodules of an R-module M is *modular*, i.e.

$$U_1 \subseteq U_3 \Rightarrow U_1 + (U_2 \cap U_3) = (U_1 + U_2) \cap U_3 \tag{11}$$

for $U_i \in V(M)$, $i = 1, 2, 3$.

Proof. Since $U_1 \subseteq U_3$ and $U_2 \cap U_3 \subseteq U_2 \subseteq U_1 + U_2$, $U_1 + (U_2 \cap U_3) \subseteq (U_1 + U_2) \cap U_3$. Conversely, let $u \in (U_1 + U_2) \cap U_3$. Then $u = u_1 + u_2$, $u_i \in U_i$, $i = 1, 2$. It follows that $u_2 = u - u_1 \in U_3$ since $U_1 \subseteq U_3$. Therefore $u = u_1 + u_2$ with $u_2 \in U_2 \cap U_3$ and $u \in U_1 + (U_2 \cap U_3)$. ∎

Definition 3. Let ϕ be a one-one mapping of the lattice V onto the lattice V'. ϕ is an (lattice) *isomorphism* of V onto V' if ϕ preserves the operations $\cup$ and $\cap$, i.e. if

$$(a \cup b)\,\phi = a\phi \cup b\phi \quad \text{and} \quad (a \cap b)\,\phi = a\phi \cap b\phi \quad \text{for all } a, b \in V \tag{12_1}$$

ϕ is an *antiisomorphism* if

$$(a \cup b)\,\phi = a\phi \cap b\phi \quad \text{and} \quad (a \cap b)\,\phi = a\phi \cup b\phi \quad \text{for all } a, b \in V \tag{12_2}$$

We showed in Theorem 1 that $\phi\colon B \to (0 : B)$ is a one-one mapping of the lattice V of left ideals B of R onto the lattice V' of K-subspaces of a faithful, irreducible R-module M of finite dimension over its centralizer K. Furthermore, ϕ reverses inclusion. We state the following corollary to Theorem 1 in the new terminology:

Theorem 3. Hypotheses: Let M be a faithful, irreducible R-module of finite dimension over its centralizer K. Conclusion: The lattice of left ideals of R is antiisomorphic to the lattice of K-subspaces of M under the mapping

$$\phi\colon B \to (0\colon B), \qquad B \text{ a left ideal in } R$$

The inverse of ϕ is the mapping

$$\psi\colon U \to (0\colon U), \qquad U \text{ a subspace of } M \quad \blacksquare$$

Theorems II.1.5 and I.3.2 show that a primitive ring with a faithful and irreducible module M of finite dimension n over its centralizer K is isomorphic to the ring of $n \times n$ matrices over K. It is natural, therefore, to try to characterize those primitive rings with modules of finite dimension over their centralizers in terms of some internal property of the rings. Theorems 1 and 3 of this section give us a clue as to the property required. Theorem 1 tells us that in these rings, every descending chain of left ideals

$$B_1 \supset B_2 \supset \cdots \tag{13}$$

stops after a finite number of steps, i.e. for some n, $B_n = B_{n+1} = \cdots$, since the corresponding ascending chain of subspaces $U_1 \subset U_2 \subset \cdots$ stops after finitely many steps by the finite dimensionality of the vector space. A ring with the property that every descending chain of left ideals stops after finitely many steps is called a *ring with descending chain condition* (abbreviated, dcc) on *left ideals*. The dcc on left ideals is equivalent to the *minimal condition on left ideals* as defined here:

Definition 4. A ring R has the *minimal condition on left* ideals if every nonempty family $\mathcal{F}$ of left ideals of R possesses a minimal left ideal *in* $\mathcal{F}$, i.e. if there exists *in* $\mathcal{F}$ a left ideal L_0 such that $L \in \mathcal{F}$ and $L \subseteq L_0$ implies $L = L_0$. Such a ring is called *artinian.*

Proof of the equivalence. Let $\mathcal{F}$ be a nonempty family of left ideals having no minimal member. Then there is a descending chain of left ideals of infinite length since for each left ideal $L_n \in \mathcal{F}$, there exists at least one $L_{n+1} \in \mathcal{F}$ such that $L_{n+1} \subset L_n$. Conversely, a descending chain of infinite length forms a family with no minimal member.

It should be noted that the minimal left ideal L_0 of Definition 4 must belong to $\mathcal{F}$. Otherwise (0) would always be a minimal left ideal for any family of left ideals in any ring [in fact, (0) is the *minimum* left ideal; i.e. it is contained in all left ideals].

Our discussion above shows that if $\dim_K(M) < \infty$, $\operatorname{Hom}_K(M, M)$ is artinian. The converse is also true.

Theorem 4. Let R be an artinian ring, M a faithful, irreducible R-module and K the centralizer of M. Then $\dim_K(M) = n$ is finite and $n + 1$ is the number of left ideals in a maximal chain, i.e. a chain which cannot be refined by the interpolation of further left ideals.

Proof. Let $\{u_i \mid i \in I\}$ be a basis of M. If I is infinite, $\{u_i \mid i \in I\}$ contains a countable subset $\{u_j \mid j = 1, 2, ...\}$. By Theorem II.1.3 applied to the left ideal $B = R$, there exists $a_n \in R$ such that

$$a_n u_1 = \cdots = a_n u_{n-1} = 0, \qquad a_n u_n \neq 0$$

The left ideals

$$L_n = \{a \mid a \in R, \quad au_1 = \cdots = au_n = 0\}, \qquad n = 1, 2, ...$$

form an infinite chain, contradicting the fact that R is artinian. The last assertion of the theorem follows from Theorem 3. ▮

Theorem II.1.4 enables us to state the following corollary to Theorem 4:

Theorem 5. A primitive artinian ring is isomorphic to a full ring $\mathrm{Hom}_K(M, M)$ of linear transformations of a finite-dimensional vector space M over a division ring K. Hence, by Theorem I.3.2, a primitive artinian ring is also isomorphic to a full ring of matrices $K_{n \times n}$, $n = \dim_K(M)$. ▮

Since we have already shown the converse, we have the following characterization:

Theorem 6. A ring is primitive artinian $\Leftrightarrow$ it is a ring of $n \times n$ matrices over some division ring K. ▮

As we shall show in Section 5, we may also characterize the primitive rings in the class of artinian rings as those which are simple in the following sense:

Definition 5. A ring R is *simple* if

α. $R^2 \neq (0)$ and
β. R has no ideals other than (0) and R

It is convenient to exclude from our investigations those rings in which the product of any two elements is 0. Such a ring R is of little interest to us since its additive structure plays the only significant role and the ideals are precisely the subgroups of $(R, +)$. If $(R, +)$ has (0) and itself as the only two subgroups, then $(R, +)$ must be cyclic of prime order, i.e. $(R, +)$ must be the additive group of $\mathbb{Z}/(p)$ for some prime p (cf. Example β in Section I.1).

In anticipation of our investigation of simple rings, we shall introduce some concepts of a general nature in Section 4.

4. Homomorphisms of semigroups, rings, and modules; relations

Let H be a set. An *equivalence relation* ρ on H is a relation (cf. Definition I.3.2) which gives rise to a decomposition of H into mutually disjoint equivalence classes (ρ-classes). The class containing a is defined by

$$\bar{a} = \{h \mid h \in H, \quad h\rho a\}$$

This means that ρ satisfies the following conditions

$$a\rho a \qquad \text{(reflexivity of } \rho) \tag{1a}$$

$$a\rho b \Leftrightarrow b\rho a \qquad \text{(symmetry of } \rho) \tag{1b}$$

$$a\rho b \quad \text{and} \quad b\rho c \Rightarrow a\rho c \qquad \text{(transitivity of } \rho) \tag{1c}$$

If H is a semigroup under $\circ$, a congruence relation on H is an equivalence relation on H which is compatible with the algebraic structure of H:

A congruence relation ρ on the semigroup H is an equivalence relation on H for which

$$a_1\rho a_2 \Rightarrow (a_1 \circ b)\,\rho(a_2 \circ b) \quad \text{and} \quad (b \circ a_1)\,\rho(b \circ a_2) \quad \text{for all } a_1, a_2, b \in H \tag{2}$$

If ρ is a congruence relation on H, we may define a semigroup operation $\circ$ on the set H/ρ of ρ-classes of H by

$$\bar{a} \circ \bar{b} = \overline{a \circ b} \tag{3}$$

From (2) it follows that if $a_1\rho a$ and $b_1\rho b$

$$(a_1 \circ b_1)\,\rho(a \circ b)$$

i.e. definition (3) is independent of the choice of representative of the classes. The $\circ$ operation in H/ρ is clearly associative since the $\circ$ operation in H is.

The mapping

$$\phi\colon a \to \bar{a}, \qquad a \in H \tag{4}$$

is an epimorphism of the semigroup H onto the semigroup H/ρ since $\phi a \circ \phi b = \bar{a} \circ \bar{b} = \overline{a \circ b} = \phi(a \circ b)$. We call it the *natural epimorphism* of H onto H/ρ.

Conversely, if ϕ is an epimorphism of the semigroup H onto the semigroup H' we obtain a decomposition of H into disjoint subsets $\bar{a}, \bar{b}, \ldots$, where

$$\bar{a} = \{h \mid h \in H, \quad \phi h = \phi a\}$$

The relation defined on H by

$$a\rho_\phi b \qquad \text{if} \quad \phi a = \phi b \tag{5}$$

is an equivalence relation. Since $\phi(a \circ b) = \phi a \circ \phi b$, it is also a congruence relation.

We therefore have a correspondence between congruence relations and epimorphisms of H, namely, each epimorphism gives rise to a congruence relation and each congruence relation gives rise to an epimorphism.

The inverse images of $\phi a \in H'$ all belong to the same ρ_ϕ-class and therefore the mapping

$$\bar{a} \to \phi a \qquad \text{for all} \quad \bar{a} \in H/\rho_\phi$$

is an isomorphism of H/ρ_ϕ onto H'.

Similar results hold for rings provided we take into account both operations and define a congruence relation on R to be compatible with the additive as well as the multiplicative structure. Therefore, ρ is a congruence relation on R if

$$a_1\rho a_2 \Rightarrow (a_1 \cdot b)\,\rho(a_2 \cdot b) \qquad \text{and} \qquad (b \cdot a_1)\,\rho(b \cdot a_2) \tag{6a}$$

and

$$a_1\rho a_2 \Rightarrow (a_1 + b)\,\rho(a_2 + b) \tag{6b}$$

There is a very close connection between the ideals of a ring R, the congruence relations on R and the epimorphisms of R. Let ϕ be an epimorphism of R onto R'. We define the kernel of ϕ by

$$\ker \phi = \{a \mid a \in R, \quad \phi a = 0\} = \{a \mid a\rho_\phi 0\} \tag{7}$$

If $a, b \in \ker \phi$, $\phi(a - b) = \phi a - \phi b = 0$ and $\phi(x \cdot a) = \phi(x) \cdot \phi(a) = 0 = \phi(a \cdot x)$ for all $x \in R$. This shows that $\ker \phi$ is an ideal in R. Conversely, every ideal $\mathfrak{A}$ in R defines a congruence relation α by

$$a\alpha b \qquad \text{if} \quad a - b \in \mathfrak{A}$$

and gives rise to the quotient ring $R/\mathfrak{A}$.

The *null relation* (only equal elements are congruent) and the *all relation* (any two elements are congruent) are called the trivial relations on R. Corresponding to these we have the zero ideal (0) and R, respectively. We have proved the following:

Theorem 1. Let $R^2 \neq (0)$, R a ring. Then the following are equivalent:

1. R is simple.
2. The null and all relations are the only congruence relations on R.
3. A homomorphism of R is either a monomorphism or the zero homomorphism, i.e. the mapping which sends every element of R to zero. ▮

Similar results may be proved for R-modules. To this end, let us recall Definition I.4.3: Let M and M' be two R-modules. An R-homomorphism of M into M' is an additive map ϕ from $(M, +)$ to $(M', +)$ satisfying

$$\phi(am) = a\phi m, \qquad a \in R, \quad m \in M \tag{8}$$

Analogously we define R-module monomorphisms, epimorphisms, etc. Such a mapping ϕ not only preserves addition but also the structure of M as an R-module. *The kernel of ϕ is the set*

$$\ker \phi = \{m \mid m \in M, \quad \phi m = 0\}$$

Since $\phi(am) = a\phi m$, $\ker \phi$ *is an R-submodule of M.* Conversely, every R-submodule U of M is the kernel of an R-homomorphism, namely the natural epimorphism defined by

$$\chi\colon m \to m + U = \{m + u \mid u \in U\} \qquad \text{for} \quad m \in M$$

The *coset* $\bar{m} = m + U$ is an element of the *factor module* M/U. Addition and multiplication by elements of R are defined by

$$\bar{m}_1 + \bar{m}_2 = \overline{m_1 + m_2} \qquad \text{and} \qquad a\bar{m} = \overline{am}$$

By definition a relation on a set H is a subset of the Cartesian product $H \times H = \{(h_1, h_2) \mid h_i \in H, i = 1, 2\}$. Since the subsets of any given set form a partial order under set theoretic inclusion $\subseteq$, this is also the case for the set of all relations on H. Call this set P. P contains, in particular, the null relation $\nu = \{(h, h) \mid h \in H\}$ and the all relation $\alpha = H \times H$. We say the relation ρ_1 is finer than the relation ρ_2 on H if $\rho_1 \subseteq \rho_2$. In this case we also say that ρ_2 is coarser than ρ_1. In terms of elements from H this means that $h_1\rho_1 h_2 \Rightarrow h_1\rho_2 h_2$.

The set of equivalence relations on H form a partial order contained in P. This set contains ν and α. Now given a set ρ_i, $i \in I$ of equivalence relations on H, the set theoretic intersection

$$\delta = \bigcap_{i \in I} \rho_i$$

is also an equivalence relation on H. Therefore, given an arbitrary relation σ

on H there exists a uniquely determined finest equivalence relation on H which is coarser than σ, namely

$$\bigcap \{\rho \mid \rho \text{ is an equivalence relation on } H \quad \text{and} \quad \sigma \subseteq \rho\}$$

This set is not empty since α belongs to it.

If H is endowed with some algebraic structure (e.g. if H is a semigroup) then the set theoretic intersection of congruence relations ρ_i, $i \in I$ is also a congruence relation. The all relation is still a congruence relation on H. Therefore, as in the case of equivalence relations, given an arbitrary relation σ on H there exists a uniquely determined congruence relation on H which is finer than any congruence relation coarser than σ, namely

$$\bigcap \{\rho \mid \rho \text{ is a congruence relation on } H \quad \text{and} \quad \sigma \subseteq \rho\}$$

Since the set P of all relations on H consists of the set of all subsets of $H \times H$, P forms a distributive lattice. In general, however, this is not the case for the set of equivalence relations. It is true, as we have seen above, that the intersection of two equivalence relations is an equivalence relation and that to given equivalence relations ρ_1 and ρ_2 there exists an equivalence relation finer than any equivalence relation coarser than the set-theoretic union $\rho_1 \sqcup \rho_2$, namely,

$$\rho_1 \cup \rho_2 = \bigcap \{\rho \mid \rho \text{ is an equivalence relation} \quad \text{and} \quad \rho_1 \sqcup \rho_2 \subseteq \rho\}$$

Nevertheless, in general $\rho_1 \cup \rho_2$ is different from $\rho_1 \sqcup \rho_2$. The set of equivalence relations form a lattice under $\cap$ and $\cup$. However, this lattice is not necessarily a sublattice of P since in general $\cup$ and $\sqcup$ do not coincide. Similar remarks apply to congruence relations in the case H is endowed with an algebraic structure.

If $H = M$, an R-module, there is a one-one correspondence between the congruence relations ρ on M and the R-submodules of M; for let ρ be a congruence relation on M and let the kernel $\ker \rho$ of this relation be the pre-image of the 0 element of M under the natural epimorphism of H to H/ρ associated with ρ. Then $\ker(\rho_1 \cap \rho_2) = \ker \rho_1 \cap \ker \rho_2$ and $\ker(\rho_1 \cup \rho_2) = \ker \rho_1 + \ker \rho_2$. The analogous result holds for congruence relations on rings. In this case $\ker \rho$ is an ideal. It is to be noted, however, that not all congruence relations on a semigroup are uniquely determined by their kernels.

5. Simple rings with minimal left ideals

We shall first characterize the simple rings in the class of artinian rings:

Theorem 1 (Artin). An artinian ring R is primitive if and only if R is simple.

Proof. 1. By Theorem II.3.5, we may take $R = K_{n \times n}$. The $n \times n$ matrices C_{ij} with 1 in the (i, j)th entry and 0 elsewhere form a basis of $K_{n \times n}$ over K. Let

$$a = \sum_{j,k=1}^{n} C_{jk}\alpha_{jk} \neq 0, \qquad \alpha_{jk} \in K \tag{1}$$

For some pair (j_0, k_0), $\alpha_{j_0k_0} \neq 0$ and any ideal $\mathfrak{A} \subseteq K_{n \times n}$ containing a also contains

$$C_{il} = C_{ij_0}a \cdot C_{k_0l}\alpha_{j_0k_0}^{-1}$$

Clearly $\mathfrak{A}$ contains every linear combination of such elements and hence $\mathfrak{A} \supseteq R$. Therefore R is simple.

2. To prove the converse we weaken the hypotheses and prove a more general result. Assume therefore that R is simple and possesses at least one minimal left ideal L in the family of left ideals different from (0). We shall show that L, considered as a submodule of ${}_RR$, is faithful and irreducible.

Clearly the only submodules of L are (0) and L. The irreducibility of L is a consequence of (2) below since $L^2 \neq (0)$ implies $RL \neq (0)$. To prove L is faithful, consider the set

$$(0 : L) = \{a \mid a \in R, \quad aL = (0)\}$$

$RaL = (0)$ if $a \in (0 : L)$ and since $RL \subseteq L$, $aRL = 0$ showing that $(0 : L)$ is a two-sided ideal. By the simplicity of R, either $(0 : L) = (0)$ or $(0 : L) = R$. If the latter, then $RL = (0)$ and $L^2 = (0)$ also. This is a contradiction since

$$R \text{ simple} \Rightarrow L^2 \neq (0), \qquad L \text{ a nonzero left ideal} \tag{2}$$

Proof of (2). $L + LR$ is a two-sided ideal since

$$R \cdot (L + LR) \subseteq L + LR \qquad \text{and} \qquad (L + LR)\, R \subseteq L + LR$$

Assume $L^2 = 0$. Then

$$R^2 = (L + LR)^2 = L^2 + LR + LRL + LRLR \subseteq L^2 + L^2R = (0)$$

contradicting the simplicity of R. From (2) it follows that $RL \neq 0$ and therefore $(0 : L) = (0)$. Hence L is a faithful, irreducible R-module and R is primitive. Since an artinian ring has at least one minimal left ideal, Theorem 1 is proved. ∎

However, we have proved something stronger, namely:

Theorem 2. A simple ring with at least one minimal left ideal L is primitive and L itself is a faithful irreducible R-module. ▮

We may characterize such subrings of $\mathrm{Hom}_K(M, M)$ with the help of the following:

Definition 1. $a \in \mathrm{Hom}_K(M, M)$ has *finite rank* if $\dim_K(aM)$ is finite.

Theorem 3. The set of elements of finite rank in $\mathrm{Hom}_K(M, M)$ form an ideal $\mathfrak{A}$ in $\mathrm{Hom}_K(M, M)$.

Proof. A linearly independent set of elements of M may be mapped by a K-linear mapping to a linearly dependent set. Therefore we have the following inequalities:

$$\begin{aligned} \dim_K(a - b)\, M &\leqslant \dim_K aM + \dim_K bM \\ \dim_K(ac)\, M &= \dim_K a(cM) \leqslant \dim_K aM \\ \dim_K(ca)\, M &= \dim_K c(aM) \leqslant \dim_K aM \end{aligned}$$

for $a, b \in \mathfrak{A}$ and $c \in \mathrm{Hom}_K(M, M)$. ▮

Corollary 4. Let R be a subring of $\mathrm{Hom}_K(M, M)$. Let $\mathfrak{A}$ be as above. Then $R \cap \mathfrak{A}$ is an ideal in R.

Proof. Let $a, b \in R \cap \mathfrak{A}$ and $c \in R$. Then $a - b \in R \cap \mathfrak{A}$. Furthermore, since $\mathfrak{A}$ is an ideal in $\mathrm{Hom}_K(M, M)$, $ca, ac \in R \cap \mathfrak{A}$. ▮

The elements in $\mathrm{Hom}_K(M, M)$ of finite rank may be represented (as in Section I.3) by so-called row-finite matrices, i.e. matrices with at most finitely many nonzero rows. This is a stronger condition than the one discussed in Section I.3, where each column had only finitely many entries different from zero.

In the characterization of simple rings with at least one minimal left ideal as primitive rings all of whose elements are of finite rank (cf. Theorems 6 and 7), we make use of "idempotent elements." For a better understanding of the significance of these elements, we introduce the following notion. A *projection* of the K-vector space M onto the K-subspace U of M is a K-linear transformation E of M onto U which leaves each element of U fixed. In symbols

$$E \in \mathrm{Hom}_K(M, M), \qquad EM = U, \qquad Eu = u, \qquad \text{for all} \quad u \in U \tag{3}$$

It follows that

$$Em = E^2m \qquad \text{for all} \quad m \in M$$

and hence $E = E^2$. Conversely, if $E \in \mathrm{Hom}_K(M, M)$ and $E^2 = E$, then E is a projection of M onto EM since for each $u = Em \in EM$,

$$Eu = E^2m = Em = u$$

Projections are closely related to the decomposition of vector spaces as sums of subspaces. Let $E = E^2$ and let

$$(0 : E) = \{m \mid Em = 0\}$$

Then $M = (0 : E) + EM$, $(0) = (0 : E) \cap EM$, i.e. M is the sum of two subspaces whose intersection is (0). When this is the case we say M is the *direct sum* of the subspaces and write

$$M = (0 : E) \oplus EM \qquad \text{as a} \quad K\text{-space} \tag{4}$$

Since $(Em)\,k = E(mk)$, EM and $(0 : E)$ are subspaces. Let $m \in EM \cap (0 : E)$. Then m is both fixed and annihilated by E and hence is zero. Since

$$m = Em + (m - Em) \qquad \text{for} \quad m \in M \tag{5}$$

the sum of the two subspaces is M.

We now make the following:

Definition 2. An element e of a ring R is *idempotent* if $e^2 = e$.

Theorem 5. Let M be a vector space over the division ring K. An element $E \in \mathrm{Hom}_K(M, M)$ is a projection of M onto EM if and only if E is idempotent. In this case M is the direct sum of the two subspaces EM and $(0 : E)$. ∎

Theorem 6 (Jacobson). Let R be a simple ring and L a minimal left ideal in R. Let K be the centralizer of the R-module L. Then R is a dense subring of $\mathrm{Hom}_K(L, L)$ all of whose elements are of finite rank. Furthermore, there exists an idempotent $e \in R$ which generates L, i.e. for which $L = Re$.

Proof. By the simplicity of R and (2), it follows that $L^2 \neq (0)$. Since L is minimal and the left ideal L^2 is contained in L, $L^2 = L$. Hence there exists $l_0 \in L$ such that $Ll_0 \neq (0)$. Ll_0 is also a left ideal contained in L. Therefore $Ll_0 = L$. It follows that there exists $e \in L$ such that $el_0 = l_0$ and hence $e^2l_0 = el_0$. Thus $e^2 - e \in L \cap (0 : l_0)$. But $L \cap (0 : l_0)$ is a left ideal contained in L and is therefore either the whole of L or (0). If $L \cap (0 : l_0) = L$, then

$L \subseteq (0: l_0)$, a contradiction since $Ll_0 \neq (0)$. Therefore $L \cap (0 : l_0) = (0)$ and $e^2 = e$. Now $Re \subseteq L$ and $Re \neq (0)$ since $0 \neq e = e^2 \in Re$. Hence $Re = L$. The idempotent e has finite rank. In fact

$$\dim_K(eL) = 1 \tag{6}$$

We prove (6) by contradiction. Assume there are two linearly independent elements, el', el'' in eL. By Theorem 2, R induces a dense ring of K-linear transformations on the R-module L. Hence there exists $a \in R$ such that

$$ael' = 0, \qquad ael'' \neq 0 \tag{7}$$

The left ideal

$$L \cap (0 : l')$$

contains the element $ae \neq 0$. Therefore $L \cap (0 : l') = L$ showing that $Ll' = 0$. This contradicts our assumption that el' is a member of a linearly independent set and therefore different from 0. We have proved (6) and also Theorem 6 since, by Corollary 4, the elements of finite rank in R form an ideal and $e \neq 0$ is in this ideal. ▮

Theorem 6 and the following Theorem 7 characterize the simple rings with at least one minimal left ideal.

Theorem 7. A dense ring of linear transformations of finite rank of a K-space M over a division ring K is simple and possesses a minimal left ideal.

Proof. 1. Let $\mathfrak{A} \neq (0)$ be an ideal in R and $c \neq 0$ an arbitrary element of R. The basic idea of the proof is to find an idempotent $e \in \mathfrak{A}$ which projects the faithful, irreducible R-module M onto the finite-dimensional subspace cM. It then follows that

$$ecm = cm \qquad \text{for all} \quad m \in M$$

and hence $c = ec \in \mathfrak{A}$ showing that $\mathfrak{A} = R$.

2. Let $x \neq 0$ and $a \neq 0$ be two elements of M and $\mathfrak{A}$, respectively, and let $\{y_1, \ldots, y_n\}$ be a K-basis of aM. Then there exists $x_1 \in M$ such that $ax_1 = y_1$ and by the density of R, there exists $b \in R$ such that

$$by_1 = x_1, \qquad by_2 = \cdots = by_n = 0$$

Therefore

$$baM = x_1K \qquad \text{and} \qquad bax_1 = x_1$$

Since $x \neq 0$ and $x_1 \neq 0$, there exist $c_1, c_2 \in R$ such that

$$c_1x = x_1, \qquad c_2x_1 = x$$

Form

$$e = c_2bac_1 \in \mathfrak{A}$$

Then

$$c_2bac_1M \subseteq c_2baM = c_2x_1K = xK$$

and each element xk, $k \in K$, is left fixed by e. Therefore e projects M onto xK and $e \in \mathfrak{A}$.

3. At this point we may prove the existence of a minimal left ideal in R. By Theorem 5

$$M = eM \oplus (0 : e) \tag{8}$$

Since $eM = xK$, a subspace of minimal positive dimension, it is natural to conjecture that the annihilator of the maximal subspace $(0 : e)$, i.e. $L = (0 : (0 : e))$, is a minimal left ideal. To prove this, let l_1 and $l_2 \in L$, $l_1 \neq 0$. Since the R-module M is faithful, it follows that $l_1M \neq 0$. l_1 annihilates $(0 : e)$ and therefore by (8)

$$0 \neq l_1M = l_1eM = l_1xK$$

The density of R implies the existence of $a \in R$ such that

$$al_1x = l_2x$$

The element $al_1 - l_2$ annihilates $xK = eM$ as well as $(0 : e)$ and therefore, by (8), annihilates M. Hence $al_1 = l_2$ proving that if $l_1 \in L$, $l_1 \neq 0$, then $Rl_1 = L$. L is therefore irreducible as a left R-module, or, in other words, L is a minimal left ideal.

4. Now let $c \in R$. If $\dim_K cM = 1$, we have shown in Part 2 that there exists an idempotent $e \in \mathfrak{A}$ such that $eM = cM$. Let U be a subspace of M and let $\dim_K U = n$. Assume inductively that, given an arbitrary $(n-1)$-dimensional subspace U'', there exists $e'' \in \mathfrak{A}$ such that e'' projects M onto U''. U contains a one-dimensional subspace U' and by Part 2, there exists an idempotent e' which projects M onto U': $U' = e'M$. Consider

$$U'' = U \cap (0 : e')$$

a subspace of U. Clearly $U'' \cap U' = (0)$. Furthermore

$$U = U'' \oplus U' \tag{9}$$

since each element of U is expressible as

$$u = e'u + (u - e'u)$$

and

$$e'(u - e'u) = 0$$

But $\dim_K U'' = \dim_K U - \dim_K U' = n - 1$ and therefore by induction there exists an idempotent $e'' \in \mathfrak{A}$ which projects M onto U''. Consider now the element

$$e = e' + e'' - e''e'$$

contained in $\mathfrak{A}$. We claim that e projects M onto U. Proof:

$$eM \subseteq e'M + e''M + e''e'M = U' \oplus U''$$

and

$$ex = e'x + e''x - e''e'x = e'x = x \qquad \text{for all} \quad x \in U'$$
$$ex = e'x + e''x - e''e'x = e''x = x \qquad \text{for all} \quad x \in U'' \subset (0 : e')$$

We have shown that for each finite dimensional subspace U of M we can find a projection $e \in \mathfrak{A}$ such that $eM = U$. By the remarks made in Part 1, the theorem is proved. ▮

The above generalization of simple artinian rings is not only of interest in the field of algebra but also in functional analysis.

6. Isomorphism theorems

We have already remarked in Theorem II.5.1 that every minimal left ideal L of a simple ring R is a faithful irreducible R-module. We now ask if there is any relationship between L and the other faithful, irreducible R-modules. The following theorem gives us some information in this direction.

Theorem 1. Let R be a primitive ring with a minimal left ideal L. Then each faithful, irreducible R-module M is isomorphic to the left R-module L.

Proof. An isomorphism of an R-module M onto an R-module M' (cf. Definition II.4.1) is an isomorphism ϕ of the additive group $(M, +)$ onto $(M', +)$ which satisfies the additional condition

$$\phi(am) = a\phi(m) \qquad \text{for all} \quad m \in M \quad \text{and} \quad a \in R$$

Since $L \neq (0)$ and M is faithful, there exists an element $u \in M$ such that $Lu \neq (0)$. $RL \subseteq L$ and therefore Lu is an R-submodule of M which, by the irreducibility of M, is the whole of M. Consider the mapping

$$\phi: l \to lu \qquad \text{for all} \quad l \in L$$

Since $(al)\,u = a(lu)$, ϕ is an R-module epimorphism of L onto $M = Lu$. The kernel

$$\ker \phi = \{l \mid l \in L, \quad lu = 0\}$$

is a left ideal contained in L. By the minimality of L and since $Lu \neq (0)$, $\ker \phi = (0)$. ▮

This theorem shows that any two faithful, irreducible R-modules are isomorphic. The same result also holds for the centralizers of two faithful, irreducible R-modules.

Theorem 2. Let R be a primitive ring with a minimal left ideal L. Let M and M' be faithful, irreducible R-modules. Then the centralizer K of M is isomorphic to the centralizer K' of M'.

Proof. By Theorem 1, M' and M are isomorphic as R-modules under some isomorphism σ. Let $k \in K$. We must assign to k an endomorphism $k^\sigma \in K'$. We define k^σ by

$$k^\sigma\colon m' \to \sigma((\sigma^{-1}m')\,k) \qquad \text{for all} \quad m' \in M \tag{1}$$

Clearly k^σ is additive. Since k is in the centralizer of M and σ is an R-module isomorphism,

$$\begin{aligned}(am')\,k^\sigma &= \sigma((\sigma^{-1}(am'))\,k) = \sigma((a(\sigma^{-1}m'))\,k)\\ &= \sigma(a(\sigma^{-1}m')\,k)\\ &= a\sigma((\sigma^{-1}m')\,k)\\ &= a(m'k^\sigma)\end{aligned}$$

Therefore $k^\sigma \in \operatorname{Hom}_R(M', M') = K'$. By using the R-module homomorphism $\tau = \sigma^{-1}$ of M' to M, we may show that each element of K' occurs as the image of exactly one element $k \in K$:

$$k' \to k'^\tau \to k'^{\tau\sigma} = k' \qquad \text{for all} \quad k' \in K'$$

Thus $\phi\colon k \to k^\sigma$ is a one-one mapping of K onto K'. The additivity of ϕ follows from the additivity of the R-module isomorphism σ. This may be seen by substituting $k_1 + k_2$ for k in (1). ϕ is also multiplicative since

$$\begin{aligned}(m'k^\sigma)\,\lambda^\sigma &= [\sigma((\sigma^{-1}m')\,k)]\,\lambda^\sigma = \sigma((\sigma^{-1}[\sigma((\sigma^{-1}m')\,k)])\,\lambda)\\ &= \sigma(((\sigma^{-1}m')\,k)\,\lambda) = \sigma((\sigma^{-1}m'(k \cdot \lambda))\\ &= m'(k \cdot \lambda)^\sigma\end{aligned}$$

This shows that the mapping $k \to k^\sigma$ for $k \in K$ is an isomorphism of the division ring K onto the division ring K'. ▮

If we put $m' = \sigma m$ in (1) for a suitable $m \in M$, we obtain

$$\sigma(mk) = (\sigma m)\, k^{\sigma} \qquad \text{for} \quad m \in M \quad \text{and} \quad k \in K \tag{2}$$

This shows that in general the R-module isomorphism σ of M onto M' is not K-linear but only K-semilinear in the following sense:

Definition 1. Let M and M' be vector spaces over the division rings K and K', respectively. Let $k \to k^*$ be an isomorphism of K onto K'. A *semilinear* map ψ of M into M' is a homomorphism of $(M, +)$ to $(M', +)$ together with an isomorphism $k \to k^*$ of K onto K^* such that

$$\psi(mk) = \psi(m)\, k^*$$

The isomorphism $k \to k^*$ is said to be associated with ψ.

So far we have considered different modules over the same ring R. If R and R' are isomorphic rings under some isomorphism S, we may define a semilinear map from M onto M'. This map will naturally depend on S.

Theorem 3. Let M and M' be faithful, irreducible modules over the rings R and R', respectively. Let the division rings $K = \mathrm{Hom}_R(M, M)$ and $K' = \mathrm{Hom}_{R'}(M', M')$ be their centralizers. Let each ring contain a minimal left ideal and let

$$S\colon a \to a^S, \qquad a \in R$$

be an isomorphism of R onto R'. Then there exists a one-one semilinear map σ of the K-space M onto the K'-space M' which is related to S in the following way:

$$a^S = \sigma a \sigma^{-1} \qquad \text{for} \quad a \in R \tag{3}$$

i.e.

$$\sigma(am) = a^S \sigma m \qquad \text{for} \quad a \in R, \quad m \in M \tag{4}$$

Proof. M' may be made into a faithful, irreducible R-module by defining

$$a \circ m' = a^S m' \qquad \text{for} \quad a \in R, \quad m' \in M' \tag{5}$$

By Theorem 1 there exists an R-module isomorphism σ of the R-module M onto the R-module M'. Therefore

$$\sigma(am) = a \circ \sigma m = a^S \sigma m \qquad \text{for all} \quad m \in M$$

and (4) is proved. If we set $\sigma m = m'$, we get

$$\sigma a \sigma^{-1} m' = a^S m' \qquad \text{for all} \quad m' \in M'$$

showing $\sigma a \sigma^{-1} = a^S$. σ is a semilinear map of the K-space M onto the K'-space M' since the centralizer of the R-module M' is the same as the centralizer K' of M' as an R'-module by (5). ▌

Conversely, if σ is a one-one semilinear map of the K-space M onto the K'-space M', we may define an isomorphism of the ring $\text{Hom}_K(M, M)$ onto the ring $\text{Hom}_{K'}(M', M')$ by

$$S: a \to \sigma a \sigma^{-1} \qquad \text{for} \quad a \in \text{Hom}_K(M, M)$$

By Theorem 1, the faithful, irreducible modules over a primitive ring R containing a minimal left ideal L are all isomorphic to L. In this case, Theorem 2 shows that the centralizers of these modules are also uniquely determined up to isomorphism.

These facts suggest that we should try to describe the centralizer in terms of the internal structure of the ring. Let e be an idempotent generating the minimal left ideal L. We showed in Theorem II.5.6 that

$$\dim_K(eL) = 1 \qquad \text{where} \quad K = \text{Hom}_R(L, L)$$

This equality gives us a clue as to how to proceed. From $L = Re$ it follows that $eL = eRe$ and hence $\dim_K(eRe) = 1$.

Theorem 4. Let e be an idempotent in a ring R. Then the centralizer K of the left R-module Re is isomorphic to the ring eRe under the mapping

$$\phi: k \to ek, \qquad k \in K \tag{6}$$

Proof. Since $(ae)\,k = a(ek)$ for all $a \in R$, k is uniquely determined by the image of e; ϕ is therefore a one-one mapping of K into Re. Now $ek = re$ for some $r \in R$. But $ek = (e \cdot e)\,k = e(ek) = ere$ and therefore $ek \in eRe$. To show that ϕ is onto, let $ere \in eRe$. Then the mapping

$$\psi: ae \to ae \cdot ere \qquad \text{for} \quad a \in R \tag{7}$$

is an element of $\text{Hom}_R(re, Re)$ since by (7)

$$x(ae) = (xa)\,e \to xae \cdot ere = x(ae \cdot ere)$$

The image of e under ψ is ere showing that ϕ is onto. We have still to show that ϕ is a homomorphism. Let $k_i \in K$ and $ek_i = er_ie$. Then

$$\begin{aligned}
\phi(k_1 + k_2) &= e(k_1 + k_2) = ek_1 + ek_2 = er_1e + er_2e \\
&= \phi k_1 + \phi k_2 \\
\phi(k_1k_2) &= (ek_1)\,k_2 = (er_1e)\,k_2 \\
&= (er_1e)(ek_2) = (er_1e)(er_2e) = (\phi k_1)(\phi k_2)
\end{aligned}$$ ▌

Theorem 5. A minimal left ideal L in a primitive ring R is generated by an idempotent e. R is a dense ring of linear transformations of the vector space Re over the division ring eRe.

Proof. In the proof of Theorem II.5.6 we showed that L is generated by an idempotent if $L^2 \neq (0)$. We therefore prove $L^2 \neq (0)$.

1. *The product of two ideals $A \neq 0$, $B \neq 0$ is different from zero.* Proof: $AM = M$ and $BM = M$ if M is a faithful, irreducible R-module. Therefore

$$(AB)\,M = A(BM) = M \neq (0)$$

2. *R contains no nilpotent left ideals $L_1 \neq (0)$*, i.e. no left ideal $L_1 \neq (0)$ such that $L_1{}^n = (0)$ for some natural number n (cf. Definition III.1.3). Proof: $L_1{}^n = (0)$ implies

$$(L_1R)^{n+1} = L_1(RL_1)^n\, F \subseteq L_1^{n+1}R = (0)$$

Therefore by Part 1 it follows that the ideal $L_1R = (0)$. But the set

$$C = \{a \mid a \in R, \quad aR = 0\}$$

is not zero since $C \supseteq L_1$. Therefore C and R are two ideals with 0 product, contradicting Part 1.

3. Hence $L^2 \neq (0)$ *and $L = Re$ for some idempotent e.* By Theorem 1 L is a faithful, irreducible R-module and by Theorem 4 its centralizer K is the ring eRe. ∎

7. Modular maximal left ideals

In Section 6 we showed that all faithful, irreducible modules over a primitive ring R containing a minimal left ideal L are isomorphic to L. If we do not know of the existence of minimal left ideals in R but assume instead that R contains an identity 1, we may first prove the existence of maximal left ideals as follows: Let $\mathfrak{F}$ be the family of left ideals not containing 1 and let

$$\cdots \subseteq L_i \subseteq L_k \subseteq \cdots$$

be a chain of members of $\mathfrak{F}$. Then the set-theoretic union $\bigcup L_i$ does not contain 1 and hence $\bigcup L_i \in \mathfrak{F}$. Clearly $\bigcup L_i$ is an upper bound of the chain and therefore by Zorn's Lemma (cf. Section I.3), there exists a maximal left ideal $L \neq R$ in $\mathfrak{F}$. *Every ring with identity therefore contains a maximal left ideal.* The maximality of L implies that the factor module ${}_RR/L$ is an irreducible R-module since there is a one-one correspondence between the submodules of ${}_R R$ containing L and the submodules of ${}_R R/L$. If all irreducible R-modules were isomorphic, ${}_R R/L$ would be a representative of the isomorphism class. If, however, R does not contain minimal left ideals, the irreducible R-modules need not be isomorphic to one another. Nevertheless, we shall show that each irreducible R-module is isomorphic to ${}_R R/L$ for some maximal left ideal L. Furthermore, we shall see that we may drop the assumption that R contain 1 and replace it by the weaker hypothesis that R contain an element e which is a *right identity modulo L*, i.e. an element such that

$$x - xe \in L \qquad \text{for all} \quad x \in R \tag{1}$$

Thus, let M be an irreducible R-module and $u \neq 0$ an element of M such that $Ru \neq 0$. Then $Ru = M$ and the mapping

$$a + (0 : u) \rightarrow au \qquad \text{for all} \quad a \in R \tag{2}$$

is an R-module isomorphism of the factor module ${}_R R/(0 : u)$ onto M as may be verified without difficulty. From $Ru = M$ it follows that there exists $e \in R$ such that

$$eu = u \tag{3}$$

This implies $xeu = xu$ and hence

$$x - xe \in (0 : u) \qquad \text{for all} \quad x \in R \tag{4}$$

The element e is therefore a right identity modulo $(0 : u)$. Furthermore, $(0 : u)$ is a maximal left ideal in R since any left ideal L' properly containing $(0 : u)$ is mapped under the isomorphism in (2) onto the whole of M and hence $L' = R$. The left ideal $(0 : u)$ is a maximal modular left ideal in the sense of:

Definition 1. A left ideal L of a ring R is *modular* if there exists an element e which is a right identity modulo L, i.e. an element $e \in R$ such that

$$x \equiv xe (\text{mod } L) \qquad \text{for all} \quad x \in R \tag{5}$$

or, equivalently, such that

$$\{x - xe \mid x \in R\} \subseteq L \tag{6}$$

If R has an identity, every left ideal in R is modular. Corresponding to a given element $e \in R$ there is a modular left ideal L with e as right identity modulo L, namely $L = \{x - xe \mid x \in R\}$.

Theorem 1. Every modular left ideal L different from R is contained in a modular, maximal left ideal L'.

Proof. Let e be a right identity modulo L and $\mathcal{O}$ the union of the members of the chain

$$L \subseteq \cdots \subseteq L_i \subseteq L_k \subseteq \cdots \tag{7}$$

of ideals containing L but not containing e. Then $e \notin \mathcal{O}$. By Zorn's lemma there exists a left ideal L' in R containing L but not e and maximal subject to these conditions. L' is also maximal in R since any left ideal L'' properly containing L' contains e and hence R. ▌

The connection between modular, maximal left ideals and irreducible modules is established in the following:

Theorem 2. The R-module M is irreducible if and only if there exists a modular maximal left ideal L such that R/L and M are isomorphic.

Proof. The necessity has already been shown above. Let L be a modular, maximal left ideal in R. By the maximality of L the R-module R/L contains only itself and (0) as submodules. We must show that $R(R/L) \neq \bar{0}$, i.e. $R^2 \not\subseteq L$. Let e be a right identity modulo L. Then $\{x - xe \mid x \in R\} \subseteq L$. If $Re \subseteq L$, then $x = xe + (x - xe) \in L$ for all $x \in R$ showing $R \subseteq L$, a contradiction. Therefore $R^2 \not\subseteq L$. ▌

Let L be a left ideal in R. The element $c \in R$ induces, by left multiplication, the 0-map on R/L if and only if c belongs to the ideal

$$(L\colon R) = \{c \mid c \in R, \quad cR \subseteq L\}$$

This set is a (two-sided) ideal since $caR \subseteq cR \subseteq L$ and $acR \subseteq aL \subseteq L$. R/L is then a faithful module over the factor ring $R/(L : R)$ as we show in the following more general theorem:

Theorem 3. 1. Let M be an R-module and $\mathfrak{A}$ an ideal contained in the ideal $(0 : M)$. Then M is an $R/\mathfrak{A}$-module.

2. M is a faithful module over $R/\mathfrak{A}$ if and only if $\mathfrak{A} = (0 : M)$.

Proof. 1. Define multiplication by elements of $R/\mathfrak{A}$ by

$$(x + \mathfrak{A})\, m = xm, \qquad x \in R, \quad m \in M \tag{8}$$

Since $\mathfrak{A} \subseteq (0 : M)$, the multiplication is well defined. The module properties are clear.

2. If $\mathfrak{A} = (0 : M)$ and $x \notin (0 : M)$, there exists $m_0 \in M$ such that

$$(x + (0 : M))\, m_0 = xm_0 \neq 0$$

3. If $\mathfrak{A} \subset (0 : M)$, let $b \in (0 : M)$, $b \notin \mathfrak{A}$. Then $b + \mathfrak{A}$ is not the zero element of $R/\mathfrak{A}$ but $(b + \mathfrak{A})\, M = 0$. Therefore M is not a faithful $R/\mathfrak{A}$-module.

It follows that $(L : R)$ is the smallest ideal $\mathfrak{A}$ in R such that R/L is a faithful $R/\mathfrak{A}$-module. ▮

Definition 2. An ideal P in a ring R is *primitive* if R/P is a primitive ring. In this terminology, our discussion above and Theorem 2 show that $(L : R)$ is a primitive ideal if and only if L is a modular maximal left ideal. This proves the following:

Theorem 4. 1. The ideal P in a ring R is primitive if and only if there exists a modular maximal left ideal L in R such that $P = (L : R)$.

2. A ring R is primitive if and only if there exists a modular maximal left ideal L in R such that $(0) = (L : R)$. If such an L exists, R/L is a faithful, irreducible R-module. ▮

Theorem 5. Every division ring K is a primitive ring.

Proof. K itself is a faithful, irreducible R-module. ▮

Theorem 6. A commutative ring R is primitive if and only if it is a field.

Proof. The sufficiency has already been established in Theorem 5. To prove the necessity we first prove

Theorem 7. If L is a modular left ideal, $(L : R)$ is the largest ideal of R contained in L.

Proof. Let e be a right identity modulo L. If $a \in (L : R) = \{c \mid cR \subseteq L\}$, then $ae \in L$. But $a - ae \in L$. Therefore $a = ae + (a - ae) \in L$ and $(L : R) \subseteq L$. Let $\mathfrak{A}$ be an ideal of R contained in L. Since $\mathfrak{A}R \subseteq \mathfrak{A} \subseteq L$, $\mathfrak{A} \subseteq (L : R)$.

Now if R is a commutative primitive ring and L a modular maximal left ideal such that $(L : R) = (0)$, $L = (L : R)$ by Theorem 7 and hence $L = (0)$. Therefore if e is a right identity modulo L, e is a (two-sided) identity in R. The set $W' = \{ay \mid y \in R\}$ is an ideal in R and $W' \neq (0)$ since $ae = a$.

Therefore $W' = R$ showing that the equation $ay = b$ has at least one solution in R. ▮

As a corollary to Theorem 7 we may state:

Theorem 8. A simple ring with identity is primitive.

Proof. The identity 1 of the simple ring R is a right identity modulo every left ideal of R. By Zorn's lemma there exists a left ideal L maximal in the family of left ideals not containing 1. Clearly L is also maximal in R. The irreducible R-module R/L is also faithful. For, $x \cdot (R/L) = \bar{0}$ implies $xR \subseteq L$ and consequently $RxR \subseteq L$ showing that $x = 0$. ▮

8. Primitive algebras

Results analogous to those proved in the preceding sections of this chapter may also be proved for algebras over a commutative ring Φ.

Definition 1. Let R be an algebra over Φ. Let M be a module over R considered as a ring. Then M is an *algebra-module* over R if M is also a right Φ-module such that

$$(am)\,\alpha = a(m\alpha) = (a\alpha)\,m, \qquad m1 = m \tag{1}$$

for all $a \in R$, $m \in M$, and $\alpha \in \Phi$.

Definition 2. An endomorphism C of an algebra-module M over an algebra R is an endomorphism of M, considered as a module over the ring R, such that

$$(m\alpha)\,C = (mC)\alpha \qquad \text{for all} \quad m \in M \quad \text{and} \quad \alpha \in \Phi \tag{2}$$

If $1 \in R$, $1\alpha \in R$ for all $\alpha \in \Phi$. Therefore every ring-module M over R is also an algebra-module and every endomorphism of M considered as a ring-module is also an endomorphism of M considered as an algebra-module. The last assertion is valid if we drop the hypothesis that $1 \in R$ and merely assume that $RM = M$.

Theorem 1. Let M be an algebra-module over the algebra R. Let $RM = M$. Then every endomorphism of M considered as a ring-module is also an endomorphism of M considered as an algebra-module, and conversely.

Proof. The converse is clear. Since $M = RM$, each element $m \in M$ is expressible in the form

$$m = \sum_{1 \leqslant i \leqslant n} a_i m_i \,, \qquad a_i \in R, \quad m_i \in M$$

Let C be an endomorphism of M considered as a ring-module. By (1)

$$\begin{aligned}(m\alpha)C &= \sum ((a_i m_i)\alpha)C = \sum ((a_i \alpha) m_i)C \\ &= \sum (a_i \alpha)(m_i C) = \sum (a_i(m_i C))\alpha \\ &= \sum ((a_i m_i)C)\alpha = (mC)\alpha \quad \blacksquare\end{aligned}$$

In order to extend the theorems of Section 7 to cover the case that R is an algebra, we make the following:

Definition 3. A left ideal L in the Φ-algebra R is an algebra-submodule of ${}_R R$ as defined in Definition 1. Analogously we define right and two-sided ideals in an algebra R.

Theorem 2. Let R be an algebra over Φ.

1. A left ideal L in the algebra R is modular with e as right identity modulo L if and only if L, considered as a left ideal in the ring R, is modular.
2. A modular left ideal L of the algebra R is maximal if and only if L is maximal as a left ideal in the ring R.

Proof. 1. The condition $\{x - xe\} \subseteq L$ is independent of the Φ-structure of R

2. Let L be a maximal left ideal of the ring R and let $\{x - xe \mid x \in R\} \subseteq L$. Assume $L\alpha \nsubseteq L$ for some $\alpha \in \Phi$. Then

$$L + L\alpha = R$$

Hence

$$e = l_1 + l_2\alpha \qquad \text{for} \quad l_i \in L$$

Therefore by Definition I.4.2(5) of an algebra it follows that

$$e^2 = el_1 + el_2\alpha = el_1 + (e\alpha)\, l_2$$

But $e - e^2 \in L$ and $e^2 \in L$ imply $e \in L$ and hence $L = R$, a contradiction. Therefore $L\Phi \subseteq L$ and L is a left ideal of R considered as an algebra. A modular maximal left ideal of the algebra R is a modular left ideal of the ring R. By Zorn's lemma L is therefore contained in a modular left ideal L' of the ring R. As we have already shown, $L'\Phi \subseteq L'$ and therefore $L = L'$. $\blacksquare$

Based on this result we may prove the following theorem which says essentially that there is no distinction between irreducible ring- and algebra-modules.

Theorem 3. 1. Let R be an algebra over Φ and M an irreducible ring-module over R. We may define multiplication of elements of M by elements of Φ in such a way that M becomes an irreducible algebra-module.

2. An R-module M is irreducible as a ring-module if and only if it is irreducible as an algebra-module.

Proof. 1. Since $Ru = M$ for some element $u \in M$, every $m \in M$ is expressible as

$$m = au, \qquad a \in R$$

We may then define

$$m\alpha = (au)\,\alpha = (a\alpha)u \qquad \text{for} \quad a \in R, \quad \alpha \in \Phi \tag{3}$$

If the mapping $m \to m\alpha$ is well defined, we have

$$(au)(\alpha \cdot \beta) = [a\alpha\beta]\,u = ([a\alpha]\,u)\,\beta = ((au)\,\alpha)\beta$$

First, however, we must show that $au = a'u$ implies $(a\alpha)\,u = (a'\alpha)u$ for $a, a' \in R$. It is sufficient to show

$$au = 0 \qquad \text{implies} \qquad (a\alpha)\,u = 0 \qquad \text{for} \quad \alpha \in \Phi, \quad a \in R \tag{4}$$

Since $M = Ru$, $R/(0{:}\,u)$ is isomorphic as a ring-module to M under the mapping

$$a + (0{:}\,u) \to au$$

Therefore $(0 : u)$ is a modular maximal left ideal of the ring R [cf. Eq. II.7(4)]. By the preceding theorem, $(0 : u)$ is also a left ideal in the algebra R, i.e. $a \in (0 : u)$ implies $a\alpha \in (0 : u)$. This proves (4).

2. If M is irreducible as a ring-module, clearly it is irreducible as an algebra-module. Conversely, let M be an irreducible algebra-module and let $0 \neq u \in M$. Since $R\Phi = R$, the ring-module $Ru = R\Phi u = Ru\Phi$ is also an algebra-module. Therefore $Ru = 0$ or $Ru = M$. Suppose $Ru = 0$. Consider the set

$$M_0 = \{m \mid Rm = 0\}$$

This is clearly an algebra-submodule of M and $M_0 \neq (0)$ since $u \in M_0$. Therefore $M_0 = M$, implying $RM = (0)$, a contradiction. Therefore $Ru = M$ proving the irreducibility of M as a ring-module. ∎

Theorem 4. Let R be an algebra over Φ. The ideal $\mathfrak{A}$ of the algebra R is primitive if and only if it is primitive considered as an ideal of the ring R.

Proof. By Section II.7 $\mathfrak{A}$ is primitive if and only if $\mathfrak{A} = (0 : M)$ for some irreducible R-module M. By Theorem 3, M is irreducible as a ring-module if and only if it is irreducible as an algebra-module and hence the theorem is proved. ▮

Since the ring $\mathbb{Z}$ of integers is a commutative ring with 1, every ring R and every R-module may be considered as an algebra and an algebra-module (over $\mathbb{Z}$), respectively.

Rings with a Faithful Family of Irreducible Modules

CHAPTER III

1. The Radical of a Ring

Not all rings possess irreducible modules. For example, consider the ring R consisting of two elements 0 and a with addition and multiplication defined by

$$a + a = 0, \qquad a^2 = 0$$

$L = (0)$ is the only proper left ideal in R. There is, however, no right identity modulo L (cf. Theorem II.7.4). Even if R does possess irreducible modules, there need not exist a faithful one and therefore R need not be primitive.

In this chapter we shall investigate those rings possessing a family of irreducible modules which is faithful as a whole, i.e. a family of irreducible modules M_α such that no nonzero element of R is contained in every $(0 : M_\alpha)$. We first introduce the concept of the Jacobson radical of a ring.

Definition 1 (Jacobson). The (Jacobson) *radical* J of the ring R is the intersection of all ideals of the form $(0: M)$, where M is an irreducible R-module:

$$J = \bigcap (0: M), \qquad M \text{ an irreducible } R\text{-module} \tag{1}$$

If R has no irreducible R-modules, we set

$$J = \bigcap \varnothing = R$$

R is *semisimple* if $J = (0)$.

Theorem 1. The radical of the factor ring R/J is zero.

Proof. Every irreducible R-module M is an irreducible R/J-module since $J \subseteq (0 : M)$. Since $\bigcap (0 : M) = J$, $\bigcap (\overline{0 : M}) = \bigcap \{\bar{a} \mid \bar{a} \in R/J, \bar{a}M = 0\} = (\bar{0})$.
The family $\mathcal{F}$ of irreducible R-modules is therefore a faithful family for the factor ring R/J provided $\mathcal{F} \neq \varnothing$. ∎

Just as in the case of primitive rings, we may define the radical of a ring internally, i.e. without involving the notion of an R-module.

Theorem 2. The radical of a ring is the intersection of its modular, maximal left ideals.

Proof. By Theorem II.7.2, there is a correspondence between the isomorphism classes of irreducible R-modules and the modular maximal left ideals L of R defined by

$$M \to R/L \tag{2}$$

From our discussion preceding Definition II.7.1, the modular, maximal left ideals are precisely those of the form

$$L = (0 : u)$$

where u is a nonzero element of some irreducible R-module. Since

$$\bigcap_{0 \neq u \in M} (0 : u) = (0 : M)$$

it follows that

$$J = \bigcap_{M} (0 : M) = \bigcap_{M} \bigcap_{0 \neq u \in M} (0 : u) = \bigcap L. \quad ∎$$

We may also describe the radical of a ring as the sum of all left ideals which consist entirely of "quasiregular" elements. To introduce the concept of quasiregularity we proceed as follows: Let $z \notin J$. Then there exists an irreducible R-module M_0 and an element $0 \neq u \in M_0$ such that $zu \neq 0$. By the irreducibility of M_0 it follows that $Rzu = M_0$. Hence there exists $a \in R$ such that

$$azu = u \tag{3}$$

First assume that R contains an identity 1. Then (3) may be written in the form

$$(1 - az)\,u = 0$$

This implies that $1 - az$ is not a right *unit* in R, i.e. there is no element $b \in R$ such that

$$b \cdot (1 - az) = 1$$

otherwise $u = b(1 - az)\,u = 0$. Therefore if $z \notin J$ there is an element az in the left ideal Rz such that $1 - az$ is not a right unit. In other words, the set

$$\{z \mid 1 - az \text{ is a right unit for all } az \in Rz\}$$

is contained in J.

In a ring without identity it is meaningless to talk about units. However, if we substitute the notion of quasiregularity we obtain, by essentially the same reasoning as in the preceding paragraph, the analogous conclusion, namely

$$\{z \mid rz \text{ is quasiregular for all } r \in R\} \subseteq J$$

To motivate the definition of quasiregularity let us for a moment assume that R has an identity. If $1 - y$, $y \in R$, is a right unit, there exists $b \in R$ such that $b(1 - y) = 1$. This element b may be expressed as $1 - y'$ for some $y' \in R$. Therefore there exists $y' \in R$ such that

$$(1 - y')(1 - y) = 1$$

Multiplying this out and canceling 1 from both sides we get

$$-y - y' + y'y = 0.$$

Hence, for a given y, $1 - y$ is a right unit if and only if there exists y' such that $y + y' - y'y = 0$.

Definition 2. The element $z \in R$ is *quasiregular* (qr) if there exists $y' \in R$ such that

$$y + y' - y'y = 0 \tag{4}$$

A left ideal is quasiregular if all its elements are quasiregular.

Now suppose R is an arbitrary ring (with or without identity). Let $z \in J$. There exists an irreducible R-module M_0 and an element $0 \neq u \in M_0$ such that $azu = u$ for some $a \in R$. Suppose now that the left ideal Rz is qr. Then, in particular, there exists $z' \in R$ such that

$$az + z' - z'az = 0$$

But since $u = azu = 0$, we have

$$0 = (u - azu) - z'(u - azu) = u - (az + z' - z'az)\,u = u$$

a contradiction. Therefore

$$\{z \mid z \in R, \quad Rz \text{ qr}\} \subseteq J \tag{5}$$

To show the reverse inclusion we use Theorem 2. If $z \in R$ is not qr then the modular left ideal $L = \{x - xz \mid x \in R\} \neq R$. For otherwise we would have $z' \in R$ such that $z' - z'z = -z$ and z would be qr. The modular left ideal L is contained in a modular maximal left ideal L'. If $z \in J$, then $z \in L'$ by Theorem 2 and since $L \subseteq L'$, then $L' = R$, a contradiction. Therefore $z \notin J$. Now if the left ideal Rz is not qr, there exists $a \in R$ such that az is not qr. Therefore $az \notin J$ as we have just shown. Since J is an ideal this implies $z \notin J$. We have proved

Theorem 3. The radical of a ring is a qr ideal containing all qr left ideals. ▮

Theorem 4. The element $y \in R$ is qr if and only if the left ideal

$$L = \{x - xy \mid x \in R\} = R$$

Proof. If $L = R$, there exists $x \in R$ such that $x - xy = -y$. Conversely, if y is qr, $x - xy = -y$ for some $x \in R$ and therefore $y \in L$. But $y \in L$ implies $Ry \subseteq L$ and hence $R \subseteq L$. ▮

Not every qr element $z \in R$ is in the radical J. In the ring of integers the only qr elements z are those for which

$$(1 - z')(1 - z) = 1$$

for some $z' \in \mathbb{Z}$. Therefore $z = 0$ or $z = 2$. The ideal generated by 2 consists of all the even integers. However, 2 is the only one which is qr showing that $2 \notin J$. Therefore the radical of $\mathbb{Z}$ is zero.

From $y \in J$ it follows that $y + y' - y'y = 0$. Since $y' = -y + y'y \in J$, y' is also qr. Let $y' + y'' - y''y' = 0$. Then

$$\begin{aligned} y'' &= y'' + (y' + y - y'y) - y''(y' + y - y'y) \\ &= y + (y' + y'' - y''y') - (y' + y'' - y''y')\,y = y \end{aligned}$$

Hence $y' + y - yy' = 0$. Therefore each element of J is also "right quasi-regular" and we obtain the same radical if we use right modules and substitute "right" for "left" throughout. It is important to note, however, that not all rings primitive on the left are also primitive on the right as Bergmann [1964] showed. It should be further noted that Jacobson uses right modules in his book [1956].

Sometimes it is difficult to determine whether an element is qr. Theorem 5 provides us with a sufficient condition for an element to be qr in terms of the following notion:

Definition 3. An element $z \in R$ is *nilpotent* if there exists a natural number n such that $z^n = 0$. The left ideal L is *nil* if all its elements are nilpotent. The *left ideal L is nilpotent* if there exists a natural number n such that the product of any n elements of L is zero, i.e. $L^n = (0)$.

Theorem 5. 1. Every nilpotent element $z \in R$ is qr.
2. Every nil left ideal is contained in the radical J of R.

Proof. If $z^n = 0$, $z + z' - z'z = 0$ for $z' = -z - z^2 - \cdots - z^{n-1}$. Therefore every left ideal consisting of nilpotent elements is contained in J. ∎

In an artinian ring the converse of Theorem 5.2 holds.

Theorem 6. Let R be artinian. Then the radical of R is nilpotent.

Proof. Consider the descending chain

$$J \supseteq J^2 \supseteq \cdots$$

Since R is artinian, there exists a natural number n such that

$$J^n = J^{n+1} = \cdots = J^{2n} = J^{2n+1} = \cdots \tag{6}$$

Assume $J^n \neq (0)$. Then by (6) J^n is contained in the family

$$\mathfrak{F} = \{L \mid L \text{ is a left ideal in } R, \quad L \subseteq J^n, \quad J^n L \neq (0)\}$$

Hence $\mathfrak{F} \neq \varnothing$. Let L_m be minimal in $\mathfrak{F}$. There exists $l \in L_m$ such that $J^n l \neq 0$. Since $J^n = J^{2n}$, $J^n l$ is a member of $\mathfrak{F}$. Furthermore $J^n l \subseteq L_m$ and therefore $J^n l = L_m$. Hence there exists $z \in J^n \subseteq J$ such that

$$zl = l \tag{7}$$

Since z is qr, there exists $z' \in R$ such that $0 = z + z' - z'z$. This implies, by (7), that

$$0 = l \cdot lz - z'(l - lz) = l - (z + z' - z'z)\, l = l$$

a contradiction since $J^n l \neq (0)$. Therefore $J^n = (0)$. ∎

In a commutative ring R every nilpotent element generates a nilpotent ideal. However, it should be observed that the absence of nonzero nilpotent elements does not imply that J is zero.

2. Semisimple rings as subdirect sums of primitive rings

If the radical

$$J = \bigcap_{M \text{ irreducible}} (0 : M)$$

of a ring R is zero, the family of irreducible R-modules is faithful as a whole. We may of course restrict ourselves to a family of pairwise nonisomorphic irreducible modules since

$$M \approx M' \quad \text{implies} \quad (0 : M) = (0 : M') \tag{1}$$

Thus let

$$\mathcal{F} = \{M_i \mid i \in I\} \tag{2}$$

be a family of pairwise nonisomorphic irreducible R-modules, so chosen that any irreducible R-module is isomorphic to some member of $\mathcal{F}$. Then

$$J = \bigcap_{i \in I} (0 : M_i) \tag{3}$$

Let i be an arbitrary fixed element of I. Let

$$a(i) = a + (0 : M_i) \in R/(0 : M_i) \tag{4}$$

The mapping

$$a \to a(i), \qquad a \in R, \qquad i \text{ fixed} \tag{5}$$

is clearly an epimorphism of R onto the primitive ring $R_i = R/(0 \colon M_i)$.

In general it is not possible to recover R from R_i in any easy way. However, we can embed R in the so-called direct sum of the R_j (cf. Definition 1 below). Again, as in the case of primitive rings, the embedding need not be an epimorphism.

Definition 1. Let $\{\mathfrak{A}_i \mid i \in I\}$ be a family of rings. The *complete direct sum*, $\sum_c \mathfrak{A}_i$, consists of the set of functions

$$a \colon i \to a(i) \in \mathfrak{A}_i\,, \qquad i \in I \tag{6}$$

with I as domain and their ranges in the set theoretical union of the $\mathfrak{A}_i$, $i \in I$. The sum and the product are defined by

$$a + b \colon i = a(i) + b(i), \qquad ab(i) = a(i)\, b(i) \tag{7}$$

Clearly $\sum_c \mathfrak{A}_i$ is a ring. The operations in this ring depend on the operations in the $\mathfrak{A}_i$. Given an arbitrary choice of $a_i \in \mathfrak{A}_i$, $i \in I$, there exists exactly one element $a \in \sum_c \mathfrak{A}_i$ such that

$$a(i) = a_i \qquad \text{for} \quad i \in I$$

Here (4) implies that the mapping which assigns to each element $a \in R$ the element

$$i \to a(i) = a + (0 : M_i) \qquad \text{for} \quad i \in I \tag{8}$$

of the complete direct sum $\sum_c R_i$ of primitive rings R_i, is a homomorphism. Furthermore, since $J = (0)$, the mapping is a monomorphism of R into $\sum_c R_i$.

This monomorphism is not in general an epimorphism. For example, consider the ring $\mathbb{Z}$ of integers. Let $\{p_i \mid i \in I\}$ denote the set of primes and (p_i) the ideal generated by p_i. Take $\{\mathbb{Z}/(p_i) \mid i \in I\}$ as a complete set of representatives of the isomorphism classes of irreducible $\mathbb{Z}$-modules. $J = (0)$ and the mapping

$$a \to (i \to a + (p_i)) \tag{9}$$

is a monomorphism of $\mathbb{Z}$ into the complete direct sum $\sum_c \mathbb{Z}/(p_i)$ of fields. However, for given $a_i \in \mathbb{Z}$ it is not always possible to solve the congruences

$$a \equiv a_i \bmod p_i$$

simultaneously.

The image R' of R in $\sum_c R_i$ is therefore in general only a subring of $\sum_c R_i$. We may, nevertheless say a little more: Each element a_i of each subring R_i occurs as the image of i under some $a \in R'$. This implies that R is isomorphic to a subdirect sum of the R_i in the sense of the following:

Definition 2. A *subdirect sum* $\sum_s \mathfrak{A}_i$ of the rings $\mathfrak{A}_i$, $i \in I$, is a subring of the complete direct sum $\sum_c \mathfrak{A}_i$ satisfying the condition that for each $i_0 \in I$ and each $a_{i_0} \in R_{i_0}$ there exists $a \in \sum_s \mathfrak{A}_i$ such that

$$a(i_0) = a_{i_0}$$

We have proved the following:

Theorem 1. A semisimple ring R is isomorphic to a subdirect sum of primitive rings R_i. ▌

Let $\mathfrak{A}$ be a subdirect sum of rings $\mathfrak{A}_i$, $i \in I$. Then the mapping $a \to a(i)$ for $a \in \mathfrak{A}$ is an epimorphism of the ring $\mathfrak{A}$ onto the ring $\mathfrak{A}_i$ with kernel the ideal $\mathfrak{U}_i = \{a \in \mathfrak{A} \mid a(i) = 0\}$ in $\mathfrak{A}$. Since $\mathfrak{A}$ is embedded in the complete direct sum $\sum_c \oplus \mathfrak{A}_i$ of the $\mathfrak{A}_i$, it follows that $\bigcap \{\mathfrak{A}_i \mid i \in I\} = (0)$. Conversely, given a ring R and a family $\{\mathfrak{A}_i \mid i \in I\}$ of ideals of R such that

$$\bigcap \{\mathfrak{A}_i \mid i \in I\} = (0),$$

then R is isomorphic to a subdirect sum of the factor rings $R/\mathfrak{A}_i$ under the mapping $a \to (a + \mathfrak{A}_i \mid i \in I)$ for $a \in R$.

Once we have shown that a ring is isomorphic to a subdirect sum of rings R_i, we are faced with the problem of describing more exactly how this subdirect sum is contained in the complete direct sum. Subdirect sums which are dense in $\sum_c R_i$ relative to the topology defined below play a particularly important role.

Definition 3. 1. Let $\mathfrak{A}$ be a ring. Define a topology on $\mathfrak{A}$ by choosing as a neighborhood basis for each $a \in \mathfrak{A}$ the family of sets consisting only of the subset $\{a\}$. The resulting topology is called the *discrete topology*.

2. Let $\mathfrak{A}_i$, $i \in I$ be a family of rings with the discrete topology. We topologize $\sum_c \mathfrak{A}_i$ by means of the so-called *product topology* of the discrete topologies on the individual $\mathfrak{A}_i$'s. This product topology is defined by taking as neighborhood basis of $a \in \sum_c \mathfrak{A}_i$ the sets

$$U(a) = \left\{ b \,\middle|\, b \in \sum\nolimits_c \mathfrak{A}_i, \quad b(i_j) = a(i_j), \quad j = 1, \ldots, n \right\}$$

$U(a)$ consists of those functions b on I which take on the same values as a on a pre-assigned finite subset $\{i_1, \ldots, i_n\}$ of I.

If I is finite, the product topology is none other than the discrete topology on $\sum_c \mathfrak{A}_i$. It is easily shown that the families $U(a)$ defined above satisfy conditions U_1, U_2, and U_3 of Section II.2 and that $\sum_c \mathfrak{A}_i$ is a topological ring under the product topology. The following theorem is an immediate consequence of the definition:

Theorem 2. A subdirect sum $\mathfrak{A}$ of topological rings $\mathfrak{A}_i$ with the discrete topology is dense in $\sum_c \mathfrak{A}_i$ if and only if for each finite subset $\{i_1, \ldots, i_n\}$ of I and for each corresponding set of elements $a_{i_k} \in \mathfrak{A}_{i_k}$, $k = 1, \ldots, n$, there exists an element $a \in \mathfrak{A}$ such that

$$a(i_k) = a_{i_k}, \qquad k = 1, \ldots, n$$

$\mathfrak{A}$ is then called a *dense subdirect sum*.

If I is finite, $\mathfrak{A}$ is dense in $\sum_c \mathfrak{A}_i$ if and only if $\mathfrak{A} = \sum_c \mathfrak{A}_i$. ∎

Theorem 3. Let $\mathfrak{A} = \sum_s \mathfrak{A}_i$ and let

$$A_i = \{a \mid a \in \mathfrak{A}, \quad a(i) = 0\}$$

be the kernels of the epimorphisms $a \to a(i)$ of $\mathfrak{A}$ onto $\mathfrak{A}_i$. Then $\mathfrak{A}$ is dense in $\sum_c \mathfrak{A}_i$ if and only if for each finite subset $\{i_1, ..., i_r\}$ of I,

$$\mathfrak{A} = A_{i_1} + \bigcap_{2 \leqslant k \leqslant n} A_{i_k} \tag{10}$$

Proof. 1. Let $\mathfrak{A}$ be dense. For given $a \in \mathfrak{A}$ and $\{i_1, ..., i_n\} \subseteq I$, there exists $b \in \mathfrak{A}$ such that $b(i_1) = a(i_1)$ and $b(i_k) = 0$, $k = 2, ..., n$. Since $(a - b)(i_1) = 0$,

$$a - b \in A_{i_1} \quad \text{and} \quad b \in \bigcap_{k=2}^{n} A_{i_k}$$

Therefore (10) is proved.

2. Suppose conversely that (10) is satisfied. Let a_{i_j} be an arbitrary element in $\mathfrak{A}_{i_j}$, $j = 1, ..., n$. Since $\mathfrak{A}$ is a subdirect sum, there exists at least one element $a_1 \in \mathfrak{A}$ such that

$$a_1(i_1) = a_{i_1} \tag{11}$$

By (10),

$$a_1 = b_1 + d_1 \quad \text{where} \quad b_1 \in A_{i_1} \quad \text{and} \quad d_1 \in \bigcap_{2 \leqslant k \leqslant n} A_{i_k} \tag{12}$$

But $(a_1 - b_1)(i_1) = 0$. Therefore $d_1(i_1) = a_{i_1}$ and $d_1(i_j) = 0$, $j = 2, ..., n$. Similarly for each $2 \leqslant m \leqslant n$, there exists $d_m \in \mathfrak{A}$ such that $d_m(i_m) = a_{i_m}$ and $d_m(i_j) = 0$ if $j \neq m$. Hence

$$(d_1 + d_2 + \cdots + d_n)(i_j) = a_{i_j}, \qquad j = 1, ..., n \tag{13}$$

and $\mathfrak{A}$ is dense. ▌ Later on in Theorem III.3.6 we shall show that condition (10) may be simplified.

Let $\{\mathfrak{A}_i \mid i \in I\}$ be a family of pairwise nonisomorphic simple rings and let $\mathfrak{A} = \sum_s \mathfrak{A}_i$. We show that (10) is satisfied in this case: Since $\mathfrak{A}/A_i \approx \mathfrak{A}_i$, the A_i are distinct maximal ideals in $\mathfrak{A}$. Furthermore, $\mathfrak{A}^2 \nsubseteq A_i$ since $\mathfrak{A}_i^2 \neq (0)$. To show (10) is satisfied it is sufficient to prove

$$\bigcap_{2 \leqslant k \leqslant n} A_{i_k} \nsubseteq A_{i_1} \tag{14}$$

since the A_i's are maximal. (14) will follow if we can show

$$A_{i_2} \cdots A_{i_n} \nsubseteq A_{i_1} \tag{15}$$

since $A_{i_2} \cdots A_{i_n} \subseteq \bigcap_{2 \leqslant k \leqslant n} A_{i_k}$. When $n = 2$, $A_{i_2} \neq A_{i_1}$ and A_{i_2}, A_{i_1} maximal imply $A_{i_2} \nsubseteq A_{i_1}$. Assume inductively that $A_{i_2} \cdots A_{i_{n-1}} \nsubseteq A_{i_1}$. Now $\mathfrak{A}^2 \nsubseteq A_{i_1}$ and therefore

$$\begin{aligned}\mathfrak{A} &= \mathfrak{A}^2 + A_{i_1} \\ &= (A_{i_1} + [A_{i_2} \cdots A_{i_{n-1}}])(A_{i_1} + A_{i_n}) + A_{i_1} \\ &= A_{i_2} A_{i_3} \cdots A_{i_n} + A_{i_1}\end{aligned}$$

proving that $A_{i_2} \cdots A_{i_n} \nsubseteq A_{i_1}$. By induction (15) follows and we have

Theorem 4. A subdirect sum of pairwise nonisomorphic simple rings is dense. ▮

By the same method we may prove:

Theorem 5. If a semisimple ring R is a subdirect sum of simple ring R_i, $i \in I$, then there exists a subset Λ of I such that R is a dense subdirect sum of simple rings R_λ, $\lambda \in \Lambda$.

Proof. By hypothesis

$$(0) = \bigcap_{i \in I} A_i, \qquad A_i = (0 : M_i), \qquad \text{and} \qquad R/A_i \text{ simple} \tag{16}$$

If an ideal A_i appears more than once we may discard all repetitions and retain only one index λ for which $A_\lambda = A_i$. Let Λ denote the set of all such indices. Under these circumstances, $(0) = \bigcap_{\lambda \in \Lambda} A_\lambda$ and (15) is still valid with Λ replacing I. Theorem 5 therefore follows in the same way as Theorem 4. ▮

If in addition the simple rings $\mathfrak{A}_i$ are primitive, as is the case, for example, when $\mathfrak{A}_i$ contains a minimal left ideal (Theorem II.5.2), we may prove (15) using the concept of a prime ideal.

Definition 4. The ideal P in R is *prime* if

$$BC \subseteq P \qquad \text{implies either} \quad B \subseteq P \quad \text{or} \quad C \subseteq P$$

where B and C are arbitrary ideals in R.

Theorem 6. A primitive ideal P is prime.

Proof. Let $P = (0 : M)$, M an irreducible R-module. If $C \nsubseteq P$, there exists $0 \neq u \in M$ such that $Cu \neq 0$. Therefore $Cu = M$. If $B \not\subseteq P$, there exists $v \in M$ and that $Bv \neq 0$. Hence $B \cdot Cu \neq 0$ and $BC \not\subseteq P$. ▮

Now if $\mathcal{A}_{i_1}$ is a primitive ring, A_{i_1} is a primitive ideal by Definition II.7.2 and therefore if

$$A_{i_2} \cdots A_{i_n} \subseteq A_{i_1}$$

it follows that at least one A_{i_k}, $k \geqslant 2$ must be contained in A_{i_1}. By maximality, $A_{i_k} = A_{i_1}$, a contradiction, proving (15).

By Theorem II.7.6 a commutative ring is primitive if and only if it is a field. Therefore Theorem 5 implies the following:

Theorem 7. A semisimple commutative ring is a dense subdirect sum of fields. ▮

3. Semisimple artinian rings

If a ring R is a dense subdirect sum of finitely many rings $R_1, ..., R_n$, then

$$R = \sum_c R_i$$

since, given n elements $a_i \in R_i$, $i = 1, ..., n$, there exists $a \in R$ such that $a(i) = a_i$, $i = 1, ..., n$. Suppose now that R is semisimple artinian. Semisimplicity implies there exists a family of primitive ideals $\{A_i \mid i \in I\}$ such that $(0) = \bigcap_{i \in I} A_i$. Since R is artinian on left ideals, it is, in particular, artinian on ideals. Therefore, we may choose a finite subset of the A_i's, say $A_1, ..., A_q$, such that

$$(0) = A_1 \cap A_2 \cap \cdots \cap A_q \tag{1}$$

Furthermore, we may assume that the set $\{A_1, ..., A_q\}$ is *irredundant*, i.e. no member of the set may be omitted without affecting the intersection. We shall call (1) a representation of (0) as the intersection of an irredundant set of primitive ideals. R/A_i is artinian since there is a one-one correspondence between the left ideals of R/A_i and the left ideals of R containing A_i. By Theorems II.5.1 and II.3.5, R/A_i is a simple ring isomorphic to the ring of $n_i \times n_i$ matrices over a division ring K^i. By Theorem III.2.4 and the irredundancy of (1) R is a dense subdirect sum and hence

$$R = \sum_c R/A_i = R/A_1 \oplus \cdots \oplus R/A_q$$

Each R/A_i is simple artinian and therefore isomorphic to $K^i_{n(i)\times n_i}$. This proves the following structure theorem first proved by Wedderburn for finite-dimensional algebras over their ground fields and later generalized by Artin to artinian rings (The proof of the uniqueness of the direct summands will be given in Theorem 3):

Theorem 1 (Wedderburn–Artin). A semisimple artinian ring R is a direct sum of finitely many simple artinian rings R_i each isomorphic to a ring of $n_i \times n_i$ matrices over a division ring K^i. The n_i and K^i are uniquely determined up to order and isomorphism (cf. Theorem 3). ▌

Of course if the artinian ring R is an algebra over the commutative ring Φ this theorem is also valid.

This fundamental structure theorem of classical ring theory has been proved in the much more general context of dense subdirect sums of primitive rings. If we are only interested in arriving at the result, there is a much shorter proof of Theorem 1.

By Theorem II.3.5 we have only to show that an artinian ring with no nonzero nilpotent left ideals is the complete direct sum of finitely many primitive artinian rings. This may be done as follows:

α. A minimal left ideal L in R is an irreducible R-module since $RL = (0)$ implies $L^2 = (0)$, a contradiction. Furthermore, L is a faithful, irreducible module over the ring $R/(0 : L)$, where

$$(0 : L) = \{a \mid a \in R, \quad aL = 0\} \tag{2}$$

$R/(0 : L)$ is therefore a primitive ring.

β. LR is an ideal different from zero since $L^2 \neq (0)$. Let

$$D = LR \cap (0 : L)$$

Then

$$D^2 \subseteq (0 : L) \cdot LR = (0)$$

and therefore $D = (0)$ since R contains no nonzero nilpotent left ideals. Hence

$$LR \cap (0 : L) = (0) \tag{3}$$

γ. $LR + (0 : L)$ is an ideal in R. Suppose

$$LR + (0 : L) \subset R$$

Then $(LR + (0 : L))/(0 : L) \subset R/(0 : L)$, since the left ideals of $R/(0 : L)$ are in one-one correspondence with the left ideals of R containing $(0 : L)$. $R/(0 : L)$

is primitive as was shown in α. Therefore by Theorem II.5.1 $R/(0 : L)$ is a simple artinian ring showing that

$$(LR + (0 : L))/(0 : L)$$

is the zero submodule of $R/(0 : L)$. Thus

$$LR + (0 : L) \subseteq (0 : L) \tag{4}$$

and $LR \subseteq (0 : L)$. Therefore $LR = LR \cap (0 : L) = (0)$, contradicting $LR \neq (0)$. Hence

$$LR + (0 : L) = R \tag{5}$$

δ. By (3) and (5) every element in R is uniquely expressible as the sum of an element in LR and an element in $(0 : L)$. Since LR and $(0 : L)$ are ideals, $LR \cdot (0 : L) \subseteq LR \cap (0 : L) = (0)$. It follows that R is isomorphic to the (complete) direct sum

$$\sum_{i=1}^{2}{}_{\mathrm{c}}\, \mathfrak{A}_i$$

of the two rings $\mathfrak{A}_1 = LR$ and $\mathfrak{A}_2 = (0 : L)$. The inverse images of the $\mathfrak{A}_i$ under this isomorphism give rise to a decomposition of R:

$$R = LR \oplus (0 : L) \tag{6}$$

[Without appealing to the concept of a complete direct sum, we say that a ring R is the direct sum of the *ideals* LR and $(0 : L)$ if (3) and (5) hold].

ϵ. By (6), the ring $R_1 = LR$ is isomorphic to the primitive artinian ring $R/(0 : L)$. The ring $(0 : L)$ is again artinian without nonzero nilpotent left ideals since by (6) a left ideal in $(0 : L)$ is also a left ideal in $LR \oplus (0 : L) = R$. Apply the same reasoning as above to the semisimple artinian ring $R^{(2)} = (0 : L)$, etc. This gives rise to a descending chain

$$R = R_1 \oplus R^{(2)} \supset R^{(2)} = R_2 \oplus R^{(3)} \supset \cdots$$

By the descending chain condition, we obtain $R^{(q+1)} = (0)$ for some q. Therefore

$$R = R_1 \oplus \cdots \oplus R_q, \qquad R_i \quad \text{primitive artinian} \tag{7}$$

and Theorem 1 is proved once again except for uniqueness of summands.

This proof just given will be better understood once we have introduced the concept of a homogeneous component of ${}_RR$ in Section IV.2 and proved Theorem IV.2.3. It will be seen that LR is in fact a homogeneous component of ${}_RR$.

Theorem 2. A semisimple artinian ring has an identity.

Proof. Each ring $K_{n\times n}$ occurring in the sum has an identity and it is clear that the sum of these identities is an identity for R. ∎

In the next chapter we shall show that R may be expressed as the direct sum of minimal left ideals in many ways. This is in sharp contrast with the uniqueness of the simple direct summands occurring in (7). We now proceed with the proof of the uniqueness of the representation (7). The proof depends on Theorem 2 and the following:

Theorem 3. A ring $R \ni 1$ may be decomposed in at most one way as the direct sum of simple rings R_i, i.e. if

$$R = R_1 \oplus \cdots \oplus R_q, \qquad R_i \quad \text{simple} \tag{8}$$

and

$$R = R_1' \oplus \cdots \oplus R_s' \tag{9}$$

R_i' simple, then $q = s$ and $R_i = R'_{\pi(i)}$, $i = 1, ..., q$, π a permutation of $\{1, ..., q\}$.

Proof. Since $R_i^2 = R_i$ and $R_iR_j = (0)$ if $i \neq j$,

$$R_i = R_iR = R_i(R_1' \oplus \cdots \oplus R_s')$$

But $R_iR_k' \cap R_iR_l' \subseteq R_k' \cap R_l' = (0)$ if $k \neq l$. Therefore at most one R_iR_k', say $R_iR'_{\pi(i)}$ is different from zero. Since R_i is simple, $R_iR'_{\pi(i)} = R_i$. Hence $R_i = R'_{\pi(i)}$. Thus to each $i \in \{1, ..., q\}$ there exists a unique $\pi(i) \in \{1, ..., s\}$ such that $R_i = R'_{\pi(i)}$. Conversely, to each $j \in \{1, ..., s\}$ there exists a unique $\pi(j) \in \{1, ..., q\}$ such that $R_j' = R_{\sigma(j)}$ and the theorem is proved. ∎

Theorem 4. Let R be semisimple artinian. By Theorem 1

$$R = R_1 \oplus \cdots \oplus R_q, \qquad R_i \text{ simple}, \qquad i = 1, ..., q \tag{8}$$

Let $\mathfrak{A} \neq (0)$ be an ideal in R. Then there exists $R_{i_1}, R_{i_2}, ..., R_{i_k}$ such that

$$\mathfrak{A} = R_{i_1} \oplus \cdots \oplus R_{i_k}$$

Proof. Since $1 \in R$, $\mathfrak{A} = \mathfrak{A} \cdot R$. Now $\mathfrak{A} \cdot R_i$ is either (0) or R_i since $\mathfrak{A}R_i \subseteq \mathfrak{A} \cap R_i$ is an ideal in the simple ring R_i. $\mathfrak{A} \cdot R_i = R_i$ implies $R_i \subseteq \mathfrak{A}$ and the result follows. ∎

The information obtained in Theorem 4 on the ideals of a semisimple artinian ring R allows us to prove the following important theorem on the lattice of ideals of R [cf. Section II.3, (10_1)]:

Theorem 5. The lattice V of ideals in a semisimple artinian ring is *distributive*, i.e.

$$A \cap (B + C) = (A \cap B) + (A \cap C) \tag{9}$$

A, B, and C ideals in R.

Proof. Let $\mathfrak{A}$ be an ideal in R. By Theorem 4, there is a subset $\{i_1, ..., i_r\} \subseteq \{1, ..., q\}$ such that $A = R_{i_1} \oplus \cdots \oplus R_{i_r}$. Consider now the mapping

$$A \to \{i_1, ..., i_r\} \tag{10}$$

of the lattice of ideals of R onto the lattice of subsets of $\{1, ..., q\}$ with set theoretic inclusion as partial ordering. Clearly the mapping is a lattice isomorphism and since the lattice of subsets of $\{1, ..., q\}$ is distributive, so is V.

If V is a distributive lattice, the dual of (9) is valid:

$$A + (B \cap C) = (A + B) \cap (A + C) \tag{11}$$

We prove this as follows:

$$\begin{aligned}(A + B) \cap (A + C) &= [(A + B) \cap A] + [(A + B) \cap C] \\ &= A + [(A + B) \cap C] \\ &= A + (A \cap C) + (B \cap C) \\ &= A + (B \cap C)\end{aligned}$$

Dually, (9) follows from (11). ∎

In this context, let us examine once again the fundamental criterion for a subdirect sum to be dense (cf. Theorem III.2.3). If the lattice V of ideals in $\mathfrak{A}$ is distributive we may replace condition (10) of Theorem III.2.3 by

$$\bigcap_{2 \leqslant k \leqslant n} (A_{i_1} + A_{i_k}) = \mathfrak{A}$$

for every finite subset $\{i_1, ..., i_n\} \subseteq I$. This in turn is equivalent to requiring that

$$\mathfrak{A} = A_i + A_k, \qquad i \neq k, \quad i, k \in I$$

We have therefore proved the following:

Theorem 6. Hypothesis: Let the lattice V of ideals of the subdirect sum $\mathfrak{A} =$

$\sum_s \mathfrak{A}_i$ be distributive. Conclusion: $\mathfrak{A}$ is dense in $\sum_c \mathfrak{A}_i$ if and only if the kernels

$$A_i = \{a \mid a \in \mathfrak{A}, \quad a(i) = 0\}$$

are *comaximal*, i.e. if and only if

$$A_i + A_k = \mathfrak{A} \qquad \text{for every pair} \quad i, k \in I, \quad i \neq k. \quad \blacksquare$$

Let us apply this criterion to the ring of integers $\mathbb{Z}$. The lattice of ideals in $\mathbb{Z}$ is distributive and $\mathbb{Z}$ is a subdirect sum of fields $\mathbb{Z}/(p)$, p a prime. The kernels are the ideals (p), p a prime and clearly if $p \neq q$, $(p) + (q) = \mathbb{Z}$. Therefore $\mathbb{Z}$ is a dense subdirect sum of the fields $\mathbb{Z}/(p)$ (cf. Theorem IV.2.7).

Completely Reducible Modules

CHAPTER IV

The complete direct sum of a family M_i, $i \in I$ of R-modules is defined (as in the case of rings) to be the set of functions m from I to $\bigcup_{i \in I} M_i$ with $m(i) \in M_i$. On the other hand, the direct sum of the modules M_i, $i \in I$ is that submodule of the complete direct sum consisting of those functions m for which $m(i) = 0$ for all but finitely many $i \in I$. A module M is said to be completely reducible if it is the direct sum of irreducible modules.

The concept of complete reducibility is of particular importance for semisimple rings because of the following reason: In Theorem II.1.5 we characterized primitive rings as dense rings of linear transformations and in Theorem III.3.1 we showed that a semisimple artinian ring is a direct sum of finitely many simple rings each isomorphic to a full matrix ring over a division ring. In each case, the information provides us with a clear picture of the structure of these two classes of rings. However, the structure theorem for (Jacobson) semisimple rings (cf. Theorem III.2.1) does not provide us with as sharp a picture; in fact, as we have seen, a semisimple ring is only a subdirect sum of primitive rings and, if these primitive rings are not simple, (cf. Theorem III.2.5), the subdirect sum is not even dense in the complete direct sum. In the Jacobson density theorem for a primitive ring R, we saw that R is a dense subring of the ring $R^0(M)$ of K-linear mappings of the faithful, irreducible R-module M considered as a right K-module over its centralizer $K = \mathrm{Hom}_R(M, M)$. Following Bourbaki, we call $R^0(M)$ the double centralizer of the R-module M (see Definition IV.3.1 below). Now a semisimple ring R, as a subdirect sum of primitive rings R_i, is a subring of the complete direct sum of these rings. Each R_i has a faithful, irreducible

R_i-module M_i. Form the direct sum M of the $\mathbb{Z}$-modules M_i and turn this into an R-module by defining

$$am = \sum a_i m_i \qquad \text{for} \qquad a = (a_i \, . \, i \in I) \qquad \text{and} \qquad m = \sum m_i$$

Again, as before, we first form the centralizer K of the completely reducible R-module M and then the double centralizer $R^0(M)$ of K-linear mappings of M to itself. We shall see (Theorem IV.3.3) that we can describe the structure of $R^0(M)$ quite explicitly. Moreover, we shall be able to show that a semi-simple ring with identity is again dense in its double centralizer relative to the finite topology.

1. Direct sums of modules

Definition 1. Let R be a ring and $\{M_i \mid i \in I\}$ a family of R-modules. The *complete direct sum*

$$\textstyle\sum_c M_i \tag{1}$$

consists of the set of functions m defined on I with values $m(i) \in M_i$. Addition and the operation by elements of R are defined by

$$(m + m')(i) = m(i) + m'(i) \tag{2}$$

and

$$(am)i = a(m(i)) \qquad \text{for all} \quad a \in R \quad \text{and } m \in \textstyle\sum_c M_i \tag{3}$$

respectively.

Clearly $\sum_c M_i$ is an R-module.

Definition 2. Let R be a ring and $\{M_i \mid i \in I\}$ a family of R-modules. The (*discrete*) *direct sum*

$$\textstyle\sum \oplus M_i\,, \qquad i \in I \tag{4}$$

is the submodule of the complete direct sum consisting of those functions m which are nonzero for at most finitely many $i \in I$.

If I is a finite set, the complete and discrete direct sums coincide. It is for this reason that we used the notation of Definition 2 in Section III.3.

If all the R-modules M_i, $i \in I$ are submodules of a fixed R-module M, we shall denote the smallest R-module contained in M and containing each M_i by

$$\sum_{i \in I} M_i \tag{5}$$

It consists of sums

$$m_{i_1} + m_{i_2} + \cdots + m_{i_r} \tag{6}$$

where $m_{i_k} \in M_{i_k}$. If each element $m \in M$ is *uniquely* expressible as in (6), $\sum M_i$ is isomorphic to the direct sum of the M_i, $i \in I$ and conversely. In this case we shall write $\sum \oplus M_i$ for $\sum M_i$.

To familiarize ourselves with these concepts, let us take as an example a semisimple artinian ring.

A semisimple artinian ring is the direct sum

$$R = R_1 \oplus \cdots \oplus R_q \tag{7}$$

of finitely many simple artinian rings R_i. By Theorem II.3.3 the lattice V of left ideals L in a simple artinian ring is antiisomorphic to the lattice V' of K-subspaces U of a faithful irreducible R-module M over its centralizer $K = \mathrm{Hom}_R(M, M)$. Since $\dim_K(M) = n < \infty$, the 0-subspace may be represented as the intersection of n "hyperplanes," i.e. $(n - 1)$-dimensional K-subspaces U_i, $i = 1, ..., n$,

$$0 = U_1 \cap \cdots \cap U_n, \qquad \bigcap_{j \neq i} U_j \not\subseteq U_i, \qquad i = 1, ..., n \tag{8}$$

By the antiisomorphism, the U_i's correspond to the minimal left ideals

$$L_i = (0 : U_i), \qquad i = 1, ..., n$$

and we get

$$R = L_1 + \cdots + L_n, \qquad L_i \not\subseteq \sum_{j \neq i} L_j, \qquad i = 1, ..., n$$

and therefore $L_i \cap \sum_{j \neq i} L_j = 0$. This implies

$$_R R = L_1 \oplus \cdots \oplus L_n$$

and the left R-module $_R R$ is expressible as the direct sum of minimal left ideals. Conversely, a representation of $_R R$ as the direct sum of minimal left ideals gives rise to a representation of the 0-subspace of the K-space as the intersection of n' hyperplanes $U_i = (0 : L_i)$, $i = 1, ..., n'$. Therefore $n' = n$. Hence by (7) and Theorem II.3.5 we have the following:

Theorem 1. 1. A semisimple artinian ring R is a direct sum of minimal left ideals:

$$R = L_1 \oplus \cdots \oplus L_s \tag{9}$$

2. If R is simple artinian, the number of summands in (9) is equal to the

dimension n of a faithful irreducible R-module over its centralizer K. Hence

$$R \approx K_{n \times n}$$

where $K_{n \times n}$ is the ring of $n \times n$ matrices over K. ∎

In the following discussion we no longer require the condition $RM \neq (0)$ included in the definition of an irreducible R-module M. All we need is for M to have no submodule other than (0) and itself. We shall call these modules *minimal* R-modules.

Theorem 2. Let M be an R-module. The following are equivalent:

1. $M = \sum_{i \in I} M_i$, M_i minimal for all $i \in I$
2. $M = \sum_{\lambda \in \Lambda} \oplus M_\lambda$, M_λ minimal for all $\lambda \in \Lambda$
3. Every R-submodule U of M is a *direct summand*, i.e. there exists an R-module U' of M such that $M = U \oplus U'$. (We say the lattice of submodules of M is *complemented*.) ∎

In the proof that Part 1 implies Part 2 we may choose

$$\Lambda \subseteq I$$

We first prove a more general result:

Theorem 3. Let $M = \sum_{i \in I} M_i$, M_i minimal R-modules and let U be a submodule of M. Then there exists $\Lambda \subseteq I$ such that

$$M = U \oplus \sum_{\lambda \in \Lambda} \oplus M_\lambda \tag{10}$$

Proof. Let $\mathcal{S}$ be the family of subsets Λ of I such that the sum

$$U \oplus \sum_{\lambda \in \Lambda} \oplus M_\lambda$$

is direct. If $M_i \subseteq U$ for all $i \in I$, $U = M$ and there is nothing to prove. Otherwise there exists some M_i with $M_i \not\subseteq U$. By the minimality of M_i, $U_i \cap M_i = (0)$ and the sum $U \oplus M_i$ is direct, showing that $\mathcal{S} \neq \varnothing$. Let

$$\cdots \Lambda_i \subseteq \Lambda_k \subset \cdots \tag{$*$}$$

be a chain of subsets in $\mathcal{S}$. The union Λ_τ of these sets is in $\mathcal{S}$. For if $m \in U + \sum_{\lambda \in \Lambda_\tau} M_\lambda$ and if

$$m = u_1 + m_{i_1} + \cdots + m_{i_r}, \qquad u_1 \in U, \qquad m_{i_k} \in M_{i_k}$$

and

$$m = u_1' + m_{i_1'} + \cdots + m_{i_s'}, \qquad u_1' \in U, \qquad m_{i_k'} \in M_{i_k'}$$

are two representations of m as sums of elements from U and $\sum_{\lambda \in \Lambda_T} M_\lambda$, there exists Λ_k in the chain $(*)$ such that $\{i_1, ..., i_r, i_1', ..., i_s'\} \subseteq \Lambda_k$. Therefore $u_1 = u_1'$, $r = s$ and $m_{i_k} = m_{i_k'}$, $k = 1, ..., r$. By Zorn's lemma, there exists a maximal set Λ_0 in $\mathcal{S}$. Let $i_0 \in I$ be an arbitrary index. If $M_{i_0} \not\subseteq U \oplus \sum_{\lambda \in \Lambda_0} \oplus M_\lambda$,

$$M_{i_0} \cap \left(U \oplus \sum_{\lambda \in \Lambda_0} \oplus M_\lambda\right) = (0)$$

by minimality of M_{i_0}. Therefore

$$\left(U \oplus \sum_{\lambda \in \Lambda_0} M_\lambda\right) \oplus M_{i_0}$$

is a direct sum. But by associativity

$$\left(U \oplus \sum_{\lambda \in \Lambda_0} \oplus M_\lambda\right) + M_{i_0} = U \oplus \left(\sum_{\lambda \in \Lambda_0} M_\lambda \oplus M_{i_0}\right)$$

Contradicting the maximality of Λ_0. Therefore, for all $i \in I$,

$$M_i \subseteq U \oplus \sum_{\lambda \in \Lambda_0} \oplus M_\lambda$$

and hence

$$M = U \oplus \sum_{\lambda \in \Lambda_0} \oplus M_\lambda . \quad \blacksquare$$

Proof of Theorem 2. Part 1 implies Part 2 by Theorem 3 with $U = (0)$. Part 2 implies Part 3 again by Theorem 3. To prove that Part 3 implies Part 1 we must first show that every submodule $0 \neq U \subseteq M$ contains a minimal submodule of M. Let $0 \neq u \in U$. Then $W = \mathbb{Z}u + Ru$ is a submodule of U different from (0). The union of a chain of submodules of W not containing u is also a submodule of W not containing u. Hence by Zorn's lemma, there exists a submodule W' maximal in the family of submodules of W not containing u. Every submodule of W properly containing W' must contain u and hence W. Therefore W' is maximal in W. By Part 3, there exists $W'' \subseteq M$ such that $M = W' \oplus W''$. Let $N = W \cap W''$. Then

α. $W' \cap N = W' \cap W \cap W'' = (0)$ and

β. $W' + N = W' + (W'' \cap W) = (W' + W'') \cap W = M \cap W = W$

where the second equality in β is obtained by an application of the modular law (cf. Theorem II.3). Therefore

$$W = W' \oplus N$$

showing that N is a minimal submodule since W/W' is minimal by the maximality of W'. Now let S be the sum of all minimal R-submodules of M and let S' be a complement of S in M. Then

$$M = S \oplus S' \tag{11}$$

If $S' \neq (0)$ there exists a minimal submodule of M contained in S', contradicting (11). ∎

Definition 3. An R-module M is completely reducible if any one of the three equivalent conditions of Theorem 2 holds. A semisimple artinian ring R considered as a left R-module is completely reducible by Theorem 1. Now let M be a unital module over the semisimple artinian ring R. Since

$$M = \sum_{m \in M} RM \tag{12a}$$

and $R = L_1 \oplus \cdots \oplus L_s$ by Theorem 1,

$$M = \sum Lm \tag{12b}$$

where m ranges over M and L over the minimal left ideals of R. Since $1 \in R$, every minimal left ideal in R is irreducible. Therefore, for a fixed m and L, the R-homomorphism

$$l \to lm, \qquad l \in L \tag{13}$$

of L onto Lm is either an isomorphism (in which case Lm is irreducible) or the zero homomorphism. In the latter case we may omit Lm from (12b) and obtain a representation of M as a sum of irreducible modules. This proves the "only if" part of the following:

Theorem 4. An artinian ring with identity is semisimple if and only if each unital R-module is completely reducible. For the sufficiency it is enough to assume that $_RR$ is completely reducible. Furthermore, the irreducible R-modules are isomorphic to the minimal left ideals of R.

We still have to show that the condition is sufficient. Since R is artinian, this entails showing that there exist no nonzero nilpotent left ideals. By hypothesis, given a left ideal $L \neq (0)$, there exists L' such that $R = L \oplus L'$. In particular

$$1 = e + e' \qquad \text{with} \quad 0 \neq e \in L, \quad 0 \neq e' \in L'$$

It follows that $e = e^2 + ee'$ and hence $e - e^2 \in L' \cap L = (0)$. Therefore $e = e^2$. Since $0 \neq e = e^2 \in L$, L is not nilpotent. ∎

2. Homogeneous components of an R-module

Definition 1. Let R be a ring with identity. (Every minimal unital R-module is therefore irreducible.) An R-module U is *homogeneous of type I* if U is the direct sum of a family $\{M_\lambda \mid \lambda \in \Lambda\}$ of irreducible R-submodules M_λ each isomorphic to the irreducible module I.

Theorem 1. Let R be a ring with identity and M a completely reducible module. Let $M = \sum_{\lambda \in \Lambda} \oplus M_\lambda$ be a decomposition of M as the sum of irreducible submodules. Partition the set $\{M_\lambda \mid \lambda \in \Lambda\}$ into classes L_ρ, $\rho \in P$, of mutually isomorphic submodules. Let $H_\rho = \sum_{M_\lambda \in L_\rho} M_\lambda$. Then

$$M = \sum_{\alpha \in A} \oplus H_\alpha$$

The H_α are called the homogeneous components of M.

Proof. $M = \sum_{\rho \in P} \sum_{M_\lambda \in L_\rho} M_\lambda = \sum_{\alpha \in A} H_\alpha$. The H_α therefore generate M. The directness of the sum follows from

Theorem 2. Let R be a ring with identity. Let U be an irreducible R-submodule of the R-module

$$M = \sum \oplus M_i\,, \qquad M_i \text{ irreducible} \tag{1}$$

Then

$$U \subseteq M_{i_1} + \cdots + M_{i_r}$$

where

$$U \approx M_{i_k}\,, \qquad k = 1, \ldots, r$$

Conversely, let

$$U \subseteq M_{i_1} \oplus \cdots \oplus M_{i_r}$$

and let some element $u \in U$ be expressible as

$$u = u_{i_1} + \cdots + u_{i_r}\,, \qquad u_{i_k} \in M_{i_k}\,, \qquad k = 1, \ldots, r \tag{2}$$

with $u_{i_s} \neq 0$. Then $U \approx M_{i_s}$.

Proof. Let $0 \neq u \in U$. Then by the definition of sum,

$$u = u_{i_1} + \cdots + u_{i_r}\,, \qquad u_{i_k} \in M_{i_k}$$

with uniquely determined $u_{i_1}, ..., u_{i_r}$. Now

$$U = Ru \subseteq Ru_{i_1} + \cdots + Ru_{i_r}$$

By the irreducibility of the M_i, $Ru_{i_k} = M_{i_k}$ and therefore

$$U \subseteq M_{i_1} + \cdots + M_{i_r}$$

The mapping of U to M_{i_k} defined by

$$au \to u_{i_k}$$

is an isomorphism. [That it is well defined follows from the directness of (1).] Hence $U \approx M_{i_k}$. The converse is proved similarly. ▮

The following theorem gives an indication of the significance of the homogeneous components in the theory of completely reducible modules.

Theorem 3. Let R be a semisimple artinian ring and let

$$R = R_1 \oplus \cdots \oplus R_q \tag{3}$$

be a representation of R as a direct sum of simple artinian rings R_i.

1. The R_i are the homogeneous components of ${}_RR$ of type U_i, say. In other words, R_i is the sum of all minimal left ideals of R of type U_i and every minimal left ideal of R is contained in some R_i and is therefore of one of the types U_i.

2. Every irreducible R-module is faithful for exactly one R_j and is annihilated by the other components.

Proof. Let M be an irreducible R-module. Then $RM = M$. Therefore

$$M = R_1M + \cdots + R_qM \neq (0)$$

Hence $R_jM \neq (0)$ for some j. Since R_jM is an R-submodule, $R_jM = M$. Suppose if possible that $R_iM = M$, $i \neq j$. Then $R_iR_jM = R_jM = M$. But $R_iR_j \subseteq R_i \cap R_j = (0)$. Therefore $M = (0)$, a contradiction. Hence

$$R_jM = M \qquad \text{and} \qquad R_iM = (0), \qquad i \neq j \tag{4}$$

Moreover, M is a faithful R_j-module. For suppose for some $0 \neq a \in R_j$, $aM = (0)$. Then the set $S = \{b \mid b \in R_j, bM = (0)\} \neq (0)$. But S is an ideal and by simplicity of R_j, $S = R_j$. Therefore $R_jM = (0)$, a contradiction.

The representation ${}_RR = L_1 \oplus \cdots \oplus L_s$ as a direct sum of irreducible left ideals was obtained in Theorem IV.1.1 by the further decomposition of each R_i in (3) as the direct sum of minimal left ideals. If L is a minimal left ideal in R, there is some R_j in (3) such that

$$R_j \supseteq R_j L = L \qquad \text{by (4)}$$

Moreover, it is clear by (4) that any minimal left ideal R-isomorphic to L is contained in R_j and that R_j is the sum of isomorphic left ideals. ▮

Let M be a completely reducible R-module and let $\alpha \in K = \mathrm{Hom}_R(M, M)$. Each homogeneous component H of M is mapped into itself by α since an irreducible R-submodule U contained in H is either mapped to (0) or to a submodule isomorphic to U. By Theorem 2, $H\alpha \subseteq H$. In general, however, a submodule $T \neq (0)$ of H_1 is not mapped into itself by all $\alpha \in K$. For T contains an irreducible submodule M_1 (Theorem IV.1.3) which may be mapped isomorphically onto any irreducible submodule M_2 in H_1 by some isomorphism ϕ. This R-isomorphism ϕ may be extended to an R-homomorphism ϕ' of M into itself in the following way: As in Theorem IV.1.2, let

$$M = \sum_{k \in K} H_k \qquad \text{and} \qquad H_k = \sum_{\rho \in P_k} M_\rho, \quad M_\rho \text{ irreducible}$$

Let $W = M_2$. Then the mapping ϕ' defined by

$$\begin{aligned} H_k\phi' &= (0) && k \neq 1 \\ M_\rho\phi' &= (0) && \rho \neq 1, \quad \rho \in P_1 \\ m\phi' &= m\phi && \text{for} \quad m \in M_1 \end{aligned}$$

is an endomorphism of M. This discussion proves the following characterization of the homogeneous components of a completely reducible R-module.

Theorem 4. Let the unital R-module M be completely reducible and $K = \mathrm{Hom}_R(M, M)$ its centralizer. Then the homogeneous components of M are the smallest R-submodules H of M which are also right K-modules. ▮

The next four theorems are aimed at proving that if M is a completely reducible R-module, then M is also a completely reducible K-module where $K = \mathrm{Hom}_R(M, M)$.

Theorem 5. If R is a ring with identity and M a unital, completely reducible R-module with $M = \sum_{\nu \in \mathcal{N}} \oplus N_\nu = \sum_{\mu \in \mathcal{M}} \oplus M_\mu$, N_ν and M_μ irreducible, then card $\mathcal{M}$ = card $\mathcal{N}$.

Proof. If both sets $\mathcal{M}$ and $\mathcal{N}$ are finite, the theorem is an easy consequence of the Jordan–Hölder theorem (Theorem V.3.2). We therefore consider only the case where one of the sets, say $\mathcal{N}$, is infinite. Let $0 \neq X_\mu \in M$, $\mu \in \mathcal{M}$.

Then $X_\mu \in \sum_{\nu\in\Lambda} N_\nu$ for some finite subset Λ of $\mathcal{N}$. Since M_μ is irreducible, this implies $M_\mu \subseteq \sum_{\nu\in\Lambda} N_\nu$. By the directness of the sum, there exists a unique minimal finite subset $\Lambda(\mu)$ of $\mathcal{N}$ such that $M_\mu \subseteq \sum_{\nu\in\Lambda(\mu)} N_\nu$. This determines a mapping $\mu \to \Lambda(\mu)$. We claim that $\mathcal{N} = \bigcup_{\mu\in\mathcal{M}} \Lambda(\mu)$. Suppose $\nu_0 \notin \Lambda(\mu)$ for any $\mu \in \mathcal{M}$. We have $N_{\nu_0} \subseteq \sum_{i=1} M_{\mu_i}$ for suitable $\mu_i \in \mathcal{M}$. Hence

$$N_{\nu_0} \subseteq \sum \left\{ N_\nu \,\middle|\, \nu \in \bigcup_{i=1}^{k} \Lambda(\mu_i) \right\}$$

contrary to the directness of the sum $\sum_{\nu\in\mathcal{N}} \oplus N_\nu$. Therefore card $\mathcal{M}$ is greater than or equal to the cardinality of a set of finite subsets of $\mathcal{N}$ whose union is $\mathcal{N}$. The latter has the same cardinality as $\mathcal{N}$. Therefore card $\mathcal{M} \geqslant$ card $\mathcal{N}$. Reversing the argument we obtain card $\mathcal{N} \geqslant$ card $\mathcal{M}$ and the theorem is proved. ▮

Theorem 6. Let M be a completely reducible R-module, R a ring with identity. Then any isomorphism of an irreducible submodule M_1 into M may be extended to an automorphism, i.e. an isomorphism of M onto itself.

Proof. Let $M = \sum_{\rho\in P} \oplus H_\rho$ be the decomposition of M into its homogeneous components and suppose $M_1 \subseteq H_1$. Let σ_1 be an isomorphism of M_1 onto M_1', say. M_1' is also contained in H_1. Now by Theorem IV.1.3

$$H_1 = \Big(\sum_{\lambda\in\Lambda} \oplus M_\lambda\Big) \oplus M_1$$

and

$$H_1 = \Big(\sum_{\lambda'\in\Lambda'} \oplus M_\lambda'\Big) \oplus M_1'$$

By Theorem 5 card Λ = card Λ'. There is therefore a one-one correspondence between Λ and Λ', say $\lambda \leftrightarrow \lambda'$. Let $\sigma_\lambda\colon M_\lambda \to M_\lambda'$ be an isomorphism of M_λ onto M_λ' and let σ denote the resulting isomorphism of H_1 onto H_1. Clearly σ extends σ_1. Now define τ by

$$\begin{aligned} m_1\tau &= m_1\sigma, &\quad m_1 &\in H_1 & \\ m_i\tau &= m_i, &\quad m_i &\in H_i, &\quad i \neq 1 \end{aligned}$$

τ extends σ (and hence σ_1) and is an automorphism of M. ▮

Theorem 7. Let $1 \in R$ and let M be a unital, completely reducible R-module, K its centralizer. If $m \neq 0$ belongs to an irreducible submodule U of M, mK is an irreducible K-module.

Proof. Let $0 \neq m' \in mK$. Then there exists $\alpha \in K$ such that

$$m' = m\alpha$$

The mapping $u \to u\alpha$ of U onto $U\alpha$ is an isomorphism. By Theorem 6 this may be extended to an automorphism β of M. Therefore $m' = m\beta$ and hence $m'\beta^{-1} = m$ showing that

$$m'K = mK \quad \blacksquare$$

Theorem 8. Let $1 \in R$ and let M be a unital, completely reducible R-module. Then M is also a completely reducible K-module, $K = \mathrm{Hom}_R(M, M)$.

Proof. $M = \sum mK$ where m ranges over all nonzero m belonging to irreducible R-submodules. ▮

3. The double centralizer of a completely reducible module

Not all rings have a faithful completely reducible R-module M. When this is the case, the last theorem of the preceding section suggests that we investigate M as a right $K(= \mathrm{Hom}_R(M, M))$-module and again form $\mathrm{Hom}_K(M, M)$ with a view to determining its structure and the extent to which R "fills" the double centralizer $\mathrm{Hom}_K(M, M)$ (cf. the introduction to Chapter IV).

Definition 1. Let R be a ring and M a left R-module. Let $K = \mathrm{Hom}_R(M, M)$. The centralizer of the right K-module M is called the *double centralizer* of M and is denoted by $R^0(M)$ (or just R^0 when there is no risk of confusion). The elements of $R^0(M)$ are written on the left of M.

Of course

$$R \subseteq R^0 \qquad \text{for every faithful } R\text{-module } M$$

The following theorem provides us with a sufficient condition for an R-module isomorphism between two submodules of M to be an R^0-module isomorphism.

Theorem 1. Let the R-submodule U of M be a direct summand of M then

1. U is an R^0-submodule of M and
2. If T is an R-submodule of M, every R-homomorphism ϕ of U into T is also an R^0-homomorphism of U into T.

Proof. 1. Let $M = U \oplus U'$ and let $\pi \in K = \text{Hom}_R(M, M)$ be the homomorphism defined by

$$\pi: m = u + u' \to u \qquad \text{for} \quad u \in U, \quad u' \in U' \tag{1}$$

i.e. π is the *projection* of the R-module M onto the R-submodule U. Let $a \in R^0 = \text{Hom}_K(M, M)$. Then

$$aU = a(M\pi) = (aM)\,\pi \subseteq U$$

2. Let $\phi \in \text{Hom}_R(U, T)$ and form $\pi\phi$. Clearly

$$\pi\phi \in \text{Hom}_R(M, T) \subseteq \text{Hom}_R(M, M)$$

Hence for an arbitrary $a \in R^0$ and each $u = m\pi$ in U,

$$a(u\phi) = a(m\pi\phi) = (am)\,\pi\phi = (am\pi)\,\phi = (au)\phi \quad \blacksquare$$

If the R-module M is a direct sum of submodules each isomorphic to a module U_1 then using Theorem 1 we may show that $R^0(M)$ depends only on U and not on the number of summands occurring in the decomposition of M:

Theorem 2. Let the R-module M be a direct sum of isomorphic submodules U_i, $i \in I$. Let U_1 be one of the direct summands. By Theorem 1 U_1 is an R^0-submodule of M. Let $a \in R^0(M)$ and let a' denote the restriction of a to U_1, i.e.

$$a'u_1 = au_1 \qquad \text{for} \quad u_1 \in U_1 \tag{2}$$

Then the mapping

$$a \to a' \qquad \text{for} \quad a \in R^0(M) \tag{3}$$

is an isomorphism of $R^0(M)$ onto the double centralizer $R^0(U_1)$ of U_1.

Proof. 1. We assert that a' is in the double centralizer of U_1. Let $k_1 \in \text{Hom}_R(U_1, U_1)$. By Theorem 1 with $U = U_1 = T$, k_1 is also an R^0-homomorphism of U_1 and therefore for $a \in R^0(M)$

$$(au_1)\,k_1 = a(u_1k_1), \qquad u_1 \in U_1, \qquad k_1 \in \text{Hom}_R(U_1, U_1) \tag{4}$$

Hence $a' \in R^0(U_1)$.

2. The mapping $a \to a'$ is a monomorphism. Let

$$\phi_i: U_1 \to U_i$$

be an R-isomorphism of U_1 onto U_i and let u_i be an arbitrary element in U_i.

Let $u_i = u_1\phi_2$ for some element $u_1 \in U_1$. By Theorem 1 applied to $U = U_1$, $T = U_i$,

$$au_i = a(u_1\phi_i) = (au_1)\,\phi_i \tag{5}$$

Let a' be the zero of $R^0(U_1)$ for some $a \in R^0(M)$. Then $aU_1 = a'U_1 = 0$. By (5), $aU_i = 0$ for all $i \in I$. Since M is the sum of the U_i this implies $aM = (0)$ and hence $a = 0$. The mapping $a \to a'$ is therefore a monomorphism of

$$R^0(M) \qquad \text{into} \quad R^0(U_1)$$

3. To show that the mapping in (3) is onto we shall construct in (6) from a given element $b \in R^0(U_1)$ an element $b_i \in R^0(U_i)$ for each i using the isomorphism ϕ_i of U_1 onto U_i, $\phi_1 =$ identity. In (7) an element $a \in R^0(M)$ will be constructed from these b_i in such a way that under the mapping $a \to a'$, $a' = b$. Now for the calculations: Let $b \in R^0(U_1)$. We assert that the mapping

$$b_i\colon u_i \to (b[u_i\phi_i^{-1}])\,\phi_i \qquad \text{for} \quad u_i \in U_i \tag{6}$$

is an element of the double centralizer $R^0(U_i)$.

Proof. $k_1 \to \phi_i^{-1}k_1\phi_i$ is an isomorphism of $\mathrm{Hom}_R(U_1, U_1)$ onto $\mathrm{Hom}_R(U_i, U_i)$. Let $k_i \in \mathrm{Hom}_R(U_i, U_i)$ and let $k_i = \phi_i^{-1}k_1\phi_i$ for some $k_1 \in \mathrm{Hom}_R(U_1, U_1)$. Then by (6)

$$\begin{aligned} b_i(u_ik_i) &= (b[(u_ik_i)\,\phi_i^{-1}])\,\phi_i = (b[(u_i\phi_i^{-1}k_1\phi_i)\,\phi_i^{-1}]) \\ &= [b(u_i\phi_i^{-1}k_1)]\,\phi_i = ([b(u_i\phi_i^{-1})]\,k_1)\,\phi_i \\ &= ([(b_iu_i)\,\phi_i^{-1}]\,k_1)\phi_i = (b_iu_i)\,k_i \end{aligned}$$

For the given $b \in R^0(U_1)$ define the mapping a by

$$a \sum_{\lambda \in I} u_\lambda = \sum_{\lambda \in I} b_\lambda u_\lambda \tag{7}$$

We must show $a \in R^0(M)$. Since $M = \sum \oplus M_\lambda$ the mapping

$$\pi_i\colon \sum_{\lambda \in I} u_\lambda \to u_i$$

is a projection of the R-module M onto U_i. To a given $u \in U_i$ there exists $u_1 \in U_1$ such that $u_1\phi_i = u$. Let $k \in \mathrm{Hom}_R(M, M)$. Then by definition (6)

$$(au)\,k = ((au_1)\,\phi_i)\,k = ((b_1u_1)\,\phi_i)\,k = (bu_1\phi_i)k \tag{8}$$

On the other hand

$$a(uk) = a(u_1\phi_i k) = a \sum_\lambda (u_1\phi_i k\pi_\lambda) = \sum_\lambda b_\lambda(u_1\phi_i k\pi_\lambda)$$

and again by (6)

$$a(uk) = \sum_\lambda (b(u_1\phi_i k\pi_\lambda\phi_\lambda^{-1}))\, \phi_\lambda \tag{9}$$

Now $\phi_i k\pi_\lambda\phi_\lambda^{-1} \in \mathrm{Hom}_R(U_1, U_1)$ and therefore

$$a(uk) = \sum_\lambda (bu_1)\, \phi_i k\pi_\lambda\phi_\lambda^{-1}\phi_\lambda = (bu_1)\, \phi_i k = (au)k$$

The last equality follows from (8).

4. It is clear that the mapping in (3) is a homomorphism. ▮

If in particular R is a ring with identity and M a homogeneous R-module of type U, this theorem shows that the double centralizer of M is isomorphic to the double centralizer $R^0(U)$ of the module U. Since U is irreducible, the double centralizer of M is isomorphic to a ring of linear transformations of the vector space U over the division ring $\mathrm{Hom}_R(U, U)$. This proves the second assertion of the following:

Theorem 3. Hypotheses: Let R be a ring with identity. Let M be a completely reducible R-module and let $M = \sum_{\rho \in P} \oplus H_\rho$ be a decomposition of M into its homogeneous components H_ρ. Let H_ρ be of type U_ρ and let $R^0(M)$ and $R^0(H_\rho)$, $\rho \in P$, be the double centralizer of the R-modules M and H_ρ, respectively.

Conclusion: 1. $R^0(M)$ is isomorphic to the *complete direct sum* $\sum_{\rho \in P} R^0(H_\rho)$ of the rings $R^0(H_\rho)$.

2. $R^0(H_\rho)$ is (up to isomorphism) the ring $\mathrm{Hom}_{K_\rho}(U_\rho, U_\rho)$ of K_ρ-linear transformations of the vector space U_ρ over the division ring

$$K_\rho = \mathrm{Hom}_R(U_\rho, U_\rho)$$

Proof. Only Part 1 remains to be proved. The complete direct sum of rings was defined in Section III.2. By Theorem 1 the H_ρ are $R^0(M)$-submodules of M. Each element $a \in R^0(M)$ therefore maps H_ρ into itself. Denote the restriction of a to H_ρ by a_ρ. Then the mapping

$$a \to (a_\rho \mid \rho \in P) \qquad \text{for} \quad a \in R^0(M) \tag{10}$$

is a ring isomorphism of $R^0(M)$ into the complete direct sum of the double centralizers $R^0(H_\rho)$, $\rho \in P$. To show that the mapping in (10) is an epimorphism, let $(b_\rho \mid \rho \in P, b_\rho \in R^0(H_\rho))$ be an arbitrary element of $\sum_c R^0(H_\rho)$. Define b by

$$b\left(\sum_{\rho \in P} m_\rho\right) = \sum_\rho b_\rho m_\rho\,, \qquad m_\rho \in H_\rho \tag{11}$$

b is clearly an endomorphism of M. (Note that by the definition of a direct sum only finitely many m_ρ in (11) are different from zero.) We must show $b \in R^0(M)$. By Theorem IV.2.4 the H_ρ are K-modules, $K = \mathrm{Hom}_R(M, M)$. Hence for $k \in K$ it follows from Definition (11) that

$$b\left(\left(\sum m_\rho\right) k\right) = b\left(\sum m_\rho k\right) = \sum b_\rho(m_\rho k)$$
$$= \sum (b_\rho m_\rho)k = \left(\sum b_\rho m_\rho\right) k = \left(b \sum m_\rho\right) k$$

i.e. $b \in R^0(M)$. The image of b under the mapping in (10) is the given element $(b_\rho \mid \rho \in P)$. ▮

The precise description of the double centralizer $R^0(M)$ of a completely reducible R-module M enables us to determine the structure of its centralizer $K = \mathrm{Hom}_R(M, M)$ completely. To this end we first form the centralizer $\mathrm{Hom}_{R^0(M)}(M, M)$ of the $R^0(M)$-module M and prove that $\mathrm{Hom}_{R^0}(M, M) = \mathrm{Hom}_R(M, M)$.

Theorem 4. Let M be an R-module and $K = \mathrm{Hom}_R(M, M)$ its centralizer. Then the double centralizer $K^0(M)$ of the right K-module M is K.

Proof. Clearly $K \subseteq K^0(M)$. By definition, the centralizer of the K-module M is the double centralizer $R^0(M)$ of the R-module M. Hence for each $\eta \in K^0(M)$ and all $a \in R^0(M)$

$$(am)\,\eta = a(m\eta) \qquad \text{for} \quad m \in M$$

Therefore, in particular, the equality is valid for $a \in R$. Hence $\eta \in \mathrm{Hom}_R(M, M)$. ▮

By Theorem IV.2.8, a completely reducible R-module M is also completely reducible as a K-module, $K = \mathrm{Hom}_R(M, M)$. Therefore the last two theorems imply the following:

Theorem 5. Let R be a ring with identity and M a unital, completely reducible R-module. Then the centralizer of M is the complete direct sum of full rings of linear transformations of vector spaces U_ρ over division rings K_ρ, $\rho \in P$. ▮

In general the ring R is not equal to the double centralizer of an R-module. However, we may prove that under certain conditions R is dense in $R^0(M)$ (relative to the finite topology).

Theorem 6 (Bourbaki). Let R be a ring with identity and let M be a unital, completely reducible R-module. Then R is dense in $R^0(M)$, i.e. to each finite set of elements $x_1, ..., x_n \in M$ and to a given $a \in R^0(M)$ there exists $b \in R$ such that

$$bx_i = ax_i, \qquad i = 1, ..., n$$

Proof. By Theorem 3 the double centralizer $R^0(M)$ is the complete direct sum of the double centralizers of the homogeneous components of M. Therefore it is sufficient to prove Theorem 6 in the case that M is homogeneous. Let $T = M \oplus \cdots \oplus M$, the direct sum of n copies of M. To a given $a \in R^0(M)$ we define a mapping A of T into itself by

$$A(m_1, ..., m_n) = (am_1, ..., am_n) \tag{12}$$

$A \in R^0(M)$ since T is a homogeneous module and by Theorem 2 its double centralizer is isomorphic to $R^0(M)$ under the mapping constructed in (6) and (7). Now let X denote the R-submodule of T generated by the element $(x_1, ..., x_n)$. T is clearly a completely reducible R-module and therefore, by Theorem IV.1.2, X is a direct summand of T. By Theorem 1, X is also an $R^0(T)$-submodule of T. Therefore $A(x_1, ..., x_n) \in X$. But

$$X = \{(bx_1, ..., bx_n) \mid b \in R\}$$

Therefore there exists $b \in R$ such that

$$(bx_1, ..., bx_n) = A(x_1, ..., x_n) = (ax_1, ..., ax_n)$$

and Theorem 6 is proved. ∎

If R contains an identity and M is a faithful, irreducible R-module Theorem 4 reduces to the Jacobson density theorem on primitive rings. In this case the double centralizer is the full ring of linear transformations of the vector space M over the division ring $K = \mathrm{Hom}_R(M, M)$.

Theorem 7. If the ring R contains an identity, then ${}_RR$ is a completely reducible R-module if and only if R is artinian and semisimple.

Proof. The sufficiency of the condition has already been shown in Theorem IV.1.1. On the other hand, if ${}_RR$ is completely reducible, then

$$ {}_RR = \sum_{i \in I} \oplus L_i , \qquad L_i \text{ minimal left ideals} $$

In particular

$$ 1 = e_{i_1} + \cdots + e_{i_n} , \qquad e_{i_k} \in L_{i_k} $$

Therefore

$$ R = Re_{i_1} + \cdots + Re_{i_n} = L_{i_1} \oplus \cdots \oplus L_{i_n} $$

By Theorem IV.1.3 applied to $M = {}_RR$, every descending chain of left ideals in R stops after finitely many steps. Therefore R is artinian. On the other hand, by Theorem 6, R is the direct sum of finitely many full rings R_k of linear transformations of the vector spaces L_{i_k} over the division rings $K_k = \mathrm{Hom}_R(L_{i_k}, L_{i_k})$. This implies the simplicity of the artinian rings R_k. ∎

4. Modules with Boolean lattice of submodules

Let R be a ring with identity, M a faithful, unital, completely reducible R-module. Let

$$ M = \sum_{\rho \in P} \oplus H_\rho , \qquad H_\rho \quad \text{of type } U \tag{1} $$

be the decomposition of M into its homogeneous components. Let

$$ H_\rho = \sum_{\mu \in M} \oplus U_{\rho\mu} \tag{2} $$

be the decomposition of H_ρ into the direct sum of isomorphic irreducible submodules $U_{\rho\mu}$. By Theorem IV.3.3 the direct sum

$$ M^* = \sum_{\rho \in P} \oplus U_\rho \tag{3} $$

of mutually nonisomorphic irreducible R-modules is a faithful completely reducible R-module. The homogeneous components of M^* are the U_ρ.

By Theorem IV.1.3 the lattice $V(M^*)$ of R-submodules of M^* is *complemented*, i.e. to each $U \in V(M^*)$ there exists $V \in V(M^*)$ such that

$$ U \cap V = (0) \qquad \text{and} \qquad U + V = M^* $$

Theorem 1. Let R be a ring with identity and let M be a faithful, unital, completely reducible R-module whose decomposition is given in (1). Then the R-module M^* defined in (3) is also faithful and completely reducible. Moreover, the submodules of M^* form a Boolean lattice, i.e. a complemented lattice satisfying the distributive law:

$$A \cap (B + C) = (A \cap B) + (A \cap C), \qquad A, B, C \in V(M^*) \tag{4}$$

$V(M^*)$ is isomorphic to the lattice $\mathfrak{T}(P)$ of subsets of P with inclusion as partial ordering.

To complete the proof of Theorem 1 we first prove:

Theorem 2. Every submodule V and every factor module of a completely reducible module M is completely reducible.

Proof. Let the decomposition of M be as in (1). By Theorem IV.1.3, given a submodule V of M there exists a complementary R-submodule V'. Again by Theorem IV.1.3

$$M = V' \oplus \sum_{\lambda \in \Lambda} U_\lambda$$

But $V \approx M/V' \approx \sum_{\lambda \in \Lambda} \oplus U_\lambda$. Therefore V is completely reducible. By interchanging the roles of V and V' we get the complete reducibility of M/V. ∎

Because of Theorem 2 we now know that every R-submodule V of M^* is completely reducible. Let

$$V = \sum_{\lambda \in \Lambda} \oplus V_\lambda$$

be a representation of V as a direct sum of irreducible R-submodules V_λ. $V_\lambda \subseteq M^* = \sum_{\rho \in P} \oplus U_\rho$ and by Theorem IV.2.2 it follows that V_λ is one of the U_ρ since the homogeneous components of M^* are the U_ρ. Therefore

$$V = \sum_{\lambda \in \Lambda} V_\lambda = \sum_{\lambda \in \Lambda \subseteq P} U_\lambda$$

The mapping

$$\psi\colon V \to P_V = \{\rho \mid \rho \in P, \quad U_\rho \subseteq V\}$$

is a one-one mapping of the lattice $V(M^*)$ into the lattice $\mathfrak{T}(P)$ of subsets of P. Given $S \subseteq P$, the submodule

$$N = \sum_{\sigma \in S} \oplus U_\sigma$$

maps into S under ψ. Hence ψ is onto. Moreover ψ is a lattice isomorphism since

$$P_{V+V'} = P_V \cup P_{V'} \qquad \text{and} \qquad P_{V\cap V'} = P_V \cap P_{V'}$$

The subsets of a set P under set theoretic intersection and union form a Boolean lattice $\mathfrak{S}(P)$ and the theorem is proved. ▌

Definition 1. A ring is *distributively representable* if it has a module whose lattice of submodules is distributive. R is then said to be *distributively represented* by M. In terms of Definition 1 we have:

Theorem 3. If R is a ring (with identity) having a faithful completely reducible R-module M, then R is distributively representable by a module M^*. Moreover M^* may be chosen so that $V(M^*)$ is a Boolean lattice. ▌

Tensor Products, Fields, and Matrix Representations

CHAPTER V

In the first four chapters we dealt with the theory of semisimple rings, with particular emphasis on semisimple artinian rings. In Chapters VI–IX we shall investigate rings with radical different from zero and it is the purpose of this chapter to forge the tools necessary for the development of the theory. The reader will therefore find the following miscellaneous topics: polynomials in an arbitrary number of indeterminates, topics in field theory, and a treatment of the more elementary properties of the tensor product. In addition we shall discuss the representation of an algebra by means of matrices over the ground field while the deeper theory of tensor products will be developed in the next chapter.

1. Polynomials

The definition of the ring of polynomials in an arbitrary number of indeterminates over a commutative ring with identity may be based on the following:

Definition 1. Let S be a semigroup and Φ a commutative ring with identity. Then the *semigroup ring* $\Phi[S]$ consists of those functions

$$a\colon s \to \alpha_s \in \Phi, \qquad s \in S \tag{1}$$

defined on S with values in Φ which are zero for all but a finite number of $s \in S$. The sum and product of elements in $\Phi[S]$ are defined by

$$a + b \colon s \to \alpha_s + \beta_s \qquad \text{for} \quad s \in S \tag{2_1}$$

$$ab \colon s \to \sum_{pr=s} \alpha_p \beta_r \qquad \text{for} \quad s \in S \tag{2_2}$$

$\Phi[S]$ is a Φ-algebra in which S may be embedded by the monomorphism which sends $s_0 \in S$ to the function

$$s \to \alpha_s = \begin{cases} 1 & \text{if } s = s_0 \\ 0 & \text{if } s \neq s_0 \end{cases} \tag{3}$$

We denote this function by s_0 also. With this identification each element of $\Phi[S]$ is uniquely expressible as a sum

$$a = \sum_{s \in S} \alpha_s s \tag{4}$$

For the product we have

$$ab = \left(\sum_s \alpha_s s\right)\left(\sum_t \beta_t t\right) = \sum_{s,t} \alpha_s \beta_t st$$

$$= \sum_s \left(\sum_{pr=s} \alpha_p \beta_r\right) s \tag{5}$$

Now let $\mathfrak{X} = \{X_i \mid i \in I\}$ be a set and $S_{\mathfrak{X}}$ the free commutative semigroup with identity generated by $\mathfrak{X}$. It consists of all expressions of the form

$$X_{i_1}^{u_1} X_{i_2}^{u_2} \cdots X_{i_N}^{u_N} \tag{6}$$

where the $X_{i_j} \in \mathfrak{X}$ are finite in number and the u_i are nonnegative integers. The product of two such expressions is obtained by adding the exponents of the same symbols. The identity of $S_{\mathfrak{X}}$ is the element all of whose exponents are zero. The *degree* of the monomial in (6) is the sum $u_1 + \cdots + u_N$ of its exponents.

Definition 2. Let Φ be a commutative ring with identity. Let $\mathfrak{X} = \{X_i \mid i \in I\}$ be a set. The *ring* $\Phi[\mathfrak{X}]$ *of polynomials in the indeterminates* X_i, $i \in I$ is the semigroup ring $\Phi[S_{\mathfrak{X}}]$, where $S_{\mathfrak{X}}$ is the free commutative semigroup with identity generated by the X_i, $i \in I$.

By (4) each element $f \in \Phi[\mathfrak{X}]$ is uniquely expressible in the form

$$f = \sum \alpha_{i_1 i_2 \cdots i_N}^{(u_1 u_1 \cdots u_N)} X_{i_1}^{u_1} X_{i_2}^{u_2} \cdots X_{i_N}^{u_N} \tag{7}$$

with $\alpha \in \Phi$. The *degree of f* (deg f) is the maximum of the degrees of the monomials occurring in (7). The degree of the zero polynomial is set equal to $-\infty$; f is *normalized* if the monomial of the highest degree in f has its coefficient equal to $1 \in \phi$.

Definition 3. Let R be a ring containing in its center the ring Φ with identity. Let u be an element in R. Then the *value of the polynomial*

$$f = \alpha_0 + \alpha_1 X + \cdots + \alpha_n X^n \in \phi[X] \tag{8}$$

at $X = u$ is the element

$$f(u) = \alpha_0 + \alpha_1 u + \cdots + \alpha_n u^n \in R$$

Theorem 1. Let Φ be a ring with identity contained in the center of R. Let $a \in R$. The mapping

$$\eta_a : f \to f(a) \qquad \text{for} \quad f \in \Phi[X] \tag{9}$$

is a homomorphism of the ring $\Phi[X]$ to the ring R.

The proof is an immediate consequence of (2_1) and (2_2). ∎

Note: For polynomials f in more than one indeterminate we define the value of f at $(a) = (a_i \mid i \in I)$ analogously. However, if R is not commutative the mapping η_a is in general not multiplicative.

Theorem 2. If Φ contains no zero divisors, neither does $\Phi[X]$.

Proof. If Φ contains no zero divisors, the product of the monomials in f and g of highest degree is the monomial of highest degree in fg. ∎

Theorem 3. If the commutative ring Φ with identity has no divisors of zero then neither has any polynomial ring $\Phi[\mathfrak{X}]$ in an arbitrary number of indeterminates.

Proof. Let f and g be in $\Phi[\mathfrak{X}]$. Only a finite number of indeterminates occur in f and only a finite number in g. Suppose $\{X_1, ..., X_N\}$ is the set of indeterminates occurring either in f or g. Then the assertion of the theorem follows by induction and Theorem 2 since

$$\Phi[X_1, ..., X_n] \approx \Phi[X_1, ..., X_{n-1}][X_n]. \quad ∎$$

If Φ is a field we construct the quotient field of $\Phi[\mathfrak{X}]$ in the same way the field of rationals is constructed from the ring of integers. More generally we have:

Theorem 4. Let R be a commutative ring with identity having no zero divisors. (Such rings are called *domains*.) Define on the set of pairs (a, b), $a, b \in R$, $b \neq 0$, an equivalence relation by

$$(a, b) \sim (a_1, b_1) \qquad \text{if and only if} \quad ab_1 = ba_1 \tag{10}$$

Let a/b denote the equivalence class of (a, b). Define on the set of these equivalence classes two operations as follows:

$$\frac{a}{b} \cdot \frac{c}{d} = \frac{ac}{bd} \qquad \text{and} \qquad \frac{a}{b} + \frac{c}{d} = \frac{ad + bc}{bd} \tag{11}$$

The set of equivalence classes together with these two operations form a field Q. The mapping $a \to a/1$, $a \in R$, is an embedding of R in Q. Q is called the *quotient field of R*. ∎

The proof depends on the fact that if b and d are different from zero so is bd. It is easily shown that the definitions of the operations in (11) are independent of the representatives of the equivalence classes. If $a \neq 0$,

$$a/b \cdot b/a = 1$$

Theorem 4 will be generalized in a suitable way to noncommutative rings in Theorem XI.1.

Definition 4. If we set $R = \Phi[\mathfrak{X}]$ in Theorem 4, where Φ is a field, we obtain the field of polynomial quotients in the indeterminates $\mathfrak{X} = \{X_i \mid i \in I\}$ with coefficients in Φ. We denote this by $\Phi(\mathfrak{X})$.

Theorem 5. The ring $R = \Phi[X]$ of polynomials in one indeterminate X over a field Φ is a principal ideal domain, i.e. each ideal in $\Phi[\mathfrak{X}]$ can be generated by one element.

Proof. Let $\mathfrak{A}$ be an ideal in R and $a \neq 0$ an element of lowest degree in $\mathfrak{A}$. Let $b \in \mathfrak{A}$. Then there are polynomials q and r in $R = \Phi[X]$ such that

$$b = aq + r \tag{12}$$

where either r is the zero polynomial or the degree of r is less than the degree of a. (This is the so-called division algorithm.) Since $a, b \in \Phi[X]$,

$r = b - aq \in \mathfrak{A}$. It follows that $r = 0$ since otherwise $\deg r < \deg a$, contradicting the choice of a. Hence $\mathfrak{A}$ consists of polynomials which are multiples of a. ▮

Theorem 6. Let Φ be a field and R the principal ideal domain $\Phi[X]$. A polynomial $f \in \Phi[X]$ is said to be Φ-reducible if there are polynomials g and h in $\Phi[X]$ with $\deg g < \deg f$, $\deg h < \deg f$, such that $f = gh$. Otherwise we say f is Φ-irreducible. The following are equivalent:

1. (f) is a maximal ideal in R
2. $R/(f)$ is a field
3. f is Φ-irreducible.

Proof. The equivalence of Parts 1 and 2 is a consequence of Theorems II.7.4–II.7.7. Calculating modulo (f), $f = gh$ implies $\bar{g}\bar{h} = \bar{0}$. Therefore Part 2 implies Part 3. If (f) is not a maximal ideal then there exists $g \in \Phi[X]$ with $(f) \subset (g) \subset (1) = R$. Hence $f = gh$, where h is a polynomial of positive degree, showing that f is Φ-reducible. Therefore Part 3 implies Part 1. ▮

The division algorithm introduced in (12) allows us to compute the greatest common divisor of f and g as follows: If g is not a divisor of f in $K[X]$, K a field, there exist polynomials q_i and r_i in $K[X]$ such that $\deg r_2 < \deg g$ and $\deg r_{i+1} < \deg r_i$, and satisfying

$$\begin{aligned} f &= gq_1 + r_2 \\ g &= r_2 q_2 + r_3 \\ r_2 &= r_3 q_3 + r_4 \\ &\ \vdots \\ r_{N-3} &= r_{N-2} q_{N-2} + r_{N-1} \\ r_{N-2} &= r_{N-1} q_{N-1} + r_N \\ r_{N-1} &= r_N q_N \end{aligned} \tag{13}$$

This process (the so-called Euclidean algorithm) ends after a finite number of steps since the degree of the remainders r_i decreases at each step. Now r_N is a divisor of r_{N-1} in $K[X]$, hence also of r_{N-2} and climbing up step by step, r_N is a divisor of both f and g. On the other hand any divisor of both f and g in $K[X]$ is also a divisor of r_2 and hence of r_3. Again, this time climbing down step by step, we have that any common divisor of both f and g is a divisor of r_N. This common divisor r_N is unique up to unit factors and by (13) is expressible in the form

$$r_N = uf + vg \qquad \text{with} \quad u, v \in K[X] \tag{14}$$

2. Fields

Let K be a division ring with identity e. Then the mapping

$$\sigma\colon z \to ze, \qquad z \in \mathbb{Z}$$

is a homomorphism of the ring $\mathbb{Z}$ of integers into the ring K. Since K has no zero divisors, the kernel of σ is either (0) or a principal ideal (p) generated by a prime p. The following definition therefore makes sense:

Definition 1. The *characteristic* char K of a division ring K is 0 if K contains a subfield Φ isomorphic to the rationals $\mathbb{Q}$. Char K is p if K contains a subfield Φ isomorphic to the prime field $C_p = \mathbb{Z}/(p)$. In either case Φ is the smallest subfield of K. It is the intersection of all subfields of K.

In the following let K be a field.

Definition 2. Let K be a subfield of the field L and let $\alpha \in L$. The element α is *transcendental over* K (K-transcendental), if the set $\{1, \alpha, \alpha^2, ...\}$ is K-linearly independent. If $\{1, \alpha, \alpha^2, ...\}$ is linearly dependent over K then α is *algebraic over* K. In this case, let

$$f(X) = X^n + a_1X^{n-1} + \cdots + a_n$$

be the polynomial of least degree with leading coefficient 1 for which $f(\alpha) = 0$. Then f is called the *minimal polynomial* of α over K. If α is transcendental over K, set the minimal polynomial of α over K equal to the zero polynomial.

The minimal polynomial f of α over K is uniquely determined. In fact, if f_1 and f_2 are minimal polynomials of α over K, then by the division algorithm there exist q_i, r_i such that

$$f_1 = f_2q_1 + r_1$$

and

$$f_2 = f_1q_2 + r_2$$

But $f_1(\alpha) = f_2(\alpha) = r_1(\alpha) = r_2(\alpha) = 0$. Hence $f_1 = f_2q_1$ and $f_2 = f_1q_2$. Therefore if $f_1 \neq 0$, $q_1q_2 = 1$ showing that f_1 and f_2 differ by a unit. In the definition above we took the minimal polynomial to be monic, i.e. with leading coefficient 1. Hence $q_1 = q_2 = 1$. Therefore $f_1 = f_2$.

Theorem 1. Let K be a subfield of a field L and let $\alpha \in L$ be algebraic over K with f as minimal polynomial. Then the smallest subfield $K(\alpha)$ of L containing K and α is isomorphic to $K[X]/(f)$ and f is irreducible.

Proof. By Theorem V.1.1,

$$\eta_\alpha : g \to g(\alpha), \qquad g \in K[X] \tag{1}$$

is an epimorphism of $K[X]$ onto $K[\alpha]$. The kernel is the principal ideal (f) in $K[X]$ generated by f. Since L has no zero divisors, f is K-irreducible and hence by Theorem V.1.6, $K[X]/(f)$ is a field. This shows $K(\alpha) = K[\alpha]$. ∎

Conversely, let f be a given K-irreducible polynomial. Then $K[X]/(f)$ is a field L and the mapping

$$j: k \to kX^0 + (f), \qquad k \in K$$

is an embedding of K in L. Identify the image of K in L with K. For $g \in K[X]$, let $\bar{g}$ be the residue class of g modulo (f). Then the polynomial $f \in K[X]$ evaluated at $\bar{X}$ is

$$f(\bar{X}) = \overline{f(X)} = \bar{0}$$

Hence f has a root $\bar{X}$ in L and L is generated by $\bar{X}$ over K. This proves:

Theorem 2. Let f be a polynomial in X irreducible over K. Then, up to isomorphism, there is a uniquely determined extension L of K containing a root α of f and generated by α over K, i.e. $L = K(\alpha)$. Moreover, $L \approx K[X]/(f)$. ∎

If the polynomial $f \in L[X]$ has a root α in L then $X - \alpha$ is a factor of f in $L[X]$ since by the division algorithm, $f = (X - \alpha)\, g + r$, $\deg r \leqslant 0$ and hence $r(\alpha) = 0$ showing that $r = 0$. Now let f_1 be an L-irreducible factor of g in $L[X]$. f_1 exists since either g itself is irreducible or g can be factored into two factors of lower degree than $\deg g$ in $L[X]$. Applying Theorem 2 to f_1 we obtain an extension field $L(\alpha_2) \approx L[X]/(f_1)$ of L. $L(\alpha_2)$ is of course an extension of K. Proceeding in this fashion we eventually obtain an extension field E of K over which f factors completely into linear factors:

$$f = (X - \alpha_1)(X - \alpha_2) \cdots (X - \alpha_n) \qquad \text{in} \quad E[X]$$

The smallest field contained in E and containing K and $\alpha_1, \ldots, \alpha_n$ is called the *splitting field* of f over K. In the following we prove that the splitting field of a polynomial is unique up to isomorphism.

Theorem 3. Let ϕ be an isomorphism of the field K onto the field K^*. Then the mapping

$$\phi' : a_0 + a_1 X + \cdots + a_t X^t \to (\phi a_0) + (\phi a_1)\, X + \cdots + (\phi a_t)\, X^t \tag{2}$$

is an isomorphism of $K[X]$ onto $K^*[X]$. Let $f(x) \in K[X]$ and $f^* = \phi' f$ and let E be a splitting field of f over K and E^* a splitting field of f^* over K^*. Then the isomorphism ϕ of K onto K^* can be extended to an isomorphism ψ of E onto E^*, i.e.

$$\psi k = \phi k \qquad \text{for all} \quad k \in K$$

Proof. Clearly, ϕ' is an isomorphism. If $E = K$, $E^* = K^*$ since

$$\sum_i (\phi a_i)(\phi a)^i = \phi \sum a_i \alpha^i = \phi 0 = 0$$

whenever $f(\alpha) = 0$. If $f = f_1 \cdots f_s$ is a decomposition of f into K-irreducible factors, then $f^* = f_1^* \cdots f_s^*$, where $f_i^* = \phi' f_i$, $i = 1, \ldots, s$ and the f_i^* are also K^*-irreducible. Suppose $\deg f_1 \geqslant 2$, $f_1(\alpha_1) = 0$ and $f_1^*(\alpha_1^*) = 0$, where $\alpha_1 \in E$ and $\alpha_1 \in E^*$. By Theorem 2 the fields $K(\alpha_1) \subseteq E$ and $K^*(\alpha_1^*) \subseteq E^*$ are isomorphic to $K[X]/(f_1)$ and $K^*[X]/(f_1^*)$, respectively. By (2), $K[X]/(f_1) \approx K^*[X]/(f_1^*)$ under the isomorphism

$$\psi_1 : g(\alpha_1) \rightarrow g^*(\alpha_1^*) \tag{3}$$

where $g^* = \phi' g$. ψ_1 is therefore an extension of ϕ. Next suppose there are $k > 0$ roots α_i of f in E but not in K. We make the inductive hypothesis that for every polynomial with coefficients in a field Φ which has fewer than k roots outside Φ in a splitting field Ψ containing Φ, we can extend every isomorphism of Φ to an isomorphism of the splitting field Ψ.

Now E and E^* are clearly splitting fields for f and $\phi' f$ over $K(\alpha_1)$ and $K^*(\alpha_1^*)$, respectively. But f has fewer than k roots in E outside $K(\alpha_1)$. By induction, there exists an isomorphism τ_1 of E onto E^* which is an extension of ϕ. ∎

If we set $K = K^*$ in the above we have that the splitting field of a polynomial is unique up to isomorphism.

To each polynomial $f \in K[X]$ there exists an extension field L of K such that f factors completely into linear factors over L. For example, the splitting field of f will do for L. The same result is true if instead of just one polynomial we consider a finite family $\mathfrak{F}$ of polynomials in $K[X]$. In this case we merely form the product of the polynomials in $\mathfrak{F}$ and construct the splitting field of this product.

For infinite families of polynomials we prove the following:

Theorem 4 (Steinitz). Every field can be embedded in an *algebraically closed* field L, i.e. a field over which every polynomial in $K[X]$ of positive degree splits completely into linear factors over L. L is also *algebraically closed* over L, i.e. every polynomial of positive degree in $L[X]$ splits completely into linear factors in $L[X]$.

Proof. Let $\mathfrak{F} = \{f\}$ be an arbitrary family of polynomials in $K[X]$. Denote the degree of f by $|f|$. Let

$$R = K[Y_{fi} \mid 1 \leqslant i \leqslant |f|, f \in \mathfrak{F}] \tag{4}$$

R is the ring of polynomials in the indeterminates Y_{fi}, where we have $|f|$ of them for each $f \in \mathfrak{F}$. In particular R contains the elementary symmetric polynomials

$$S_{fi} = \sum_{1 \leqslant j_1 \leqslant \cdots \leqslant j_i \leqslant |f|} Y_{fj_1} \cdots Y_{fj_i}, \qquad 1 \leqslant i \leqslant |f| \tag{5}$$

For a fixed f these polynomials may be defined by

$$\sum_{0 \leqslant i \leqslant |f|} (-1)^i S_{fi} X^{|f|-i} = \prod_{1 \leqslant k \leqslant |f|} (X - Y_{fk}) \tag{6}$$

where we set $S_{f0} = 1$. Now let

$$f(x) = \sum_{0 \leqslant i \leqslant |f|} (-1)^i a_{fi} X^{|f|-1}, \qquad a_{fi} \in K \tag{7}$$

Suppose that the f have been normalized so that $a_{f0} = 1$. Let $\mathfrak{A}$ be the ideal generated by the polynomials $S_{fi} - a_{fi}$, $0 \leqslant i \leqslant |f|$, $f \in \mathfrak{F}$. It consists of polynomials of the form

$$\sum_{i,f} g_{fi}[S_{fi} - a_{fi}], \qquad g_{fi} \in R \tag{8}$$

where the sum extends over finitely many i and f. This ideal is not the whole of R. In fact the only element in common with both $\mathfrak{A}$ and K is 0. Otherwise, suppose $0 \neq k \in \mathfrak{A} \cap K$. Then

$$k = \sum_{1 \leqslant j \leqslant N} g_{f_j i_j}[S_{f_j i_j} - a_{f_j i_j}] \tag{9}$$

In the splitting field L_N of the polynomial $f_1(X) f_2(X) \cdots f_N(X)$ over K, let $\alpha_{f_j 1}, \ldots, \alpha_{f_j |f_j|}$ be the roots of $f_j(X)$, $j = 1, \ldots, N$. Then in L_N, by the very definition (6) of the elementary symmetric functions we have

$$a_{f_j i_j} = S_{f_j i_j}(\alpha_{f_j 1}, \ldots, \alpha_{f_j |f_j|})$$

Substituting α_{fj_i} for Y_{fj_i} in (9) we have $k = 0$, a contradiction. Since R contains an identity there is a maximal ideal $\mathfrak{A}_\omega$ in R containing $\mathfrak{A}$. The factor ring $L = R/\mathfrak{A}_\omega$ is a field since it is a commutative simple ring with identity. Set $\bar{Y}_{fi} = Y_{fi} + \mathfrak{A}_\omega$. Since $s_{fi} - a_{fi} \in \mathfrak{A}_\omega$ and using (6) we have the following equality in $L[X]$:

$$f(X) = \sum_{0 \leqslant i \leqslant |f|} (-1)^i a_{fi} X^{|f|-i} = \prod_{1 \leqslant k \leqslant |f|} (X - \bar{Y}_{fk})$$

for all $f \in \mathfrak{F}$. The field K can be embedded in L since $\mathfrak{A}_\omega \cap K = 0$ (otherwise $\mathfrak{A}_\omega = R$). The fact that L is algebraically closed over L is a consequence of the following: Let x be a root of the polynomial

$$\lambda_0 + \lambda_1 X + \cdots + \lambda_n X^n \in L[X]$$

and set $A = K(\lambda_0, ..., \lambda_n)$. Then $\dim_K A(\alpha) \leqslant n \dim_K A < \infty$ since $\lambda_0, ..., \lambda_n$ are algebraic over K. Hence α is algebraic over K and is therefore already in L. ∎

Theorem 3 enables us to prove an important theorem about finite fields, i.e. fields containing only a finite number of elements. Clearly a finite field K cannot contain a copy of the rationals. Hence by Theorem 1 it must contain a copy of $C_p = \mathbb{Z}/(p)$, p a prime. K is a finite-dimensional vector space over C_p and therefore the number of elements in K, or the *order* of K, $o(K)$, is a power of p: $o(K) = p^n$.

The multiplicative subgroup $K^\times$ of nonzero elements of K contains $p^n - 1$ elements. Now the order of every subgroup U of a group G is a divisor of the order of the group $o(G)$ as can be seen by decomposing the group G into its left cosets aU. In particular the order of an element a of a group (i.e. the order of the subgroup generated by a) is a divisor of $p^n - 1$. Hence each element of $K^\times$ satisfies the equation

$$X^{p^n-1} = 1$$

Therefore every element of K is a root of the polynomial

$$X^{p^n} - X \in C_p[X] \tag{10}$$

On the other hand, a polynomial $f(X)$ of degree N has at most N roots $\alpha_1, ..., \alpha_N$ in a field as may be seen by an application of the division algorithm. Therefore the finite commutative field K consists of the roots of the polynomial in (10) and is therefore the splitting field of this polynomial over C_p. By Theorem 3, K is uniquely determined up to isomorphism. We have therefore proved:

Theorem 5. Every finite field K is of p power order, p a prime and is the splitting field of the polynomial $X^{p^n} - X$, where $o(K) = p^n$. Moreover, for a given p^n, p a prime, there is essentially only one field of order p^n. Finite fields are called *Galois fields*. ∎

We may easily show that given p^n, p a prime, there is a finite field of that order. This will not be needed in the sequel. However, later on we shall need to know that every finite field contains a *primitive element*, i.e. an element θ such that

$$K = C_p(\theta) \tag{11}$$

where by $C_p(\theta)$ we mean the smallest subfield of K containing C_p and θ. We prove this result by proving:

Theorem 6. The multiplicative group $K^\times$ of a finite field K is cyclic. Any generator θ of $K^\times$ is called a *primitive element* of K over C_p. ▮

We first prove the following:

Lemma. Let G be an abelian group and let a and b be elements of G of finite order, $o(a) = \alpha$ and $o(b) = \beta$. Then there is an element c in G whose order is the least common multiple (lcm) of α and β:

$$o(c) = \operatorname{lcm}\{\alpha, \beta\} \tag{12}$$

Proof. Let G be written multiplicatively and let e be the identity.

1. Let (α, λ) denote the greatest common divisor (gcd) of α and λ. Then

$$\alpha = (\alpha, \lambda)\, \alpha' \qquad \text{and} \qquad \lambda = (\alpha, \lambda)\, \lambda' \qquad \text{with} \quad (\alpha', \lambda') = 1$$

Therefore

$$a^\lambda = e \Leftrightarrow \alpha \mid \lambda\xi \Leftrightarrow \alpha' \mid \lambda'\xi \Leftrightarrow \alpha' \mid \xi$$

(Here by "$\alpha \mid \beta$" we mean "α divides β.") Therefore

$$o(\alpha^\lambda) = \alpha/(\alpha, \lambda) \tag{13}$$

2. Suppose $(\alpha, \beta) = 1$. Since G is commutative, $c^{\alpha\beta} = e$, where $c = ab$. Hence $\gamma = o(c) \mid \alpha \cdot \beta$. On the other hand by Part 1

$$\gamma/(\gamma, \alpha) = o(c^\alpha) = o(a^\alpha b^\beta) = o(b^\alpha) = \beta/(\beta, \alpha) = \beta \tag{14}$$

and hence $\beta \mid \gamma$. Similarly $\alpha \mid \gamma$. Since $(\alpha, \beta) = 1$, $\alpha\beta \mid \gamma$.

3. Let $\alpha = o(a) = \prod_i p_i^{\rho_i}$ and $\beta = o(b) = \prod_i p_i^{\sigma_i}$ with ρ_i, $\sigma_i \in 0 \cup N$, p_i prime and $p_i \neq p_k$. Then setting $\alpha_i = \alpha p_i^{-\rho_i}$, $\beta_i = \beta \cdot p_i^{-\sigma_i}$, a^{α_i} and b^{β_i} have orders $p_i^{\rho_i}$ and $p_i^{\sigma_i}$, respectively. Thus one of these two elements, say c_i has order $p_i^{\max(\rho_i, \sigma_i)}$. By Part 2 the product of these elements c_i has order the lcm of α and β. ▮

Proof of Theorem 6. Let $\theta \in K^\times$ be an element of maximal order and let k_0

be any other element of $K^\times$. Then $o(k_0) \mid o(\theta)$ since otherwise by the lemma we could find an element of order greater than $o(\theta)$. Hence every element of $K^\times$ is a root of the polynomial $X^{o(\theta)} - 1 \in C_p[X]$. This implies that $K^\times$ contains at most $o(\theta)$ elements. Therefore $o(K^\times) = o(\theta)$ and θ generates $K^\times$. ∎

Let L be a finite-dimensional extension of K. We may prove the existence of a primitive element in L if L is a separable extension of K in the sense of the following:

Definition 1. Let K be a field. 1. A *K-irreducible polynomial* $f(X)$ in $K[X]$ is *separable over* K if $f(X)$ has no repeated linear factors in any splitting field of f.

2. $f(X)$ in $K[X]$ is separable over K if its K-irreducible factors are separable.

3. An *element* α in $L \supseteq K$ is separable over K if α is algebraic over K and the minimal polynomial of α over K is separable over K.

4. An *extension* L of K with $\dim_K L < \infty$ is *separable over* K if each element of L is separable over K.

If $\dim_K L = n < \infty$ we say L is finite over K. In this case all elements of L are algebraic over K since if $\alpha \in L$, $\{1, \alpha, \alpha^2, ..., \alpha^n\}$ is linearly dependent over K.

Theorem 7. Let L be a separable extension of K. Then there exists a (separable) primitive element θ in L, i.e. $L = K(\theta)$ for some $\theta \in L$.

Proof. If K contains a finite number of elements, a finite-dimensional extension L of K also contains only a finite number of elements. In this case the existence of θ is guaranteed by Theorem 6. Moreover, every $\alpha \in L$, is a root of

$$X^{p^n} - X \in C_p[X] \tag{15}$$

If $X - \alpha$ is a repeated factor of (15), then

$$X^{p^n} - X = (X - \alpha)^2 h(X), \qquad h(X) \in L[X]$$

Differentiating formally with respect to X we have

$$-1X^0 = p^n X^{p^n - 1} - 1 = 2(X - \alpha)h(X) + (X - \alpha)^2 h'(X)$$

showing that $X - \alpha$ is a factor of $-1X^0$. This contradiction establishes that $X^{p^n} - X$ has no repeated linear factors in any extension field. But the minimal polynomial of α over C_p divides $X^{p^n} - X$ and therefore it is separable. (*Note*: It is easily shown that the ordinary product rule of the calculus for the differentiation of the product of two functions holds also for the formal differentiation of the product of two polynomials.)

2. Suppose now that K contains infinitely many elements. Since $\dim_K L < \infty$, L is obtained by the successive adjunction to K of a finite number of algebraic elements. Therefore the theorem will follow by induction if we can show the existence of a primitive element when $L = K(\alpha, \beta)$. Let $f(X)$ and $g(X)$ be the minimal polynomials of α and β, respectively over K. Let M be the splitting field of $f(X)\, g(X)$ over L and let $\alpha = \alpha_1, ..., \alpha_r$ and $\beta = \beta_1, ..., \beta_s$ be the roots of f and g in M, respectively. Since α and β are separable both the α_i and the β_i are distinct. Since K contains infinitely many elements there exists $c \in K$ such that

$$\alpha_i + c\beta_j \neq \alpha + c\beta \qquad \text{for all} \quad (i, j) \neq (1, 1) \tag{16}$$

We show that $\theta = \alpha + c\beta$ is the element we are looking for: Clearly $K(\theta) \subseteq L$. The polynomial

$$f_1(X) = f(\theta - cX) \in K(\theta)[X]$$

has β as a root since $f(\alpha) = 0$. f_1 and g have no further common roots in M since, if

$$g(\beta_j) = 0 \qquad \text{and} \qquad f_1(\beta_j) = f(\theta - c\beta_j) = 0$$

then $\theta - c\beta_j = \alpha_{i_0}$ for some i_0, contradicting (16). Hence the greatest common divisor of $f_1(X)$ and $g(X)$ in M is $X - \beta$. But, using the fact that the greatest common divisor of f_1 and g in $K(\theta)[X]$ is expressible in the form $uf_1 + vg$ (cf. Section 1, Eq. (14)), it follows that $X - \beta \in K(\theta)[X]$. Therefore $\beta \in K(\theta)$. But $c\beta$ and $\alpha + c\beta \in K(\theta)$. Hence $\alpha + c\beta - c\beta = \alpha \in K(\theta)$ and therefore $L = K(\alpha, \beta) \subseteq K(\theta) \subseteq L$. Since L is separable over K, θ is a separable element.

Not all extensions are separable. As a counterexample, consider the following: Let $K = C_p(Y)$, the quotient field of $C_p[Y]$. Let L be the splitting field of $X^p - Y \in K[X]$ and let α be a root of this polynomial in L. Thus

$$X^p - Y = X^p - \alpha^p = (X - \alpha)^p \qquad \text{in} \quad L[X] \tag{17}$$

where the last equality follows from the binomial theorem and the fact that all the binomial coefficients $\binom{p}{n}$ except for $\binom{p}{0}$ and $\binom{p}{p}$ are divisible by p, the characteristic of L. Now α does not lie in K; for if $\alpha \in K$, then $\alpha = f(Y)[g(Y)]^{-1}$, hence $\alpha^p = Y = f(Y)^p\, g(Y)^{-p}$ and $p \deg f - p \deg g = p[\deg f - \deg g] = 1$, a contradiction. Now let $h(X)$ be a K-irreducible factor of $X^p - Y$ with leading coefficient 1. Then $X^p - Y$ is a power of $h(X)$:

For if $j(X)$ is an irreducible factor of $X^p - Y$ different from $h(X)$, $j(X)$ and $h(X)$ are relatively prime. Hence $1X^0 = a(X)\,h(X) + b(X)\,j(X)$ by Section1, Eq. (14). This, however, leads to a contradiction since by (17) both h and j have α as a root. Hence

$$X^p - Y = (h(X))^m$$

and therefore $p = m \deg h$. It follows that since $\alpha \notin K$, $m = 1$ and $X^p - Y$ is K-irreducible. By (17), α is an inseparable element over K and hence L is not a separable extension of K.

Theorem 8. Let K be a field of characteristic 0. Then every polynomial in $K[X]$ is separable and therefore every finite-dimensional extension of K is separable.

Proof. Suppose $f(X) \in K[X]$ has a repeated factor in its splitting field L, say

$$f(X) = (X - \alpha)^2 g(X) \qquad \text{in} \quad L[X]$$

By formal differentiation we have

$$f'(X) = 2(X - \alpha)\,g(X) + (X - \alpha)^2 g'(X)$$

The polynomials f and f' in $K[X]$ have the common factor $(X - \alpha)$ and hence f and f' are not relatively prime. By Section 1, Eq. (14), the greatest common divisor of f and f' already lies in $K[X]$. Since f is irreducible over K, this implies that f is a factor of f'. But $\deg f' = \deg f - 1$. Therefore f' is the zero polynomial. Since char $K = 0$, $f' = 0$ implies $f = cX^0$, $c \in K$, a contradiction to the fact that $\deg f > 0$. ∎

3. Tensor products

Let M be a left R-module and $\Phi = \mathrm{Hom}_R(M, M)$ its centralizer. Then M is an (R, Φ)-bimodule in the sense of the following:

Definition 1. Let R and Φ be rings and let M be a right Φ-module as well as a left R-module. We call M an *(R, Φ)-bimodule* if in addition

$$(am)\,\alpha = a(m\alpha) \qquad \text{for all} \quad a \in R, \quad m \in M, \quad \text{and} \quad \alpha \in \Phi$$

Let M be an (R, Φ)-bimodule and let Λ be a ring containing Φ. We now ask if it is possible to construct an (R, Λ)-bimodule containing an isomorphic copy of M. In the particular case that Φ is a field we may proceed as follows: Let $\{b_i \mid i \in I\}$ be a Φ-basis. Form the set S of formal sums $\sum b_i \otimes \lambda_i$,

$\lambda_i \in \Lambda$ and $\lambda_i = 0$ for all but a finite number of $i \in I$. Two such expressions are defined to be equal if the corresponding λ_i are equal. We define addition componentwise on S thus making it into an abelian group. For $\lambda \in \Lambda$ define

$$\left(\sum b_i \otimes \lambda_i\right) \lambda = \sum b_i \otimes \lambda_i \lambda$$

and for $a \in R$ define

$$a \sum_i b_i \otimes \lambda_i = \sum_k b_k \otimes \sum_i \alpha_{k_i} \lambda_i$$

where

$$ab_i = \sum_{k \in I} b_k \alpha_{k_i}, \qquad \alpha_{k_i} \in \Phi$$

S together with the operations defined above is an (R, Λ)-bimodule denoted by $M \otimes \Lambda$. The mapping

$$\sum b_i \alpha_i \to \sum b_i \otimes \alpha_i, \qquad \alpha_i \in \Phi$$

is an embedding of the (R, Φ)-bimodule into the (R, Λ)-bimodule $M \otimes \Lambda$. We shall prove this and the independence of the construction of the choice of basis in Theorem 2. Indeed, the approach we choose to take does not appeal to the existence of a basis and moreover we can do away with the assumption that Φ is a field.

Definition 2. Let M' be a right and M a left unital module over the ring Φ with identity. Denote the elements of M' by m' and those of M by m. The set of all expressions of the form

$$\sum_{1 \leqslant i \leqslant n} z_i(m_i', m_i), \qquad n \in \mathbb{N}, \quad z_i \in \mathbb{Z}$$

together with componentwise addition is the free abelian group $\mathfrak{F}$ with generators $\{(m', m) \mid m' \in M', m \in M\}$. Let U be the subgroup of $\mathfrak{F}$ generated by all elements of the form

$$\begin{gathered} (m_1' + m_2', m) - (m_1', m) - (m_2', m) \\ (m', m_1 + m_2) - (m', m_1) - (m', m_2) \\ (m'\alpha, m) - (m', \alpha m) \end{gathered} \tag{1}$$

where $m_i' \in M'$, $m_i \in M$ and $\alpha \in \Phi$. Denote the coset $(m', m) + U \in \mathfrak{F}/U$ by $m' \otimes m$. The factor group $\mathfrak{F}/U$ is called the *tensor product of M' and M over Φ* and is denoted by $M' \otimes_\Phi M$. $M' \otimes_\Phi M$ consists of finite sums

$$\sum_{1 \leqslant i \leqslant r} m_i' \otimes m_i, \qquad m_i' \in M', \quad m_i \in M$$

Because of (1), we have

$$\begin{aligned} (m_1' + m_2') \otimes m &= m_1' \otimes m + m_2' \otimes m \\ m' \otimes (m_1 + m_2) &= m' \otimes m_1 + m' \otimes m_2 \\ (m'\alpha) \otimes m &= m' \otimes (\alpha m) \end{aligned} \tag{2}$$

In the context of relations the tensor product is the factor group of $\mathfrak{F}$ by that relation T on $\mathfrak{F}$ which is the finest congruence relation coarser than the relation defined by the pairs,

$$\begin{gathered} ((m_1 + m_2, m), (m_1, m) + (m_2, m)) \\ ((m, m_1 + m_2), (m, m_1) + (m, m_2)) \\ ((m\alpha, m), (m, \alpha m)) \end{gathered} \tag{3}$$

This "minimality" of the relation T leads to a characterization of the tensor product which will be frequently used later on.

Definition 3. Let Φ be a ring with identity and let ${}_\Phi M$ be a left and M_Φ' a right unital Φ-module. Let N be an additively written abelian group. Let η be a mapping from the Cartesian product $M' \times M$ to N. The mapping η is said to be *balanced with respect to* Φ if it satisfies the following:

$$\begin{aligned} \eta(m_1' + m_2', m) &= \eta(m_1', m) + \eta(m_2', m) \\ \eta(m', m_1 + m_2) &= \eta(m', m_1) + \eta(m', m_2) \\ \eta(m'\alpha, m) &= \eta(m', \alpha m), \ \alpha \in \Phi \end{aligned} \tag{4}$$

The canonical map

$$\pi\colon (m', m) \to m' \otimes m \tag{5}$$

from the Cartesian product $M' \times M$ to the tensor product $M' \otimes {}_\Phi M$ is an example of such a map.

The following theorem states that every balanced map from $M' \times M$ to an abelian group can be "factored through" $M' \otimes {}_\Phi M$.

Theorem 1. Let Φ be a ring with identity and let M_Φ' and ${}_\Phi M$ be right and left Φ-modules, respectively. Let N be an additively written abelian group and η a balanced map from $M' \times M$ to N. Then there is a unique additive mapping $\sigma\colon M' \otimes {}_\Phi M \to N$ making the following diagram commutative:

$$\begin{array}{ccc} M' \times M & \xrightarrow{\pi} & M' \otimes {}_\Phi M \\ & \searrow^{\eta} & \downarrow{\sigma} \\ & & N \end{array}$$

Here π is the mapping defined in (5). (Note: by the diagram being commutative we mean $\sigma\pi = \eta$.)

Proof. Since we want $\sigma\pi$ to be the same mapping as η, and since the elements $m' \otimes m$ generate $M' \otimes_\Phi M$, the only possible additive candidate for σ is the mapping defined by

$$\sigma: \sum_i m_i' \otimes m_i \to \sum_i \eta(m_i', m_i) \tag{6}$$

We must therefore prove that σ is well-defined, i.e. we must show that

$$\sum x_i' \otimes x_i = \sum y_j' \otimes y_j \tag{7}$$

implies

$$\sum \eta(x_i', x_i) = \sum \eta(y_j', y_j) \tag{8}$$

The set of pairs of elements of $\mathcal{F}$

$$\left\{\left(\sum (x_i', x_i), \sum (y_j', y_j)\right) \middle| \sum \eta(x_i', x_i) = \sum \eta(y_j', y_j)\right\} \tag{9}$$

is a relation T^* on $\mathcal{F}$. Let T' be the finest congruence relation on $\mathcal{F}$ which is coarser than T^*. Since η is balanced, $T' \supseteq T$, where T is the congruence relation induced by the pairs in (3). But T is the defining relation of the tensor product $M' \otimes_\Phi M = \mathcal{F}/T$ and in the language of relations (7) becomes $\sum (x_i', x_i) T \sum (y_j', y_j)$. Hence the equality in (7) implies the equality in (8) and σ is well defined. ∎

Let M and M_1 be right Φ-modules and N and N_1 left Φ-modules. If $A: M \to M_1$ and $B \to B: N \to N_1$ are Φ-homomorphisms then the mapping

$$(m, n) \to mA \otimes nB \tag{10}$$

is a balanced map from $M \times N$ to the tensor product $M_1 \otimes_\Phi N_1$. By Theorem 1 there exists a homomorphism denoted by $A \otimes B$ from $M \otimes N$ to $M_1 \otimes N_1$. If in addition A and B are epimorphisms we have the following:

Theorem 2. Let M, M_1, N and N_1 be as in the previous paragraph. Let $A: M \to M_1$ and $B: N \to N_1$ be epimorphisms with kernels U and W, respectively. Then $A \otimes B$ is an epimorphism of the additive group $M \otimes N$ onto $M_1 \otimes N_1$ such that

$$A \otimes B: m \otimes n \to mA \otimes nB \tag{11}$$

Moreover, ker $A \otimes B$ is the subgroup S of $M \otimes N$ generated by the elements $u \otimes n$, $u \in U$, $n \in N$ and $m \otimes w$, $m \in M$, $w \in W$.

Proof. The mapping $\eta: (m, n) \to mA \otimes nB$ is balanced. Hence by Theorem 1 there exists exactly one homomorphism $A \otimes B$ from $M \otimes N$ onto $\eta(M \otimes N) = M_1 \otimes N_1$ such that $(A \otimes B)\,\pi = \eta$ where $\pi: (m, n) \to m \otimes n$. Clearly ker $A \otimes B \supseteq S$. Consider now the mapping $(A \otimes B)^*$ induced by $A \otimes B$ on $M \otimes N/S$. In particular we have

$$(A \otimes B)^*: \; m \otimes n + S \to mA \otimes nB$$

We show $(A \otimes B)^*$ is an isomorphism of the additive group $M \otimes N/S$ onto $M_1 \otimes N_1$. The mapping $\zeta: (mA, nB) \to m \otimes n + S$ is well defined: For if $m_1 A = m_2 A$ and $n_1 A = n_2 B$, then $m_1 - m_2 \in \ker A$ and $n_1 - n_2 \in \ker B$. But

$$m_1 \otimes n_1 - m_2 \otimes n_2 = (m_1 - m_2) \otimes n_1 + m_2 \otimes (n_1 - n_2)$$

Therefore $m_1 \otimes n_1 + S = m_2 \otimes n_2 + S$ and ζ is well defined. Clearly ζ is balanced. Hence by Theorem 1 there exists a homomorphism $\tau: M_1 \otimes N_1 \to M \otimes N/S$ such that

$$\tau: mA \otimes nB \to m \otimes n + S$$

Now $\tau(A \otimes B)^*$ is the identity on $M \otimes N/S$ and $(A \otimes B)^*\,\tau$ is the identity on $M_1 \otimes N_1$. Therefore $S = \ker A \otimes B$. ▌

Theorem 2 may also be stated as follows: The mapping $(m, n) \to mA \otimes nB$ induces a group homomorphism $A \otimes B$ from $M \otimes N$ to $M_1 \otimes N_1$. If both A and B are epimorphisms so is $A \otimes B$. The analogous result does not hold in general if A and B are monomorphisms as the following example shows:

Let $M = \mathbb{Z}/2\mathbb{Z}$ be the prime field C_2 of characteristic 2, considered as a $\mathbb{Z}$-module and let A be the identity on M. Let $N = \mathbb{Z}$ and B the embedding of ${}_{\mathbb{Z}}\mathbb{Z}$ into ${}_{\mathbb{Z}}\mathbb{Q}$, $\mathbb{Q}$ the rationals. Then $M \otimes_{\mathbb{Z}} N = C_2 \otimes_{\mathbb{Z}} \mathbb{Z} \approx C_2$ under the mapping $\bar{z}_1 \otimes z_2 \to \bar{z}_1 z_2 = \overline{z_1 z_2}$. On the other hand $C_2 \otimes_{\mathbb{Z}} \mathbb{Q} = 0$ since

$$\bar{z}_1 \otimes z_2 = \overline{2z_1} \otimes \tfrac{1}{2} z_2 = 0$$

Hence $A \otimes B$ is not a monomorphism.

From this discussion it follows that in general if $N_1 \subseteq N$, $M \otimes_{\Phi} N_1$ is not isomorphically contained in $M \otimes_{\Phi} N$ and therefore $m \otimes n_1 \in M \otimes N_1$ *cannot be identified with* $m \otimes n_1 \in M \otimes N$. However, if N_1 is a direct summand of N we may prove the following:

Theorem 3. Let M be a right Φ-module and let the left Φ-module N be the direct sum of its submodules N_1 and N_2: $N = N_1 \oplus N_2$. Then $M \otimes_\Phi N \approx M \otimes_\Phi N_1 \oplus M \otimes_\Phi N_2$ under an isomorphism σ, where

$$\sigma: m \otimes (n_1 + n_2) \to (m \otimes n_1, m \otimes n_2) \tag{12}$$

Proof. Let η be the mapping defined by

$$\eta: (x, y_1 + y_2) \to (x \otimes y_1, x \otimes y_2), \qquad x \in M, \quad y_i \in N_i$$

Clearly η is a balanced map from $M \otimes N$ to the direct sum $M \otimes N_1 \oplus M \otimes N_2$. Therefore by Theorem 1 there exists a homomorphism σ of $M \otimes N$ onto $M \otimes N_1 \oplus M \otimes N_2$ satisfying $\sigma\pi = \eta$. Under this homomorphism $m \otimes (n_1 + n_2) \to (m \otimes n_1, m \otimes n_2)$. Now let π_i, $i = 1, 2$ be the canonical balanced map from $M \times N_i$ to $M \otimes N_i$ and let $\eta_i: M \times N_i \to M \otimes N$ be the mapping defined by

$$\eta_i: (m, n_i) \to m \otimes n_i \in M \otimes N_i, \qquad i = 1, 2$$

Clearly η_i is balanced and therefore by Theorem 1 there exists a homomorphism $\tau_i: M \otimes N_i \to M \otimes N$ such that $\tau_i \pi_i = \eta_i$, $i = 1, 2$. Set $\tau = \tau_1 + \tau_2$. Then τ is a homomorphism from $M \otimes N_1 \oplus M \otimes N_2$ to $M \otimes N$ such that $\tau: (m \otimes n_1, m \otimes n_2) \to m \otimes n_1 + m \otimes n_2 = m \otimes (n_1 + n_2)$. Therefore $\tau\sigma$ is the identity on $M \otimes N$ and σ is one-one. ∎

Suppose now that M is an (R, Φ)-bimodule and N a left Φ-module. We may give the free abelian group $\mathscr{F}$ with generators $(m, n) \in M \times N$ the structure of a left R-module by defining

$$a \sum (m_i, n_i) = \sum (am_i, n_i)$$

The subgroup U of $\mathscr{F}$ generated by the elements in (1) is an R-submodule of $\mathscr{F}$. To prove this it is clearly sufficient to show that if x is a generator of U, $ax \in U$. We show this when x is of the form $(m_1 + m_2, n) - (m_1, n) - (m_2, n)$ and leave the other cases to the reader:

$$\begin{aligned} &a((m_1 + m_2, n) - (m_1, n) - (m_2, n)) \\ &\quad = (a(m_1 + m_2), n) - (am_1, n) - (am_2, n) \\ &\quad = (am_1 + am_2, n) - (am_1, n) - (am_2, n) \in U \end{aligned}$$

This enables us to make the following:

Definition 4. Let M be an (R, Φ)-bimodule and N a left Φ-module. Then $M \otimes_\Phi N$ may be given the structure of a left R-module by defining

$$a \sum (m_i \otimes n_i) = \sum (am_i) \otimes n_i \tag{13}$$

If Φ is a commutative ring and M a right Φ-module we make M into a left Φ-module by defining

$$\alpha m = m\alpha, \qquad m \in M, \quad \alpha \in \Phi \tag{14}$$

It is easily checked that with this definition M is actually a (Φ, Φ)-bimodule. Therefore the tensor product of two Φ-modules over the commutative ring Φ is a Φ-module.

An algebra $\mathfrak{A}$ over the commutative ring Φ may be considered as a (Φ, Φ)-bimodule. Therefore the tensor product of two Φ-algebras is a Φ-module. Moreover, if we define multiplication as below, the tensor product is again a Φ-algebra.

Definition 5. Let $\mathfrak{A}'$ and $\mathfrak{A}$ be two algebras over the commutative ring Φ with 1. Then the tensor product $\mathfrak{A}' \otimes_\Phi \mathfrak{A}$ of the two algebras is an algebra if we define

$$\Big(\sum_i a_i' \otimes a_i\Big)\Big(\sum_j b_j' \otimes b_j\Big) = \sum_{i,j} (a_i' b_j') \otimes (a_i b_j) \tag{15}$$

To prove that this definition is independent of the representatives $\sum_i (a_i', a_i)$ and $\sum_j (b_j', b_j)$ we need only point out that because of the distributive laws in a ring, we have, for example,

$$(a'(b_1' + b_2')\, a) - (a'b_1', a) - (a'b_2', a) \in U$$

for all $a', b_i' \in \mathfrak{A}'$ and $a \in \mathfrak{A}$.

Theorem 4. Let $\mathfrak{A}$ be an algebra over the commutative ring Φ. Let Λ be a commutative ring containing Φ. Then the tensor product $\mathfrak{A} \otimes_\Phi \Lambda$ is an algebra over Λ if we define

$$\sum_i (a_i \otimes \lambda_i)\lambda = \sum_i a_i \otimes (\lambda_i \lambda) \tag{16}$$

for $a_i \in \mathfrak{A}$, $\lambda \in \Lambda$. We denote the algebra $\mathfrak{A} \otimes_\Phi \Lambda$ by $\mathfrak{A}_\Lambda$.

Proof. $\mathfrak{A} \otimes_\Phi \Lambda$ is an algebra over Φ. The operation of Λ on $\mathfrak{A} \otimes_\Phi \Lambda$ defined in (16) is independent of the choice of representative for $\sum a_i \otimes \lambda_i$.

This may be seen in exactly the same fashion as indicated above. Furthermore

$$\begin{aligned}[(a_1 \otimes \lambda_1)(a_2 \otimes \lambda_2)]\, \lambda &= (a_1 \otimes \lambda_1)[(a_2 \otimes \lambda_2)\, \lambda] \\ &= [(a_1 \otimes \lambda_1)\, \lambda](a_2 \otimes \lambda_2)\end{aligned}$$

since each of the three expressions is equal to $a_1a_2 \otimes \lambda_1\lambda_2\lambda$ because $\lambda_1\lambda\lambda_2 = \lambda_1\lambda_2\lambda$. ∎

We are now in position to solve the embedding problem posed at the beginning of this section:

Theorem 5. Let Φ be a subfield of the ring Λ with identity. Let M be an (R, Φ)-bimodule. Then M is isomorphic as an (R, Φ)-bimodule to $M \otimes_{\Phi} \Phi$ under the mapping

$$\sigma\colon m \to m \otimes 1$$

It then follows that M is isomorphically embedded in the (R, Λ)-bimodule $M \otimes_{\Phi} \Lambda$.

Proof. $\Phi 1$ is a subspace of the Φ-vector space Λ. Hence $\Lambda = \Phi \oplus \Psi$, a direct sum as a Φ-vector space. By Theorem 3 the mapping σ is a monomorphism. ∎

In a similar fashion we may prove:

Theorem 6. Let R be an algebra over the field Φ. Let Λ be a commutative ring containing Φ. Then

$$\sigma\colon a \to a \otimes 1 \qquad \text{for} \quad a \in R$$

is a monomorphism of the algebra R over Φ into the algebra $R \otimes_{\Phi} \Lambda$ over Λ. ∎

From Definition 1 of the tensor product $M' \otimes_{\Phi} M$, it follows without difficulty that if N' is a right Φ-module isomorphic to M' and N a left Φ-module isomorphic to M then $M' \otimes_{\Phi} M \approx N' \otimes_{\Phi} N$.

4. Representations by matrices

Let R be an algebra over the field Φ and let M be a left (algebra) module over R, that is to say, according to Definition II.8.1, M is an (R, Φ)-bimodule and in addition $(a\alpha)\, m = am\alpha$ for all $a \in R$, $\alpha \in \Phi$, and $m \in M$. The elements a of R induce Φ-linear mappings A of the Φ-vector space M into itself:

$$am = Am, \qquad a \in R \quad \text{and} \quad m \in M$$

These Φ-linear maps form an algebra over Φ since Φ is commutative (Theorem I.4.1). The mapping

$$a \to A \qquad \text{for} \quad a \in R$$

is therefore an algebra homomorphism.

Definition 1. Let R be an algebra over Φ. A *Φ-representation of R* is a homomorphism of the algebra R into the algebra of Φ-linear mappings of a Φ-vector space M into itself. M is called a *Φ-representation module* for this Φ-representation of R.

Definition 2. Let Δ be an extension field of the field Φ and R an algebra over Φ. A *Δ-representation of the algebra R over Φ* is a Δ-representation of the algebra $R \otimes_\Phi \Delta$ (cf. Theorem V.3.4).

The theory of representations of Φ-algebras R in extension fields Δ of Φ is therefore the same as the theory of $R \otimes_\Phi \Delta$-modules about which something will be said in Section VI.I. In this chapter we shall discuss the matrix representations of algebras. These are obtained by choosing a Φ-basis in the representation module M.

For the remainder of this section let us fix our notation: Let R denote an algebra over the field Φ and let M be a representation module for the algebra R, i.e. M is an (R, Φ)-bimodule so that $(a\alpha)\, m = am\alpha$ for all $a \in R$, $m \in M$, and $\alpha \in \Phi$. Let $\dim_\Phi M = n < \infty$ and let $\{u_1, ..., u_n\}$ be a Φ-basis of M. The action of a on M is determined by its action on the u_k:

$$au_k = \sum_{1 \leqslant i \leqslant n} u_i \alpha_{ik}, \qquad k = 1, ..., n, \quad \alpha_{ik} \in \Phi \tag{1}$$

and the mapping

$$a \to A = (\alpha_{ik}) \in \Phi_{n \times n}, \qquad a \in R \tag{2}$$

is a *matrix representation* of R in Φ, i.e. an algebra homomorphism of R into $\Phi_{n\times n}$.

Conversely let (2) be a given matrix representation of the Φ-algebra R and let M be a vector space over Φ of dimension n. Let $\{u_1, ..., u_n\}$ be a Φ-basis of M. Then M may be turned into an R-algebra module by defining

$$au_k = \sum_i u_i k_{ik}, \qquad a \in R \tag{3_1}$$

$$a \sum_k u_k \beta_k = \sum_{i,k} u_i \alpha_{ik} \beta_k, \qquad \beta_k \in \Phi \tag{3_2}$$

Since (3_1) agrees with (1), relative to the Φ-basis $\{u_1, ..., u_n\}$ of the algebra module M, we get back the original Φ-representation of R.

Suppose $\{u_1', \ldots, u_n'\}$ is another Φ-basis of M where

$$u_k' = \sum_i u_i \pi_{ik}, \qquad \pi_{ik} \in \Phi, \qquad k = 1, \ldots, n$$

Let

$$au_k' = \sum_i u_i' \alpha_{ik}', \qquad k = 1, \ldots, n$$

Then

$$\sum_i u_i' \alpha_{ik}' = \sum_{i,j} u_j \pi_{ji} \alpha_{ik} = au_k' = \sum_i au_i \pi_{ik} = \sum_{i,j} u_j \alpha_{ji} \pi_{ik}$$

Hence, relative to the new basis a corresponds to the matrix A', where

$$A' = P^{-1}AP, \qquad P = (\pi_{ik}) \tag{4}$$

Two matrix representations

$$a \to A \qquad \text{and} \qquad a \to A'$$

of the algebra R are said to be *equivalent* if there exists a matrix $P \in \Phi_{n \times n}$ such that (4) holds for all $a \in R$. The module M immediately determines a Φ-representation of R in $\mathrm{Hom}_\Phi(M, M)$. However, to obtain a matrix representation of R we must first choose a basis in M.

Let R be a Φ-algebra and let M and M' be isomorphic R-modules under the isomorphism $\psi: m \to m'$. [Thus

$$\psi(am) = a\psi(m) \qquad \text{and} \qquad \psi(m\alpha) = \psi(m)\, \alpha.]$$

Then M and M' give rise to the same equivalence classes of Φ-representations of the algebra R. In fact, if $\{u_1, \ldots, u_n\}$ is a Φ-basis of M then $\{\psi u_1, \ldots, \psi u_n\}$ is a Φ-basis of M' and

$$a\psi u_k = \psi(au_k) = \sum_i u_i \alpha_{ik} = \sum_i (\psi u_i)\, \alpha_{ik}, \qquad k = 1, \ldots, n$$

In other words:

Theorem 1. Let R be a Φ-algebra. The equivalence classes of matrix representations of R are in one-one correspondence with the isomorphism classes of R-modules which are finite dimensional over Φ. These (R, Φ)-modules are called the representation modules of R. ∎

This correspondence between matrix representations of a Φ-algebra R and representation modules enables us to apply the theory of the preceding chapters to matrix representations. We also take over the terminology introduced. Thus if R is an algebra over the field Φ we call a matrix representation of R:

1. *irreducible* if its corresponding representation module M is an irreducible R-module,
2. *completely reducible* if M is completely reducible,
3. *reducible* if M contains a submodule different from 0 and M.

Since the radical J of a ring R annihilates every irreducible module M, we may always turn an irreducible R-module M into an R/J-module. If R/J is a finite-dimensional algebra over Φ, R/J is artinian and therefore

$$R/J = \bar{\mathfrak{A}}_1 + \cdots + \bar{\mathfrak{A}}_q$$

where the $\bar{\mathfrak{A}}_i$ are ideals. By Theorem IV.2.3, M is a faithful module for exactly one $\bar{\mathfrak{A}}_j$ and is annihilated by the other $\bar{\mathfrak{A}}_i$. Therefore an irreducible matrix representation of the Φ-algebra R induces a representation of R/J which is faithful on exactly one of the simple direct summands of R/J and which annihilates all the others. This does not mean that the matrices of this representation "fill out" the whole matrix ring $\Phi_{n\times n}$ since equality may not hold in

$$\Phi \subseteq \operatorname{Hom}_R(M, M)$$

Representations in which this happens will be discussed later in the context of absolutely irreducible representations.

If the representation module M is the direct sum of two submodules each different from 0, say

$$M = U \oplus V$$

then we may complete a Φ-basis $\{u_1, ..., u_n\}$ of U to a Φ-basis of M by adjoining a Φ-basis $\{v_1, ..., v_s\}$ of V. The resulting representation matrices relative to the basis $\{u_1, ..., u_r, v_1, ..., v_z\}$ have the form

$$A = \begin{pmatrix} U & 0 \\ 0 & V \end{pmatrix}, \qquad \text{where} \quad U \in \Phi_{r\times r}, \quad V \in \Phi_{s\times s} \tag{5}$$

We may express this as follows: The representation module M of the representation $\phi: a \to A$ is the direct sum of two submodules each different from 0 if and only if there exists a matrix $P \in \Phi_{n\times n}$ such that

$$P^{-1}AP = \begin{pmatrix} U & 0 \\ 0 & V \end{pmatrix}, \qquad U \in \Phi_{r\times r}, \quad V \in \Phi_{s\times s}$$

$r > 0$, $s > 0$, $r + s = n$ for all matrices $\phi(a)$, $a \in R$.

In particular, this happens when M is completely reducible. In this case the matrices of the representation are of the form

$$P^{-1}AP = \begin{pmatrix} U_1 & & 0 \\ & \ddots & \\ 0 & & U_t \end{pmatrix} \tag{6}$$

where $a \to U_j$, $j = 1, ..., t$ is an irreducible Φ-representation of the Φ-algebra R and the square $r_j \times r_j$ matrices U_j lying along the diagonal of the $n \times n$ matrix $P^{-1}AP$ are the only submatrices of $P^{-1}AP$ having entries different from 0. Naturally, $n = r_1 + \cdots + r_t$. By Theorem IV.1.4 every R-module of a semisimple algebra R is completely reducible and therefore so is every matrix representation of R.

If M is not completely reducible let

$$U_0 = (0) \subset U_1 \subset \cdots \subset U_t = M \tag{7}$$

be a maximal chain of R-submodules U_j so that each U_j is maximal in U_{j+1}. Choose a Φ-basis of M as follows: Choose a basis of U_1 and extend successively to a basis of U_2, U_3, ..., $U_t = M$. Relative to this basis, the matrices A of the representation which M gives rise to are of the form

$$A = \begin{pmatrix} U_{11} & U_{12} & \cdots & U_{1t} \\ 0 & U_{22} & \cdots & U_{2t} \\ \vdots & \vdots & & \vdots \\ & & & U_{tt} \end{pmatrix} \tag{8}$$

where $U_{jj} = U_{jj}(a)$, $j = 1, ..., t$ are the matrices corresponding to the representation module U_j/U_{j-1} and the submatrices U_{jk}, $j < k$ lie in $\Phi_{n_j \times n_k}$ where $n_j = \dim_\Phi(U_j/U_{j-1})$. All entries under the U_{jj} are zero. The submatrices U_{11}, ..., U_{tt} occurring in (8) are uniquely determined up to order and equivalence by the representation module M. This implies that, up to equivalence and order, they are independent of the way the maximal chain in (7) was chosen. This is a consequence of the purely module-theoretic Theorem 2 below.

Definition 3. Let M be an R-module and let $U_0 \subset U_1 \subset \cdots \subset U_t$ be a chain of submodules. The *length* of this chain is defined to be the number of proper inclusions, i.e. t.

Theorem 2 (Jordan–Hölder). Let M be an R-module and suppose all chains of submodules are finite in length. Let

$$0 = U_0 \subset U_1 \subset \cdots \subset U_r = M \quad \text{and} \quad 0 = V_0 \subset \cdots \subset V_s = M \tag{9}$$

be two chains of maximal length. Then $r = s$ and there is a permutation π of the numbers 1, 2, ..., r such that

$$U_j/U_{j-1} \approx V_{\pi(j)}/V_{\pi(j)-1}$$

as R-modules. ∎

This theorem is a corollary to a much more general theorem the proof of which depends on a number of intermediate results.

Theorem 3 (isomorphism theorem for R-modules and rings). 1. Let U and V be R-submodules of the R-module M. The mappings

$$\psi: u + (U \cap V) \to u + V \qquad \text{for all} \quad u \in U$$

is an isomorphism of $U/U \cap V$ onto $(U + V)/V$.

2. The submodules X of M lying between $U \cap V$ and U are in one-one correspondence with the submodules Y lying between V and $U + V$ under the mapping

$$\psi': X \to X + V$$

ψ' is a lattice isomorphism with inverse

$$Y \to Y \cap U$$

3. Similar results hold if U is a subring and V an ideal of a ring R.

Proof of the isomorphism theorem. $u \to u + V$ is a module homomorphism of U onto $U + V/V$. The kernel is clearly $U \cap V$ and hence ψ is an isomorphism. In the same way we easily see that ψ' is a lattice isomorphism of the "interval"

$$\{X \mid U \cap V \subseteq X \subseteq U\} \qquad \text{onto} \qquad \{Y \mid V \subseteq Y \subseteq U + V\}$$

The assertion about the inverse of ψ' follows from the modular law (Theorem II.3.2) for the R-submodules of a module M

$$(X + V) \cap U = X + (V \cap U) = X$$

for all X such that $U \cap V \subseteq X \subseteq U$. Part 3 is proved in exactly the same way. ∎

The proof of the following theorem due to Zassenhaus depends on the isomorphism theorem just proved.

Theorem 4. Let $A' \subseteq A$ and $B' \subseteq B$ be submodules of M. Then the following three factor modules are isomorphic:

$$(A \cap B)/((A' \cap B) + (A \cap B')) \tag{10}$$
$$(A' + (A \cap B))/(A' + (A \cap B')) \tag{11}$$
$$(B' + (A \cap B))/(B' + (A' \cap B)) \tag{12}$$

Proof. By symmetry it is sufficient to prove (10) and (11) isomorphic. Since the submodules of M form a modular lattice,

$$\begin{aligned}(A' \cap B) + (A \cap B') &= (A \cap B') + (A' \cap (A \cap B))\\ &= ((A \cap B') + A') \cap (A \cap B)\end{aligned}$$

Furthermore,

$$A' + (A \cap B) = (A' + (A \cap B')) + (A \cap B)$$

Put $U = A \cap B$ and $V = A' + (A \cap B')$ in Theorem 3 and notice that

$$U + V = A' \cap (A \cap B) \quad \blacksquare$$

Definition 4. 1. We say that the chain of submodules $U = U_0 \subset U_1 \subset \cdots \subset U_n = W$ is a *refinement* of the chain $U = V_0 \subset \cdots \subset V_m = W$ if the terms of the first chain include all of the terms that occur in the second chain.

2. The two chains in Part 1 are said to be *equivalent* if it is possible to set up a one-one correspondence between the factors U_i/U_{i-1} and V_j/V_{j-1} such that the paired factors are isomorphic.

Theorem 5 (Schreier). Let

$$U = A_0 \subseteq A_1 \subseteq \cdots \subseteq A_m = W \tag{13}$$

and

$$U = B_0 \subseteq B_1 \subseteq \cdots \subseteq B_n = W \tag{14}$$

be two chains of submodules of the module M. Then the chains have equivalent refinements.

Proof. Let

$$A_{ij} = A_i + (A_{i+1} \cap B_j), \qquad j = 0, 1, ..., n \tag{15}$$
$$B_{ij} = B_j + (B_{j+1} \cap A_i), \qquad i = 0, ..., m \tag{16}$$

Then since (11) and (12) of Theorem 4 are isomorphic,

$$A_{i,j+1}/A_{ij} \qquad B_{i+1,j}/B_{ij} \tag{17}$$

[each factor module in (17) could be the 0-module]. $\blacksquare$

In particular if the chains in (13) and (14) are maximal (we call maximal chains *composition series*) with $U = 0$ and $W = M$, neither can be refined. Hence Theorem 2, due to Jordan and Hölder, follows.

If the R-module M has a composition series of finite length, then all chains are of finite length. Indeed, assume M has a maximal chain of length r and suppose if possible that there is a chain of infinite length from 0 to M. Then there is a chain of length $r + 1$. But the first chain cannot be refined and therefore the two chains do not have equivalent refinements, contradicting Theorem 5.

Definition 5. The *trace of the matrix* $A = (\alpha_{ik}) \in \Phi_{n\times n}$ is the sum of the diagonal elements of A:

$$\operatorname{Tr} A = \alpha_{11} + \alpha_{22} + \cdots + \alpha_{nn} \in \Phi$$

It is easily checked that if A and P are matrices in $\Phi_{n\times n}$,

$$\operatorname{Tr}(AP) = \operatorname{Tr}(PA) \tag{18}$$

In particular, therefore, if P^{-1} exists,

$$\operatorname{Tr}(P^{-1}AP) = \operatorname{Tr} A \tag{19}$$

We may therefore speak of *the trace of an element a of R in a representation of R by means of Φ-linear mappings of the corresponding representation module M to itself* without having to refer to any specific Φ-basis. We also speak of the *trace of a Φ-representation* $a \to A$ of R. By this we mean the function Tr: $R \to \Phi$ defined by

$$\operatorname{Tr}(a) = \operatorname{Tr}(A)$$

Let $a \to A$ be a completely reducible representation and suppose that a basis has been so chosen that the matrices A are of the form

$$\begin{pmatrix} U_1 & & 0 \\ \vdots & \ddots & \vdots \\ 0 & & U_t \end{pmatrix}$$

where $a \to U_j(a)$, $j = 1, \ldots, t$ is an irreducible Φ-representation of R [cf. (6)]. Let $\operatorname{Tr}_j(a)$ denote $\operatorname{Tr} U_j(a)$. Then

$$\operatorname{Tr}(a) = \sum_{1\leqslant j\leqslant t} \operatorname{Tr}_j(a), \qquad a \in R$$

Now let R be artinian and let

$$\bar{R} = R/J = \bar{\mathfrak{A}}_1 + \cdots + \bar{\mathfrak{A}}_q \tag{20}$$

be the decomposition of the semisimple artinian ring R/J into its simple components $\bar{\mathfrak{A}}_i$. The irreducible representation induced by $a \to U_j(a)$ on R/J is a faithful representation of $\bar{\mathfrak{A}}_{k(j)}$, say, and annihilates the other $\bar{\mathfrak{A}}_i$. Moreover, the identity element $\bar{e}_{k(j)}$ of $\bar{\mathfrak{A}}_{k(j)}$ is mapped onto the $n_j \times n_j$ identity matrix $U_j(e_{k(j)})$. Hence

$$\mathrm{Tr}_j(\bar{e}_{k(j)}) = n_j \qquad \text{and} \qquad \mathrm{Tr}_j(\bar{e}_i) = 0 \qquad \text{if} \quad i \neq k(j) \tag{21}$$

By Theorem 2 a completely reducible Φ-representation $\bar{a} \to A$ of the semisimple artinian ring $\bar{R} = R/J$ is determined up to equivalence once we know the number of irreducible components which are isomorphic to any given irreducible component occurring in the direct sum decomposition of the corresponding representation $\bar{R}$-module M into irreducible submodules. By Theorem IV.1.4 there is a one-one correspondence between the isomorphism classes of $\bar{R}$-modules and the isomorphism classes of minimal left ideals. But each of the simple components $\bar{\mathfrak{A}}_i$ is the sum of all mutually isomorphic minimal left ideals. There are therefore q isomorphism classes of left ideals, represented by $\bar{L}_1, \ldots, \bar{L}_q$, say. Let

$$M = U_1 \oplus \cdots \oplus U_t \tag{22}$$

and let q_i denote the number of U_k isomorphic to $\bar{L}_i$. From (21) it follows that the trace of the identity $\bar{e}_i$ of $\bar{\mathfrak{A}}_i$ relative to the representation $\bar{a} \to A$ satisfies

$$\mathrm{Tr}(\bar{e}_i) = q_i n_i$$

We may solve this equation for q_i provided the characteristic of the field Φ is either 0 or relatively prime to n_i.

Theorem 4. Let R be a semisimple artinian ring and Φ a field of characteristic 0. A matrix Φ-representation of R is determined, up to equivalence, by the traces of the matrices of the Φ-representation. ▮

5. The trace of a separable field extension

Let L be a field containing a subfield K. Then L may be considered as an algebra over K. If $\dim_K L$ is finite, ${}_L L$ is a K-representation module of L (Definition V.4.1) and we may therefore speak of the trace of this representation. We refer to it more briefly as the *trace of L over K* and denote it $\mathrm{Tr}_{L/K}$. In this section we prove a theorem on the trace of a separable extension L over K. This theorem will be used in Chapter VIII to show that a semisimple Φ-algebra remains semisimple when we extend the ground field to L provided L is a separable extension of Φ.

Let L be a field containing the subfield K and let $\dim_K L = n < \infty$. Let $\omega_0, \omega_1, ..., \omega_{n-1}$ be a basis of L over K. Let $\lambda \in L$ and let

$$\lambda \cdot \omega_j = \sum_{0 \leqslant i \leqslant n-1} \omega_i a_{ij}(\lambda) \qquad \text{where} \quad a_{ij}(\lambda) \in K \tag{1}$$

Then $\lambda \to (a_{ij}(\lambda))$ is the K-representation referred to above and

$$\mathrm{Tr}_{L/K}(\lambda) = \sum_i a_{ii}(\lambda) \tag{2}$$

By V.4, (19), $\mathrm{Tr}_{L/K}$ is independent of the choice of basis.

Theorem 1. Let L be a separable extension of K. Then there is an element $\lambda \in L$ such that $\mathrm{Tr}(\lambda) \neq 0$.

Proof. 1. By Theorem V.2.7, L has a primitive element θ over K. Let $f(X) \in K[X]$ be the minimal polynomial of θ over K. By hypothesis $f(X)$ is separable. Let $\theta = \theta_0, \theta_1, ..., \theta_{n-1}$ be the roots of $f(X)$. Adjoin $\theta_1, ..., \theta_{n-1}$ to $L = K(\theta)$ to obtain a splitting field M of f over K. The mapping

$$\tau_q\colon c_0 + c_1\theta + \cdots + c_{n-1}\theta^{n-1} \to c_0 + c_1\theta_q + \cdots + c_{n-1}\theta_q^{n-1} \tag{3}$$

where the c_i are in K, is an isomorphism of the field $L = K(\theta) = K(\theta_0)$ onto $K(\theta_q)$, $q = 0, ..., n-1$, since by Theorem V.1.2, each field $K(\theta_q)$ is isomorphic to $K[X]/(f)$ under the mapping $X + (f) \to \theta_q$, $q = 0, ..., n-1$.

Since $\tau_q\theta = \theta_q$ and $\theta = \theta_0, \theta_1, ..., \theta_{n-1}$ are pairwise distinct elements, the τ_q are also n distinct isomorphisms.

2. In (1) choose as K-basis of $L = K(\theta)$ the set $\{1, \theta, ..., \theta^{n-1}\}$ and apply $\tau_0, \tau_1, ..., \tau_{n-1}$ to (1). Since $a_{ij}(\lambda) \in K$ we get

$$(\tau_q\lambda) \cdot (\tau_q\theta^j) = \sum_{0 \leqslant i \leqslant n-1} (\tau_q\theta^i)\, a_{ij}(\lambda), \qquad j, q = 0, 1, ..., n-1 \tag{4}$$

But $\tau_q\theta = \theta_q$. Therefore

$$\tau_q(\lambda) \cdot \theta_q{}^j = \sum_{0 \leqslant i \leqslant n-1} \theta_q{}^i a_{ij}(\lambda), \qquad j, q = 0, 1, ..., n-1$$

The matrix

$$T = \begin{pmatrix} 1 & \theta_0 & \cdots & \theta_0^{n-1} \\ 1 & \theta_1 & \cdots & \theta_1^{n-1} \\ \vdots & \vdots & & \vdots \\ 1 & \theta_{n-1} & \cdots & \theta_{n-1}^{n-1} \end{pmatrix}$$

lies in $M_{n \times n}$ and the determinant of T is

$$\prod_{0 \leqslant q < r \leqslant n-1} (\theta_r - \theta_q)$$

by a theorem of Van der Monde. Hence det $T \neq 0$ and T has an inverse $T^{-1} \in M_{n \times n}$. The n^2 equations in (4) may be written as a single matrix equation:

$$\begin{pmatrix} \tau_0\lambda & \cdots & 0 \\ \vdots & \ddots & \\ 0 & & \tau_{n-1}\lambda \end{pmatrix} T = T \cdot (a_{ij}(\lambda)) \tag{5}$$

Thus multiplying on the right by T^{-1} we have

$$\begin{pmatrix} \tau_0\lambda & \cdots & 0 \\ \vdots & \ddots & \vdots \\ 0 & & \tau_{n-1}\lambda \end{pmatrix} = T \cdot (a_{ij}(\lambda))\, T^{-1} \tag{6}$$

Taking traces of both sides we get

$$\tau_0\lambda + \cdots + \tau_{n-1}\lambda = \mathrm{Tr}(T(a_{ij}(\lambda))\, T^{-1}) = \mathrm{Tr}(a_{ij}(\lambda)) = \mathrm{Tr}(\lambda) \tag{7}$$

3. Suppose now if possible that $\mathrm{Tr}(\lambda) = 0$ for all $\lambda \in L$. Then the distinct isomorphisms $\tau_0, \ldots, \tau_{n-1}$ are linearly dependent over M since

$$(\tau_0 + \tau_1 + \cdots + \tau_{n-1}) = 0 \qquad \text{for all} \quad \lambda \in L \tag{8}$$

But this is impossible as we show in the following proof due to Dedekind: Each singleton $\{\tau_\mu\}$ is linearly independent; for $\mu\tau_q(\lambda) = 0$ for some fixed $\mu \in M$ and for all $\lambda \in L$ implies $\mu = 0$ since $\tau_q(1) = 1$. Now assume that $\{\tau_0, \ldots, \tau_{n-1}\}$ is linearly dependent and let

$$\mu_1\tau_{q_1} + \cdots + \mu_t\tau_{q_t} = 0, \qquad \mu_i \in M, \quad \mu_i \neq 0 \tag{9}$$

be a nontrivial dependence of minimal length $t \geqslant 2$ between the τ. Since

$\tau_{q_1} \neq \tau_{q_t}$, there exists $\lambda_0 \in L$ such that $\tau_{q_1}\lambda_0 \neq \tau_{q_t}\lambda_0$. Applying (9) to $\lambda_0\lambda$ we get

$$\mu_1(\tau_{q_1}\lambda_0)(\tau_{q_1}\lambda) + \cdots + \mu_t(\tau_{q_t}\lambda_0) \cdot (\tau_{q_t}\lambda) = 0 \qquad \text{for all} \quad \lambda \in L \tag{10}$$

Multiplication of (9) by $\tau_{q_1}\lambda_0$ yields

$$\mu_1(\tau_{q_1}\lambda_2)(\tau_{q_1}\lambda) + \cdots + \mu_t(\tau_{q_1}\lambda_0)(\tau_{q_t}\lambda) = 0 \qquad \text{for all} \quad \lambda \in L \tag{11}$$

Finally, subtracting (11) from (10) we obtain a linear dependence of length less than t. Moreover, the dependence is nontrivial since $\mu_t[\tau_{q_t}\lambda_0 - \tau_{q_t}\lambda_0] \neq 0$. This contradicts the minimality of the length of the dependence in (9). Thus (8) cannot hold and there is some $\lambda \in L$ such that $\mathrm{Tr}(\lambda) \neq 0$. ∎

Separable Algebras

CHAPTER VI

1. Central, simple algebras

In Section V.4 we dealt with the Φ-representations of a Φ-algebra R. We shall now investigate what happens to an R-module M (and hence to the associated matrix representation) when we extend Φ to an overfield Δ.

The following example shows that a simple Φ-algebra R does not necessarily remain simple when we extend the ground field Φ to Δ. In fact we shall see that R_Δ is not even semisimple. In Theorem VIII.3.4 however, we shall prove that R_Δ is a "Frobenius algebra."

Example. Let $f(X)$ be a polynomial in X with coefficients in a field Φ. Suppose f is irreducible over Φ but not separable, i.e. f has multiple roots in some splitting field; for example, we could take Φ to be the field discussed in V.1.2(17). Let Δ be a splitting field of f over Φ. Since f is irreducible over Φ, the factor ring $R = \Phi[X](f(X))$ is a field. Let n be the degree of f. The elements of R are of the form

$$\alpha_0 + \alpha_1\bar{X} + \alpha_2\bar{X}^2 + \cdots + \alpha_{n-1}\bar{X}^{n-1}, \qquad \alpha_i \in \Phi$$

where $\bar{X} = X + (f(X))$ and $\alpha \in \Phi$ is identified with $\alpha + (f(X))$. The mapping

$$\psi: \bar{X} \otimes \delta \to \delta\bar{X}, \qquad \delta \in \Delta$$

induces an isomorphism of $R \otimes_\Phi \Delta$ onto $\Delta[X]/(f(X))$.

Proof. ψ is clearly an epimorphism. Now let $\{\delta_1, ..., \delta_N\}$ be a basis of Δ over Φ and suppose

$$\psi\colon \sum_{1\leqslant i\leqslant N} \Big(\sum_{0\leqslant k\leqslant n-1} \alpha_{ik}X^k\Big) \otimes \delta_i = \sum_k \bar{X}^k \sum_i \alpha_{ik}\delta_i = 0$$

Since $\{\bar{X}^0, ..., \bar{X}^{n-1}\}$ is linearly independent over Δ, it follows that

$$\sum_{1\leqslant i\leqslant N} \alpha_{ik}\delta_i = 0, \qquad k = 0, ..., n-1$$

But $\{\delta_1, ..., \delta_N\}$ is a Φ-basis of Δ and hence $\alpha_{ik} = 0$ for all i and k. Now if $f(X) = (X - \lambda_1)^{t_1} \cdots (X - \lambda_k)^{t_k}$ in $\Delta[X]$, let $g(X) = (X - \lambda_1) \cdots (X - \lambda_k)$. Since some $t_i \geqslant 2$, $g(X) \neq f(X)$ and hence $f(X)$ does not divide $g(X)$. Therefore $g(\bar{X}) \neq \bar{0}$. On the other hand $(g(X))^m$ is divisible by $f(X)$ for some m. Thus $\Delta[X]/(f(X))$ contains a nonzero, nilpotent element and since $\Delta[X]/(f(X)) \approx R \otimes_\Phi \Delta$, so does $R \otimes_\Phi \Delta$. By Theorem III.1.5, $R \otimes_\Phi \Delta$ is not semisimple.

In the above example the Φ-algebra R is a field and consequently, when considered as a left R-module ${}_RR$, it is irreducible. On the other hand, such is not the case with $R \otimes_\Phi \Delta$ since the radical of this ring is neither zero nor the whole of $R \otimes_\Phi \Delta$ and so is a proper $R \otimes_\Phi \Delta$-submodule of $R \otimes_\Phi \Delta$. This hints at the importance of the following concept:

Definition 1. Let R be a Φ-algebra. A left R-module M is *absolutely irreducible* if the $R \otimes_\Phi \Delta$-module $M \otimes_\Phi \Delta$ is irreducible whenever Δ is an algebraic extension of Φ, i.e. whenever Δ is an overfield of Φ consisting entirely of elements which are algebraic over Φ. A matrix Φ-representation of R is said to be *absolutely irreducible* if the associated representation module is absolutely irreducible.

Theorem 1. Hypotheses: Let the artinian ring R be a Φ-algebra and let M be an irreducible R-module.

Conclusion: 1. If

$$\mathrm{Hom}_R(M, M) \approx \Phi \tag{1}$$

M is absolutely irreducible.

2. If M is absolutely irreducible and $\dim_\Phi R < \infty$, then (1) follows.

Property (1) is of particular importance with regard to the Jacobson density theorem (Theorem II.1.4, also see Theorem 2).

Proof of 1. 1. Let $R/J = \bar{\mathfrak{A}}_1 \oplus \cdots \oplus \bar{\mathfrak{A}}_q$ be the decomposition of the semisimple artinian ring R/J as a direct sum of the simple rings $\bar{\mathfrak{A}}_i$. Then M

is a faithful $\bar{\mathfrak{A}}_i$-module for some i. Suppose, without loss of generality, M is a faithful $\bar{\mathfrak{A}}_1$-module. Hence M is isomorphic as an R-module to a minimal left ideal $\bar{L}_1$ of $\bar{\mathfrak{A}}_1$.

2. From (1) and the density theorem applied to the simple artinian ring $\bar{\mathfrak{A}}_1$, it follows that

$$\bar{\mathfrak{A}}_1 \approx \Phi_{n\times n} \qquad \text{and hence} \qquad \bar{L}_1 \approx \sum_{1\leqslant i\leqslant n} C_{i1}\Phi$$

where C_{ik} is the $n \times n$ matrix with 1 in the (i, k) position and 0 elsewhere (cf. Section II.5, Eq. (1)). Let Δ be an overfield of Φ. The mapping

$$\psi\colon C_{ik} \otimes \delta_{ik} \to \delta_{ik}C_{ik}, \qquad i, k = 1, \ldots, n$$

induces an isomorphism ψ of the algebra $\Phi_{n\times n} \otimes_\Phi \Delta$ onto $\Delta_{n\times n}$ as can be easily checked. Under this isomorphism

$$\psi \sum_i C_{i1} \otimes \Delta = \sum_i C_{i1}\Delta$$

and $\sum_i C_{i1}\Delta$ is a minimal left ideal in $\Delta_{n\times n}$. Hence it is an irreducible $R \otimes_\Phi \Delta$-module which is isomorphic to $M \otimes_\Phi \Delta$ since

$$M \otimes_\Phi \Delta \approx \bar{L}_1 \otimes_\Phi \Delta \approx \Big(\sum_i C_{i1}\Phi\Big) \otimes_\Phi \Delta \approx \sum_i C_{i1}\Delta$$

3. Conversely assume that M is absolutely irreducible and let Ω be an algebraically closed overfield of Φ. The simple Φ-algebra $\bar{\mathfrak{A}}_1$ is isomorphic to $K_{n\times n}$, where K is the centralizer of M:

$$K = \mathrm{Hom}_R(M, M)$$

By Schur's lemma, K is a division ring. It follows that

$$\bar{\mathfrak{A}}_1 \otimes_\Phi \Omega \approx K_{n\times n} \otimes_\Phi \Omega \approx (K \otimes_\Phi \Omega)_{n\times n} \tag{2}$$

We assert that $K \otimes_\Phi \Omega$ is a division ring. For suppose not. Then the Φ-algebra $K \otimes \Omega$ contains a proper left ideal Λ. Then $\sum C_{i1}\Lambda$ is a proper left ideal contained in the left ideal $\sum_{1\leqslant i\leqslant n} C_{i1}K \otimes_\Phi \Omega$ of the ring $\sum_{i,k} C_{ik}K \otimes_\Phi \Omega \approx \bar{\mathfrak{A}}_1 \otimes_\Phi \Omega$. But $M \otimes_\Phi \Omega \approx \sum_{1\leqslant i\leqslant n} C_{i1}K \otimes_\Phi \Omega$ as $\bar{\mathfrak{A}}_1 \otimes_\Phi \Omega$-modules. Therefore $M \otimes_\Phi \Omega$ is not irreducible contradicting the absolute irreducibility of the R-module M. It now follows from (2) that $\bar{\mathfrak{A}}_1 \otimes_\Phi \Omega$ is a simple algebra over Ω. From (2) we obtain

$$\dim_\Phi \bar{\mathfrak{A}}_1 = \dim_\Omega(\bar{\mathfrak{A}}_1 \otimes_\Phi \Omega) = n^2 \dim_\Omega(K \otimes_\Phi \Omega)$$

where the first equality follows from the definition of the tensor product and the second from the embedding of the field Ω in $K \otimes_{\Phi} \Omega$. Since $\dim_{\Phi} \bar{\mathfrak{A}}_1 \leqslant \dim_{\Phi} R$ and the latter is finite by hypothesis 2, it follows that $\dim_{\Omega} K \otimes_{\Phi} \Omega < \infty$. Each element $\rho \in K \otimes_{\Phi} \Omega$ is therefore the root of a polynomial with coefficients in Ω and hence lies already in Ω. This proves

$$K \otimes_{\Phi} \Omega \approx \Omega \tag{3}$$

and hence $\dim_{\Phi} K = 1$, i.e. $K = \Phi$. ▮

Applying Theorem 1 and the density theorem (Theorem II.3.1) to a simple artinian ring we have

Theorem 2. Let R be a simple Φ-algebra, $\dim_{\Phi} R < \infty$. Then R has a faithful absolutely irreducible module M if and only if R is isomorphic to $\Phi_{n \times n}$ for a suitable n. ▮

The structure of the algebra $\Phi_{n \times n}$ over the field Φ is, of course, well-known. This explains our interest in absolutely irreducible algebra modules.

The following discussion leads naturally to an important example.

Definition 2. Let R be a Φ-algebra, Φ a field. The elements a of R induce Φ-linear maps of R to itself by left and right multiplication by a:

$$\mathfrak{L}_a : x \to a \cdot x, \qquad \mathfrak{R}_a : x \to x \cdot a \qquad \text{for} \quad x \in R$$

The set of elements $\mathfrak{L}_a$ and $\mathfrak{R}_a$ with $a \in R$ does not form a subring of $\text{Hom}_{\Phi}(R, R)$ but it generates a subring, the so-called *multiplication algebra* $T(R)$. The ring R is a $T(R)$-module and the $T(R)$-submodules are precisely the ideals of R. The centralizer

$$\Gamma = \text{Hom}_{T(R)}(R, R)$$

is called the *centroid* of R.

Theorem 3. If $R^2 = R$, the centroid Γ is a commutative algebra over Φ.

Proof. Let $\gamma_1, \gamma_2 \in \Gamma$, $x, y \in R$. On the one hand

$$\begin{aligned}(x \cdot y)\,\gamma_1\gamma_2 &= ((x \cdot y)\,\gamma_1)\,\gamma_2 = (\mathfrak{L}_x y\gamma_1)\,\gamma_2 = (x(y\gamma_1))\,\gamma_2 \\ &= (\mathfrak{R}_{y\gamma_1} x)\,\gamma_2 = \mathfrak{R}_{y\gamma_1}(x\gamma_2) = (x\gamma_2)(y\gamma_1)\end{aligned}$$

while

$$\begin{aligned}(x \cdot y)\,\gamma_1\gamma_2 &= ((x \cdot y)\,\gamma_1)\,\gamma_2 = (\mathfrak{R}_y x\gamma_1)\,\gamma_2 = ((x\gamma_1)\,y)\,\gamma_2 \\ &= \mathfrak{L}_{x\gamma_1}(y\gamma_2) = (x\gamma_1) \cdot (y\gamma_2)\end{aligned}$$

It follows that $(xy)\,\gamma_1\gamma_2 = (xy)\,\gamma_2\gamma_1$. Since $R^2 = R$, $\gamma_1\gamma_2 = \gamma_2\gamma_1$. ▮

In connection with the Jacobson density theorem, the structure of simple Φ-algebras with centroid equal to the ground field is easily determined.

Definition 3. A Φ-algebra R is said to be *central* if the centroid of R is Φ.

From Theorems 1 and 2 it follows that a central simple algebra R of finite dimension over the ground field Φ is a faithful absolutely irreducible $T(R)$-module where $T(R)$ is the multiplication algebra of R. Moreover, if Ω is an algebraically closed overfield of Φ, the tensor product

$$T(R) \otimes_{\Phi} \Omega \approx \Omega_{n\times n}$$

for a suitable n.

Every ring is an algebra over its center Z. If R is a simple ring with identity, Z is a field. For if $0 \neq z \in Z$, Rz is an ideal of R different from (0). Therefore $Rz = R$ and there exists $z' \in R$ such that $z'z = zz' = 1$. But clearly $z' \in Z$ and Z is a field.

The relationship between the center and the centroid of a simple artinian ring is described in the following:

Theorem 4. The centroid Γ of a simple algebra $R = K_{n\times n}$ over Φ is $\{\mathfrak{R}_z \mid z \in Z\}$, where Z is the center of $K_{n\times n}$.

Proof. Clearly $\{\mathfrak{R}_z \mid z \in Z\} \subseteq \Gamma$. Conversely let $\gamma \in \Gamma$ and set $c = 1\gamma$. Then for all $a \in R$,

$$a\gamma = (a \cdot 1)\gamma = (\mathfrak{L}_a 1)\gamma = \mathfrak{L}_a(1\gamma) = a \cdot c$$

Hence $\gamma = \mathfrak{R}_c$. Moreover

$$a\gamma = (1 \cdot a)\gamma = (\mathfrak{R}_a 1)\gamma = \mathfrak{R}_a(1\gamma) = \mathfrak{R}_a c = ca$$

for all $a \in \mathfrak{R}$. Therefore $c \in Z$. ∎

The following theorem is of basic importance in the theory of central algebras.

Theorem 5. Let $\mathfrak{A}$ be a central, simple algebra over the field Φ. Let $\mathfrak{B}$ be an arbitrary Φ-algebra with identity 1. Then if $\mathfrak{b}$ is an ideal in $\mathfrak{B}$,

$$\mathfrak{b} = \left\{\sum_i a_i \otimes b_i \,\middle|\, a_i \in \mathfrak{A}, b_i \in \mathfrak{b}\right\}$$

is an ideal in the algebra $\mathfrak{A} \otimes_{\Phi} \mathfrak{B}$ over Φ. The mapping ψ: $\mathfrak{b} \to \mathfrak{A} \otimes_{\Phi} \mathfrak{b}$ is a lattice isomorphism of the lattice $V(\mathfrak{B})$ of ideals in $\mathfrak{B}$ onto the lattice $V(\mathfrak{A} \otimes_{\Phi} \mathfrak{B})$ of ideals in $\mathfrak{A} \otimes_{\Phi} \mathfrak{B}$.

Proof. From the very definition of multiplication in the tensor product $\mathfrak{A} \otimes_{\Phi} \mathfrak{b}$, it is clear that $\mathfrak{A} \otimes_{\Phi} \mathfrak{b}$ is an ideal of $\mathfrak{A} \otimes_{\Phi} \mathfrak{B}$ if $\mathfrak{b}$ is an ideal of $\mathfrak{B}$. Moreover, the implication

$$\mathfrak{b}_1 \subseteq \mathfrak{b}_2 \Rightarrow \mathfrak{A} \otimes \mathfrak{b}_1 \subseteq \mathfrak{A} \otimes \mathfrak{b}_2 \tag{4}$$

is trivial. To show ψ is one-one, let $\mathfrak{b}_2 \not\subseteq \mathfrak{b}_1$ and let $b_2 \in \mathfrak{b}_2$, $b_2 \notin \mathfrak{b}_1$. Then $a \otimes b_2 \notin \mathfrak{A} \otimes \mathfrak{b}_1$ if $a \neq 0$. Otherwise, let $\{u_i \mid i \in I\}$ be a Φ-basis of $\mathfrak{b}_1$ (cf. Theorem I.3.1). Then $\{b_2, u_i \mid i \in I\}$ is a linearly independent family over Φ. $a \otimes b_2 \in \mathfrak{A} \otimes \mathfrak{b}_1$ implies the existence of elements $a_1, ..., a_N$ in $\mathfrak{A}$ such that

$$a \otimes b_2 - \sum_{1 \leqslant j \leqslant N} a_j \otimes u_{i_j} = 0$$

The linear independence of $\{b_2, u_{i_1}, ..., u_{i_N}\}$ and the definition of the tensor product imply $a = a_1 = \cdots = a_N = 0$, a contradiction. Hence, if $\mathfrak{b}_2 \not\subseteq \mathfrak{b}_1$, $\psi\mathfrak{b}_2 \not\subseteq \psi\mathfrak{b}_1$. This, together with (4) proves that ψ is one-one and preserves the partial ordering $\subseteq$ of the lattices. Finally we must show ψ is onto. Let $W' = \{b \mid b \in \mathfrak{B}, a \otimes b \in W \text{ for all } a \in \mathfrak{A}\}$, where W is an ideal of $\mathfrak{A} \otimes_{\Phi} \mathfrak{B}$. The vector space W' over Φ is contained in $\mathfrak{B}$. Each $V \in W$ is expressible in the form

$$V = \sum_{1 \leqslant i \leqslant N} a_i \otimes b_i, \qquad a_i \neq 0, \qquad i = 1, ..., N \tag{5}$$

where, without loss of generality, we may assume $\{a_1, ..., a_N\}$ is linearly independent over Φ and $b_i \in \mathfrak{B}$. We now show $b_i \in W'$. $\mathfrak{A}$ is a faithful irreducible $T(\mathfrak{A})$-module and by hypothesis

$$\mathrm{Hom}_{T(\mathfrak{A})}(\mathfrak{A}, \mathfrak{A}) = \Phi$$

Hence given $a \in \mathfrak{A}$ there exists $T \in T(\mathfrak{A})$ such that

$$Ta_1 = a, \quad Ta_j = 0, \qquad j = 2, ..., N$$

Since $T \otimes \mathscr{L}_1$ is an element of $T(\mathfrak{A} \otimes_{\Phi} \mathfrak{B})$, the multiplication algebra of $\mathfrak{A} \otimes_{\Phi} \mathfrak{B}$, and since W is an ideal of $\mathfrak{A} \otimes \mathfrak{B}$, it follows from (5) that $(T \otimes \mathscr{L}_1)V = a \otimes b_1 \in W$. Thus $a \otimes b_1 \in W$ for all $a \in \mathfrak{A}$ and $b_1 \in W'$. Similarly we show $b_i \in W'$ for all i. Hence $\psi W' = W$. To complete the proof we merely observe that since

$$a \otimes vb = (a \otimes v)(1 \otimes b) \in W \qquad \text{for all} \quad a \in \mathfrak{A}, v \in W' \text{ and } b \in \mathfrak{B}$$

and

$$a \otimes bv = (1 \otimes b)(a \otimes v) \in W \qquad \text{for all} \quad a \in \mathfrak{A}, v \in W' \text{ and } b \in \mathfrak{B}$$

W' is an ideal of $\mathfrak{B}$. ∎

We now give two applications of this theorem.

Theorem 6. Hypotheses: Let $\mathfrak{A}$ and $\mathfrak{B}$ be subalgebras of Φ-algebra $\mathbb{C}$. Let $\mathfrak{A}$ be central and simple and let $\mathfrak{B}$ contain an identity. Suppose $a \cdot b = b \cdot a$ for all $a \in \mathfrak{A}$ and $b \in \mathfrak{B}$, and assume that if $b \in \mathfrak{B}$ and $\mathfrak{A} \cdot b = 0$, then $b = 0$.

Conclusion: $\mathfrak{A} \cdot \mathfrak{B}$ is isomorphic to $\mathfrak{A} \otimes_\Phi \mathfrak{B}$ under the isomorphism

$$\psi: \sum a_i \otimes b_i \to \sum a_i b_i$$

Proof. Clearly ψ is an epimorphism. Let $\ker \psi$ be the kernel. If $\ker \psi \neq 0$, then by the preceding theorem

$$\ker \psi = \mathfrak{A} \otimes \mathsf{b}$$

where $\mathsf{b} \neq 0$ is some ideal of $\mathfrak{B}$. Thus for each $b \neq 0$ in b and all $a \in \mathfrak{A}$, $ab = 0$. By hypothesis $b = 0$ for all $b \in \mathfrak{B}$, a contradiction. ∎

Theorem 7. Let $\mathfrak{A}$ be a central, simple algebra over Φ and Δ an arbitrary overfield of Φ. Then $\mathfrak{A} \otimes_\Phi \Delta$ is a simple Δ-algebra.

Proof. The lattice of ideals of $\mathfrak{A} \otimes \Delta$ is isomorphic to the lattice of ideals of Δ by Theorem 5. The latter consists only of (0) and Δ. ∎

The last theorem enables us to complete the investigation of algebras with absolutely irreducible modules begun in Theorems 1 and 2.

Definition 4. Let R be a Φ-algebra. An overfield Δ of Φ is a *splitting field of the algebra* R if each irreducible module of the Δ-algebra $R \otimes_\Phi \Delta$ is absolutely irreducible.

Theorem 8. Let Φ be the center of the simple algebra R, let $\dim_\Phi R < \infty$ and let Δ be an overfield of Φ. Then Δ is a splitting field of the Φ-algebra R if and only if

$$R \otimes_\Phi \Delta \approx \Delta_{r \times r}$$

for a suitable r. If in addition $\dim_\Phi \Delta$ is finite, $\dim_\Phi R = r^2$.

Proof. By Theorems 4 and 7, $R \otimes_\Phi \Delta$ is a simple Δ-algebra. Hence

$$R \otimes_\Phi \Delta \approx K_{r \times r}$$

where $K = \operatorname{Hom}_{R \otimes \Delta}(M, M)$, M an irreducible $R \otimes \Delta$-module by Theorem II.6.2. By Theorem 1, M is an absolutely irreducible module for the

Δ-algebra $R \otimes \Delta$ if and only if $K = \Delta$. The Φ-dimension of R is obtained as follows:

$$\dim_\Phi R \cdot \dim_\Phi \Delta = \dim_\Phi(R \otimes \Delta) = \dim_\Phi \Delta_{r\times r} = r^2 \dim_\Phi \Delta \quad \blacksquare$$

As an immediate consequence we have:

Theorem 9. Let R be a central simple Φ-algebra with $\dim_\Phi R < \infty$. Then each irreducible R-module is absolutely irreducible if and only if R is isomorphic to $\Phi_{n\times n}$ for a suitable n. $\blacksquare$

From Theorems 2 and 8 it follows that an algebraically closed field Ω containing Φ is a splitting field of the simple artinian ring R with center Φ if R has faithful absolutely irreducible R-module.

We have come across the concept of a splitting field of a field Φ on two occasions: once in Section V.1 as the splitting field or *root field* of a polynomial $f(X)$ in $\Phi[X]$ and above, in Definition 4, as the splitting field of an algebra R over Φ.

The reason for the importance of a splitting field Δ of a Φ-algebra R (if Δ exists) is due in part to the fact that R may be embedded in $R \otimes_\Phi \Delta \approx \Delta_{n\times n}$, the structure of which is well known. The embedding map is

$$a \to a \otimes 1$$

Theorem 10. 1. A simple artinian ring $R = K_{n\times n}$ is an algebra over its center Φ. If E is a maximal subfield of the division ring K, then E is a splitting field of the Φ-algebra R. Moreover, every overfield of E is a splitting field of R over Φ.

2. Let $n = 1$ in Part 1. Then $R = K$, a division ring. We know $K \otimes_\Phi E = E_{r\times r}$ for some r. Then $\dim_E K = r$. Furthermore, if $\dim_\Phi K < \infty$, then $\dim K = r^2$ and $\dim E = r$.

Proof. 1. The set theoretic union of a chain of subfields of K containing Φ

$$\Phi = \Delta_0 \subset \cdots \subset \Delta_i \subset \Delta_k \subset \cdots$$

is a subfield of K. Hence by Zorn's lemma there exists at least one maximal subfield of K.

2. A field E lying between Φ and K is maximal in the family of fields containing Φ and contained in K if and only if E contains all the elements of K which commute with E, i.e. if and only if

$$E = \{k \mid k \in K, \quad ek = ke \text{ for all } e \in E\} \tag{6}$$

Proof. Denote the set on the right of (6) by $\mathcal{E}$. Every overfield Δ of E lies in $\mathcal{E}$. Hence if (6) holds, E is maximal. On the other hand if (6) does not hold, then there is an element $z \in \mathcal{E}$, $z \notin E$. But the division ring generated by E and z is a field strictly containing E. Thus E is not maximal.

3. Let M be a faithful irreducible R-module with centralizer

$$\mathrm{Hom}_R(M, M) = K$$

and let E be a maximal subfield of K containing Φ. Since by hypothesis R is central, simple, $R \otimes_\Phi E$ is a simple E-algebra by Theorem 7. We turn M into an $R \otimes_\Phi E$-module by defining

$$\left(\sum a_i \otimes e_i\right) m = \sum_i a_i m e_i \qquad \text{for} \quad m \in M, \quad a_i \in R, \quad e_i \in E \tag{7}$$

(7) is independent of the choice of representative since

$$[(a, e + e') - (a, e) - (a, e')]m = am(e + e') - ame - ame' = 0$$
$$[(a + a', e) - (a, e) - (a', e)]m = (a + a')\, me - ame - a'me = 0$$

and

$$[(a\alpha, e) - (a, \alpha e)]m = (am\alpha)e - (am)(\alpha e) = 0 \qquad \text{for all} \quad \alpha \in \Phi$$

Furthermore,

$$(a \otimes e)[(a' \otimes e')m] = (a \otimes e)(a'me') = aa'me'e = (aa' \otimes ee')m$$

since E is commutative. By hypothesis M is an irreducible R-module and since $R = R \otimes 1 \subseteq R \otimes E$, it is also an irreducible $R \otimes E$-module. Hence, since $R \otimes E$ is simple, M is faithful as an $R \otimes E$-module. By the density theorem, $R \otimes E$ is isomorphic to $D_{r \times r}$, where

$$D = \mathrm{Hom}_{R \otimes E}(M, M) \tag{8}$$

and $r = \dim_D M$. By the first assertion in Theorem 1, it is sufficient to prove

$$E \approx D \tag{9}$$

Proof of (9). By definition 7, $1 \otimes E \approx E$ is contained in D, since for all $e \in E$

$$\left(\sum a_i \otimes e_i\right) [(1 \otimes e)m] = \sum a_i(me)\, e_i = \sum a_i m e_i e$$
$$= \left[\left(\sum a_i \otimes e_i\right) (1 \otimes e)\right] m$$

On the other hand, $D \subseteq K$ since M is an R-module and $K = \mathrm{Hom}_R(M, M)$. Let $\delta \in D$. From $m\delta e = (1 \otimes e)(m\delta) = [(1 \otimes e)m]\delta = me\delta$ it follows that $m\delta e = me\delta$ for all $e \in E$ and all $m \in M$. Thus $\delta e = e\delta$ for all $e \in E$. By the characterization of maximal subfields of K proved in 2, it follows that $\delta \in E$ and hence $D \subseteq E$.

4. That each overfield Δ' of a splitting field Δ is a splitting field follows from

$$\Delta_{r \times r} \otimes {}_{\Delta}\Delta' \approx \Delta'_{r \times r}$$

5. The assertions in Part 2 concerning the various dimensions follow from the fact that if $R = K$ is a division ring, the module $M = {}_R R$ is a faithful irreducible R-module. Thus by (8) and (9), $r = \dim_D M = \dim_D R = \dim_E R$. If in addition $\dim_\Phi R < \infty$, then $\dim_\Phi E = r$ by Theorem 8 and since $r^2 = \dim_\Phi R = \dim_E R \cdot \dim_\Phi E = r \cdot \dim_\Phi E$. ▮

To generalize the above investigation of simple artinian rings to the semisimple case we introduce the following:

Definition 5. A *Φ-algebra R is separable* over Φ if for each overfield Δ of Φ, the Δ-algebra $R \otimes {}_\Phi \Delta$ is semisimple. We note that, in particular, a separable algebra is semisimple ($\Delta = \Phi$).

Theorem 11. Let the artinian ring R be a Φ-algebra. R is separable over Φ if there is an overfield E of such that

$$R \otimes {}_\Phi E \approx E_{r_1 \times r_1} \oplus \cdots \oplus E_{r_q \times r_q} \tag{10}$$

Moreover the condition is also necessary if $\dim_\Phi R < \infty$.

Proof. 1. If R is separable, $\dim_\Phi R < \infty$ and Ω is an algebraically closed field containing Φ, the Ω-algebra $R \otimes {}_\Phi \Omega$ is separable and hence isomorphic to a direct sum

$$K^{(1)}_{r_1 \times r_1} \oplus \cdots \oplus K^{(q)}_{r_q \times r_q}$$

of simple Ω-algebras. Since Ω is algebraically closed, then by (3) in the proof of the second assertion of Theorem 1, each $K^{(j)} = \Omega$.

2. Conversely, suppose condition (10) holds. We must show that for any overfield Δ of Φ (and not only of E!), $R \otimes {}_\Phi \Delta$ is semisimple. We first construct a field Λ containing as subfields E and Δ: The mappings $e \to e \otimes 1$ and $\delta \to 1 \otimes \delta$ are embeddings of the fields E and Δ, respectively in $E \otimes {}_\Phi \Delta$. By Zorn's lemma there exists a maximal ideal A in the commutative ring

$E \otimes_\Phi \Delta$ since $E \otimes_\Phi \Delta$ contains an identity $1 \otimes 1$. The quotient ring $E \otimes_\Phi \Delta / A$ is a field Λ in which E and Δ are embedded by means of the embeddings

$$e \to e \otimes 1 + A \qquad \text{and} \qquad \delta \to 1 \otimes \delta + A$$

By Theorem V.3.6 and (10) we have

$$\begin{aligned} R \otimes_\Phi \Lambda &\approx R \otimes_\Phi (E \otimes_E \Lambda) \approx (R \otimes_\Phi E) \otimes_E \Lambda \\ &\approx (E_{r_1 \times r_1} \otimes_E \Lambda) \oplus \cdots \oplus (E_{r_q \times r_q} \otimes_E \Lambda) \\ &\approx \Lambda_{r_1 \times r_1} \oplus \cdots \oplus \Lambda_{r_q \times r_q} \end{aligned} \tag{11}$$

Hence $R \otimes_\Phi \Delta$ is semisimple. On the other hand

$$R \otimes_\Phi \Lambda \approx (R \otimes_\Phi \Delta) \otimes_\Delta \Lambda \tag{12}$$

If $R \otimes_\Phi \Delta$ is not semisimple, its radical contains a nilpotent ideal $W \neq 0$. But then $W \otimes_\Delta \Lambda$ is a nilpotent ideal in $(R \otimes_\Phi \Delta) \otimes_\Delta \Lambda \approx R \otimes_\Phi \Lambda$, contradicting (11). Therefore $R \otimes_\Phi \Delta$ is semisimple. ▮

Theorem 12. Let R be a semisimple artinian ring R which is an algebra over Φ. Let

$$R = \mathfrak{A}_1 \oplus \cdots \oplus \mathfrak{A}_q$$

be the decomposition of R as a direct sum of simple algebras $\mathfrak{A}_j$, $j = 1, \ldots, q$ and suppose the center of $\mathfrak{A}_j$ for each j is Φ. Then R is separable.

Proof. Let E be a field containing the splitting fields $E^{(j)}$ of the $\mathfrak{A}_j$, $j = 1, \ldots, q$. By Theorem 10, E is also a splitting field of each $\mathfrak{A}_j$ and hence

$$R \otimes_\Phi E \approx E_{r_1 \times r_1} \oplus \cdots \oplus E_{r_q \times r_q}$$

By Theorem 11, R is separable over Φ. ▮

It may be shown that if Δ is a separable extension of Φ in the field-theoretic sense, (this is the case, for example, when char $\Phi = 0$) then $R \otimes_\Phi \Delta$ is semisimple if R is. This result will be proved in the context of Frobenius algebras in Chapter VIII (cf. Theorem VIII.6.2). A proof which is independent of the theory of Frobenius algebras may be found in van der Waerden [1959], Section 153.

2. The commutativity of rings satisfying $x^{n(x)} = x$

It was known to Wedderburn that a finite division ring K is a field. (We shall prove this result in Theorem 3.) The multiplicative group $K^{\times} = K \backslash 0$ is finite and therefore each element $k \in K^{\times}$ has finite order, i.e. there exists a natural number m such that $k^m = 1$. If we include the zero of K we may say quite generally that

$$a^{n(a)} = a \tag{1}$$

for all $a \in K$ and suitable $n(a) \in \mathbb{N}$. Wedderburn's theorem was generalized by Jacobson to the following:

Theorem 1 (Jacobson). Let R be a ring such that for each $a \in R$ there exists a natural number $n(a) > 1$ dependent on a for which

$$a^{n(a)} = a$$

Then R is commutative. ∎

The proof of this theorem depends on a number of intermediate results. First, we may write $a^{n(a)} = a$ as $a^{n(a)-1}a = a$. It follows that $a^{n(a)-1}a^k = a^k$, $k = 1, 2, \ldots$. Taking $k = n(a) - 1$ we have

$$e = e^2 \qquad \text{where} \quad e = a^{n(a)-1}$$

Since $e \in (a)$ and $a \in (e)$, $(e) = (a)$. If R is not semisimple, the radical J contains an idempotent e different from 0. But e must then be quasiregular and there exists $b \in R$ such that $e + b - eb = 0$. Multiplying both sides by e we get $e = 0$, a contradiction. Hence R is semisimple and as such is a subdirect sum of primitive rings:

$$R \approx \sum_{P \in \mathscr{P}}{}_{s}\, R/P \tag{2}$$

where each R/P has the same property as R. We now show that a primitive ring $\bar{R}$ with the property expressed in (1) is a division ring: Let M be a faithful, irreducible $\bar{R}$-module and let K be its centralizer. If $\dim_K M > 1$, there exist two K-linearly independent elements $u_1, u_2 \in M$ and an element $\bar{a} \in \bar{R}$ such that $\bar{a}u_1 = u_2$ and $\bar{a}u_2 = 0$. This implies $\bar{a}^n u_1 = 0 \neq u_2$ for all $n \geqslant 2$, a contradiction since $\bar{a}^{n(a)} = \bar{a}$. Thus $\dim_K M = 1$ and $\bar{R} \approx K$. The commutatitivity of the subdirect sum R follows from the commutativity of the direct summands R/P. Therefore to complete the proof of Theorem 1 it is sufficient to consider the special case that R is a division ring K.

We start our investigation of this special case with the following:

Theorem 2. Let K be a division ring and let Φ be the center of K. Clearly K is a Φ-algebra. Let S_1 and S_2 be Φ-subalgebras of K which are division rings. Suppose $\dim_\Phi S_1 = \dim_\Phi S_2 < \infty$ and let

$$\phi: s_1 \to \phi s_1 \qquad \text{for} \quad s_1 \in S_1$$

be an isomorphism of S_1 onto S_2. Then there exists an element $k_0 \in K$ such that

$$\phi s_1 = k_0^{-1} s_1 k_0 \qquad \text{for all} \quad s_1 \in S_1$$

We say that ϕ may be extended to an inner *automorphism* of K.

Proof. Let K^* denote the division ring consisting of the same elements as K with the same additive structure as K but with multiplication in K^* defined by

$$a * b = b \cdot a \tag{3}$$

Both K and K^* are central over Φ. By Theorem VI.1.5 applied to $\mathcal{A} = K^*$ and $\mathcal{B} = S_1$, the tensor product $S_1 \otimes_\Phi K^*$ is a simple ring. It contains a minimal left ideal since each left ideal of $S_1 \otimes_\Phi K^*$ is a K^*-module and by the very definition of the tensor product

$$\dim_K{}^*(S_1 \otimes_\Phi K^*) = \dim_\Phi S_1 < \infty \tag{4}$$

It follows from Theorem II.6.1 that all irreducible $S_1 \otimes_\Phi K^*$-modules are isomorphic to each other. Now K itself can be turned into an $S_i \otimes K^*$-module for $i = 1, 2$ by defining

$$(s_i \otimes k_i)k = s_i k k_i\,, \qquad i = 1, 2 \tag{5}$$

and extending the definition to $S_i \otimes K^*$ by linearity. Since distributivity is built into the definition we need only check

$$\begin{aligned}[(s_i \otimes k_i)(s_i' \otimes k_i')]k &= [s_i s_i' \otimes k_i' k_i]k = s_i s_i' k k_i' k_i \\ &= (s_i \otimes k_i)[(s_i' \otimes k_i')k]\end{aligned}$$

We note that we used (3) in our calculations. On the other hand, using the isomorphism ϕ in the hypothesis of the theorem, K may be turned into an $S_1 \otimes_\Phi K^*$-module by defining

$$(s_1 \otimes k_1)k = \phi(s_1) \cdot k k_1 \qquad \text{for} \quad s_1 \otimes k_1 \in S_1 \otimes_\Phi K^* \quad \text{and} \quad k \in K \tag{6}$$

as may be easily checked. Since K is a division ring it is an irreducible $S_1 \otimes K^*$-module relative to the definitions in (5) and (6). Hence, as we remarked above, there exists an $S_1 \otimes K^*$-isomorphism ψ of the module K as defined in (5) onto the module K as defined in (6): Since ψ is an $S_1 \otimes K^*$-homomorphism,

$$\psi(s_1 k k_1) = \psi((s_1 \otimes k_1)k) = \phi(s_1)\, \psi(k)\, k_1 \tag{7}$$

for all $s_1 \in S_1$, $k \in K$, $k_1 \in K^*$. Set $k_1 = k = 1$ in (7) and we get

$$\psi(s_1) = \phi(s_1) \cdot \psi(1) \tag{8}$$

Since $s_1 \in S_1 \subseteq K$, s_1 is an element of K^*. Setting $(s_1, k, k_1) = (1, 1, s_1)$ in (7) we obtain

$$\psi(s_1) = \phi(1) \cdot \psi(1) \cdot s_1 = \psi(1) \cdot s_1 \tag{9}$$

Comparing (8) and (9) it follows that

$$\phi(s_1) = \psi(1)\, s_1 (\psi(1))^{-1} = k_0^{-1} s_1 k_0$$

where $k_0^{-1} = \psi(1)$. ▮

The proof we now give of Wedderburn's theorem is based on the previous result.

Theorem 3. A finite division ring is a field.

Proof. Let Z be the center of the finite division ring K. Since $Q \not\subseteq K$, the characteristic of K is a prime p. K therefore contains the field C_p as the smallest field in K. Assume by way of contradiction that $Z \neq K$. Since C_p consists of multiples of the identity element 1_K of K, $C_p \subseteq Z$. By Theorem VI.1.10 there is a subfield E of K containing Z which is maximal in the family of subfields of K. Furthermore, applying Theorem VI.1.10 with $\Phi = Z$ we have

$$r^2 = \dim_Z K \qquad \text{and} \qquad \dim_Z E = r$$

It follows that all maximal subfields E of K have the same dimension over C_p, namely

$$n = \dim_{C_p} D = \dim_Z E \cdot \dim_{C_p} Z = r \cdot \dim_{C_p} Z$$

The number of elements in E is therefore p^n. Now by Theorem V.1.5, finite fields of the same order are isomorphic. Since $Z \neq K$, there exist elements θ_1 and θ_2 in K such that $\theta_1\theta_2 \neq \theta_2\theta_1$. Hence there exist at least two distinct (but isomorphic) maximal subfields E_1 and E_2 of K. By Theorem 2 applied to $S_i = E_i$, $i = 1, 2$, each isomorphism ϕ of E_1 onto E_2 can be extended to an inner automorphism of K:

$$\phi e_1 = k_2^{-1} e_1 k_2 \qquad \text{for all} \quad e_1 \in E \tag{10}$$

where k_2 is a suitable element of K. Every maximal subfield E_j of K is the image under an inner automorphism of E_1 :

$$E_j = k_j E_1 k_j^{-1}, \qquad j = 1, \ldots, N, \quad N \geqslant 2 \tag{11}$$

The set theoretic union of the E_j is the whole of K since each element $k \in K$ is contained in the subfield $Z(k)$ which in turn is contained in a maximal subfield of K. If we restrict our attention to the multiplicative group $K^\times$ of nonzero elements of K, we see that (11) expresses the fact that the subgroups $E_j^\times$ are conjugate to one another, i.e.

$$E_j^\times = k_j E_1^\times k_j^{-1}, \qquad j = 1, \ldots, N, \quad N \geqslant 2 \tag{12}$$

The set theoretic union of the $E_j^\times$ if $K^\times$. We show that this leads to a contradiction: Let k' and k'' be elements in the same right coset of $E_i^\times$ in $K^\times$. Then there exists $e_1 \in E_1$ such that $k'' = k'e_1$. Hence

$$k'' E_1^\times k''^{-1} = k' e_1 E_1^\times e_1^{-1} k'^{-1} = k' E_1^\times k'^{-1}$$

showing that conjugation of $E_1^\times$ by k' and k'' gives rise to the same subgroup of $K^\times$. The number N of distinct maximal subfields E_j, $j = 1, \ldots, N$ of K is therefore at most $| K^\times : E_1^\times |$, the number of right cosets of $E_1^\times$ in $K^\times$. But the union of the $E_j^\times$ is the whole of $K^\times$ and hence the $E_j^\times$ are pairwise disjoint, a contradiction since they each contain the identity of $K^\times$. ∎

We now return to the proof of Jacobson's theorem. We recall that, from our discussion preceding Theorem 2, it is sufficient to prove the theorem in the case the ring is a division ring K. Assume therefore that K is a division ring with the property that for each $a \in K$

$$a^{n(a)} = a, \qquad n(a) \geqslant 2$$

It follows that the smallest subfield contained in the center Z of K cannot be the field of rationals Q since, for example, $2^s \neq 2$ for any $s > 1$; hence it must be $C_p = C$ where p is prime (Theorem V.2.1). Suppose now $Z \neq K$ and let $u \in K \backslash Z$. Let $f(X) = a_0 + a_1 X + \cdots + X^N$ be the monic polynomial of least degree with coefficients in Z having u as a root; $f(X)$ exists since u is a root of $X^{n(u)} - X$. Let F be the field generated by C and $a_0, \ldots, a_{N-1}$. F is a C-space of finite dimension t over C. Since the order of C is p, the

C-space F contains exactly $q = p^t$ elements. The multiplicative group $F^\times$ of F therefore contains $q - 1 = p^t - 1$ elements. Each element $f \in F^\times$ satisfies $f^{q-1} = 1$ and hence each element of F is a root of the polynomial $X^q - X$. Since K is a division ring containing F, u cannot be a root of $X^q - X$ since a polynomial of degree q in the field $F(u)$ can have at most q roots. Therefore

$$u^q \neq u \tag{13}$$

On the other hand u is a root of the polynomial $f(X)$ with coefficients in $F = C(a_0, ..., a_{N-1}) \subseteq Z$ since $f(X)$ is irreducible over Z.

$$Z[X]/(f(X)) \approx F(u) \tag{14}$$

by Theorem V.2.2. Since $q = p^t$, where p is the characteristic of K, $p \cdot 1 = 0$ in K and hence

$$f(u^q) = (f(u))^q = 0$$

By (14), the mapping

$$\phi\colon g(u) \to g(u^q) \qquad \text{for} \quad g(X) \in Z[X] \tag{15}$$

is an isomorphism of $Z(u)$ onto itself. Since $\dim_Z Z(u) = \deg f = N < \infty$, we now have the situation described in Theorem 2 with $S_1 = S_2 = Z(u)$, $\Phi = Z$, ϕ as in (15). Hence there exists an inner automorphism of K induced by some $k_0 \in K$ which extends ϕ. In particular therefore

$$\phi u = u^q = k_0^{-1} u k_0$$

and hence

$$u \cdot k_0 = k_0 \cdot u^q \tag{16}$$

This last equality implies that the division ring $C(u, k_0)$ generated over C by u and k_0 is finite. Proof: Because of (16) each element of the *ring* $C[u, k_0]$ generated by u and k_0 over C is of the form $h(u, k_0)$, where $h(X, Y)$ is a polynomial in X and Y with coefficients in C. Since $u^n = u$ and $k_0{}^t = k_0$ for suitable n and $t \in \mathbb{N}$, it follows that the degrees of X and Y in $h(X, Y)$ are at most $n - 1$ and $t - 1$, respectively. Hence since C is finite, the *ring* generated by u and k_0 over C is finite. Now if $h(u, k_0) \in C[u, k_0]$, there exists $s \in N$ such that

$$(h(u, k_0))^s = h(u, k_0)$$

Therefore $h(u, k_0)^{s-1} = 1$ and $C[u, k_0] = C(u, k_0)$. By Theorem 3 (Wedderburn), $C(u, k_0)$ is commutative. It follows from (16) that

$$k_0 u = u \cdot k_0 = k_0 u^q$$

and hence $u = u^q$, contradicting (13).

Jacobson's Theorem 1 has been generalized (cf. appendix) and has found application in areas outside ring theory, for example, in the foundations of geometry. The proof of the theorem is interesting in that use is made of the subdirect sum of rings. In general the fact that a ring S is a subdirect sum of rings R_i, $i \in I$ gives us little information about the ring S. In this case, however, the relevant property, namely commutativity, is inhereited by S.

3. The quarternions

Let $\mathbb{R}$ be the field of real numbers and $\mathbb{C}$ the field of complex numbers. Then the only finite-dimensional field extension of $\mathbb{R}$ is $\mathbb{C}$. For it is well known that $\mathbb{C}$ is the smallest algebraically closed overfield of $\mathbb{R}$ and the adjunction of a transcendental element, i.e. an indeterminate X, to $\mathbb{R}$ gives rise to an infinite-dimensional extension. Now let K be a noncommutative division ring containing $\mathbb{R}$ as center and suppose $\dim_{\mathbb{R}} K$ is finite. By Theorem V.1.10 if E is a maximal subfield of K,

$$\dim_{\mathbb{R}} K = r^2, \qquad \dim_{\mathbb{R}} E = \dim_E K = r \tag{1}$$

From the remarks made above, E is either isomorphic to $\mathbb{R}$ or to $\mathbb{C}$, i.e. $r = 1$ or $r = 2$. If $r = 1$, $\dim_{\mathbb{R}} K = 1$ and $K = \mathbb{R}$ showing that K is commutative. If $r = 2$,

$$\dim_R K = 4 \tag{2}$$

Now let $E \approx \mathbb{C}$ be a fixed maximal subfield of K. Let $\phi: e \to \bar{e}$ be the automorphism of E which maps a complex number e onto its complex conjugate $\bar{e}$. By Theorem VI.2.2 with $\Phi = \mathbb{R}$, $S_1 = S_2 = E \approx \mathbb{C}$ and $\phi: e \to \bar{e}$, there exists an element $k_0 \in K$ which induces an inner automorphism of K which coincides with ϕ on E:

$$k \to k_0 k k_0^{-1} \tag{3}$$

Let i denote the square root of -1 in E. Thus $E = \mathbb{R}(i)$, $i^2 = -1$. By (3) and since $\phi e = \bar{e}$

$$k_0 i k_0^{-1} = -i \qquad \text{and hence} \qquad k_0 i \neq i k_0 \tag{4}$$

Now since E is a maximal subfield of K, each element of k commuting with E is already in E (cf. Eq. VI.1(1)). Thus by (4),

$$\mathbb{R} \subseteq \mathbb{R} \cap E \neq E$$

This implies

$$\mathbb{R}(k_0) \cap E = \mathbb{R} \tag{5}$$

since $\dim_{\mathbb{R}} E = 2$. On the other hand it follows from (3) and $\phi^2 e = \bar{e} = e$ that

$$e = k_0^2 e k_0^{-2} \qquad \text{and hence} \qquad e k_0^2 = k_0^2 e, \quad \forall\, e \in E$$

Therefore by VI.1(1), $k_0^2 \in E$. But $k_0^2 \in \mathbb{R}(k_0)$ and hence $k_0^2 \in E \cap \mathbb{R}(k_0) = \mathbb{R}$. If $k_0^2 > 0$, $k_0 \in \mathbb{R} \subseteq E$, a contradiction. Thus there exists $\omega \in \mathbb{R}$ such that $k_0^2 = -\omega^2$. Let

$$i_1 = i \qquad \text{and} \qquad i_2 = \frac{1}{\omega} k_0$$

Then $i_2^2 = \omega^{-2} k_0^2 = -1$ and by (4),

$$i_2 i i_2^{-1} = \omega^{-1} k_0 i k_0^{-1} \omega = -i$$

showing that $i_2 i_1 = -i_1 i_2$. Finally, set $i_3 = i_1 i_2$. Then

$$i_3^2 = i_1 i_2 i_1 i_2 = -i_1^2 i_2^2 = -(-1)\cdot(-1) = -1$$

The following relationships hold between i_1, i_2, and i_3:

$$i_1^2 = i_2^2 = i_3^2 = -1$$

$$i_1 i_2 = i_3 = -i_2 i_1, \qquad i_2 i_3 = i_1 = -i_3 i_2, \qquad i_3 i_1 = i_2 = -i_1 i_3 \tag{6}$$

It follows that the set of elements of the form,

$$k = a_0 + a_1 i_1 + a_2 i_2 + a_3 i_3, \qquad a_j \in R, \qquad j = 0, 1, 2, 3 \tag{7}$$

constitute a subring K' of K. Moreover, K' is even a division ring; for if

$$\bar{k} = a_0 - a_1 i_1 - a_2 i_2 - a_3 i_3 \tag{8}$$

then

$$k\bar{k} = \bar{k}k = a_0^2 + a_1^2 + a_2^2 + a_3^2 \tag{9}$$

The equalities in (6) show that $\mathbb{R}$ is the center of K'. Hence, as in (1), $\dim_{\mathbb{R}} K' = 4$. From this and the fact that $\mathbb{R} \subseteq K' \subseteq K$ it follows that $K' = K$. In particular, $\{1, i_1, i_2, i_3\}$ generates K' over $\mathbb{R}$.

Definition 1. The ring of real *quarternions* K is an algebra over the reals $\mathbb{R}$ with basis $\{1, i_1, i_2, i_3\}$ and identity 1. Multiplication of elements in K is defined by (6).

Our discussion above proves:

Theorem 1 (Frobenius). The ring of real quaternions is the only non-commutative division ring of finite dimension over the reals. ∎

4. Algebras with segregated radical

As a final application of the results on central, simple and on separable algebras obtained in Section VI.1, we shall prove in this section a theorem of Wedderburn's which deals with the following question: If R is an algebra over the field Φ and J its radical, does there exist a subalgebra S such that R is a direct sum of S and J? If by "direct sum" we mean a direct sum as rings then we require S to be an ideal and not only a subring of R. In this case such an S very rarely exists. If however by "direct sum" we mean a direct sum as Φ-vector spaces, we can prove the existence of S if we impose the restriction that the semisimple algebra R/J be separable in the sense of Definition VI.1.5.

Theorem 1 (Wedderburn). Let R be a finite-dimensional algebra over the field Φ. Let J be the radical of R and let R/J be separable. Then there exists a subalgebra isomorphic to R/J such that R, considered as a Φ-vector space, is the direct sum of S and J.

Proof. 1. Define the exponent of the radical J ($\exp J$) as the smallest exponent e for which $J^e = 0$. If $\exp J = 1$, $J = 0$ and there is nothing to prove. We now show that the proof in the case $\exp J > 2$ can be reduced to the case $\exp J = 2$. This we do by induction on $d = \dim_\Phi J$. If $d = 0$, $J = 0$. Let $\mathfrak{A}$ denote R/J and let

$$\phi\colon a \to a + J, \qquad a \in R \tag{1a}$$

and

$$\psi\colon a \to a + J^2 \tag{1b}$$

be the natural homomorphisms of R onto $\mathfrak{A} = R/J$ and R/J^2, respectively. Let

$$\phi'\colon a + J^2 \to a + J \qquad \text{for} \quad a + J^2 \in R/J^2$$

be the natural homomorphism of R/J^2 onto $R/J = \mathfrak{A}$. The radical of R/J^2 is J/J^2 and thus $\dim (J/J^2) < d$ since otherwise $J^2 = 0$ and hence $\exp J = 2$. By the induction hypothesis on d, there exists a subalgebra $\bar{S}$ of R/J^2 with $\bar{S} \approx \mathfrak{A} = R/J$ and such that $R/J^2 = \bar{S} + J/J^2$, where the sum is a direct sum of Φ-vector spaces. Let $S = \psi^{-1}(\bar{S})$. Then S is a Φ-subalgebra of R and

$$R/J^2 = S/J^2 \oplus J/J^2$$

Since $S/J^2 \approx \mathfrak{A}$ it follows that J^2 is the radical of S. Moreover, $\dim J^2 < d$, otherwise $J = J^2 = \cdots = J^e = 0$. Again by the induction hypothesis, there exists a Φ-subalgebra T of S (and hence a subalgebra of R) such that $T \approx \mathfrak{A}$ and $S = T \oplus J^2$. It follows from

$$R/J^2 = (T + J^2)/J^2 \oplus J/J^2$$

that $R = T + J$. But $T \cap J \subseteq S \cap J \subseteq J^2$ and hence $T \cap J = T \cap J^2 = 0$, showing that $R = T \oplus J$.

2. Now let $\exp J = 2$ and let $\{b_1, ..., b_N\}$ be a set of elements in R which map to a Φ-basis $\{\bar{b}_1, ..., \bar{b}_N\}$ of $R/J = \mathfrak{A}$ under the natural homomorphism of R to $\mathfrak{A}$:

$$\bar{b}_i = \phi b_i, \qquad i = 1, ..., N$$

The mapping

$$f: \sum_i \bar{b}_i \alpha_i \to \sum b_i \alpha_i, \qquad \alpha_i \in \Phi \tag{2}$$

is a one-one Φ-linear mapping of the Φ-vector space $\mathfrak{A}$ onto the Φ-subspace L of R generated by $\{b_1, ..., b_N\}$ with ϕ as inverse. Each element $a \in R$ is uniquely expressible in the form

$$f(\bar{a}) + \omega \qquad \text{where} \quad \bar{a} \in \mathfrak{A} \quad \text{and} \quad \omega \in J \tag{3}$$

since $\phi a \equiv \phi f(\bar{a}) (\text{mod } J)$ for all $a \in R$. Since $\exp J = 2$,

$$(f(\bar{a}_1) + \omega_1)(f(\bar{a}_2) + \omega_2) = f(\bar{a}_1) \cdot f(\bar{a}_2) + \omega_1 f(\bar{a}_2) + f(\bar{a}_1) \cdot \omega_2 \tag{4}$$

If $f(\bar{a}_1) \cdot f(\bar{a}_2) = f(\bar{a}_1 \bar{a}_2)$, f is an algebra-isomorphism of $\mathfrak{A}$ onto L. In this case, by (3), $R = L \oplus J$ and $L \approx \mathfrak{A}$ proving the theorem. If, however, $f(\bar{a}_1) \cdot f(\bar{a}_2) \neq f(\bar{a}_1 \cdot \bar{a}_2)$, we must alter the $f(\bar{a})$ by the addition of elements $p(\bar{a})$ from J in such a way as to make the mapping $f + p$ multiplicative. Thus we must find $p \in \text{Hom}\_(\mathfrak{A}, J)$ such that

$$(f(\bar{a}_1) + p(\bar{a}_1)) \cdot (f(\bar{a}_2) + p(\bar{a}_2)) = f(\bar{a}_1 \cdot \bar{a}_2) + p(\bar{a}_1 \cdot \bar{a}_2)$$

Because of (4), this is equivalent to finding $p \in \text{Hom}_\Phi(\mathfrak{A}, J)$ such that

$$f(\bar{a}_1)\, p(\bar{a}_2) - p(\bar{a}_1 \cdot \bar{a}_2) + p(\bar{a}_1) f(\bar{a}_2) = h(\bar{a}_1 \cdot \bar{a}_2) \qquad \text{for all} \quad \bar{a}_1, \bar{a}_2 \in \mathfrak{A} \tag{5}$$

where

$$h(\bar{a}_1 \cdot \bar{a}_2) = f(\bar{a}_1) f(\bar{a}_2) - f(\bar{a}_1 \bar{a}_2) \tag{6}$$

The existence of such a p is guaranteed by the fact that R/J is separable.

3. By Theorem VI.1.11 there exists an overfield E of Φ such that

$$R/J \otimes_\Phi E = E_{r_1 \times r_1} \oplus \cdots \oplus E_{r_q \times r_q} \tag{7}$$

In this third part of the proof we show that if the assertion of the theorem holds for $R \otimes_{\Phi} E$ it also holds for R. $J \otimes_{\Phi} E$ is a nilpotent ideal in $R \otimes_{\Phi} E$ since J is nilpotent. Hence $J \otimes_{\Phi} E$ is contained in the radical of $R \otimes_{\Phi} E$. On the other hand,

$$\phi'' \colon \ \sum a_j \otimes e_j \to \sum (\phi a_j) \otimes e_j \qquad \text{for} \quad a_j \in R \quad \text{and} \quad e_j \in E$$

is a homomorphism of the algebra $R \otimes_{\Phi} E$ onto the algebra $(R/J) \otimes_{\Phi} E$. The kernel of ϕ'' is $J \otimes_{\Phi} E$. By hypothesis R/J is separable over Φ and therefore $R/J \otimes_{\Phi} E$ is semisimple. This implies that the radical of $R \otimes_{\Phi} E$ is contained in $J \otimes_{\Phi} E$. Hence the radical of $R \otimes_{\Phi} E$ is $J \otimes_{\Phi} E$. The Φ-spaces $\mathfrak{A} \otimes_{\Phi} E$ and $L \otimes_{\Phi} E$ may be turned into E-spaces by defining

$$\left[\sum a_j \otimes e_j\right] e = \sum a_j \otimes (e_j e) \qquad \text{for} \quad a_j \in \mathfrak{A} \quad \text{and} \quad e_j\,, e \in E \tag{8}$$

$\mathfrak{A} \otimes_{\Phi} E$ is moreover an algebra over E since E is commutative. The mapping $f \in \mathrm{Hom}_{\Phi}(\mathfrak{A}, L)$ defined in (2) induces $f' \in \mathrm{Hom}_E(\mathfrak{A} \otimes E, L \otimes E)$ defined by

$$f' \colon \sum \bar{b}_i \otimes e_i \to \sum (f\bar{b}_i) \otimes e_i\,, \qquad e_i \in E$$

The set $\{\bar{b}_1 \otimes 1, \ldots, \bar{b}_N \otimes 1\}$ is an E-basis of $\mathfrak{A} \otimes E$ and $f'\phi''$ is the identity on $L \otimes_{\Phi} E$. This shows that the mappings ϕ'' and f' play the same roles with respect to the E-algebra $R \otimes_{\Phi} E$ and the corresponding semisimple algebra $\mathfrak{A} \otimes E$ as do ϕ and f with respect to the Φ-algebra R. In the fourth part of this proof we shall show that if E is chosen so that (7) holds, there exists $p' \in \mathrm{Hom}_E(\mathfrak{A} \otimes E, J \otimes E)$ satisfying the analog of (5), namely

$$f'(\bar{A}_1)\, p'(\bar{A}_2) - p'(\bar{A}_1 \bar{A}_2) + p'(\bar{A}_1) f'(\bar{A}_2) = h'(\bar{A}_1\,, \bar{A}_2) \tag{5'}$$

for all $\bar{A}_1\,, \bar{A}_2 \in \mathfrak{A} \otimes E$. Of course, h' is the analog of h defined in (6):

$$h'(\bar{A}_1\,, \bar{A}_2) = f'(\bar{A}_1) \cdot f'(\bar{A}_2) - f'(\bar{A}_1 \cdot \bar{A}_2) \tag{6'}$$

Since by definition $f'(\bar{a} \otimes 1) = f(\bar{a}) \otimes 1$

$$h'(\bar{a}_1 \otimes 1, \bar{a}_2 \otimes 1) = h'(\bar{a}_1\,, \bar{a}_2) \otimes 1 \qquad \text{for all} \quad \bar{a}_1\,, \bar{a}_2 \in \mathfrak{A}$$

We now show how, given a p' satisfying (5′), we may construct $p \in \mathrm{Hom}_{\Phi}(\mathfrak{A}, J)$ satisfying (5): Relative to the E-basis $\{\bar{b}_1 \otimes 1, \ldots, \bar{b}_N \otimes 1\}$ the E-linear mapping p' gives rise to a system of linear forms with coefficients in E. On substitution of these linear forms in (5′) we obtain a system of defining linear equations over E for the coefficients of the linear forms. Restricting our attention to $\mathfrak{A} \otimes 1$ which we identify with $\mathfrak{A}$, we see that f' restricts to f, h' to h, (5′) becomes (5) and the system of linear equations satisfied by the coefficients of the linear forms reduces to a system of linear equations over Φ. Since we are assuming that p' exists, these equations can be solved in E. But since these solutions are expressible in terms of determinants with entries from Φ, they can be solved in Φ. Hence there exists $p \in \mathrm{Hom}_\Phi(\mathfrak{A}, J)$ satisfying (5).

4. This shows that we can restrict our attention to the case that the separable Φ-algebra $R/J = \mathfrak{A}$ is isomorphic to

$$\Phi_{r_1 \times r_1} \oplus \cdots \oplus \Phi_{r_q \times r_q} \tag{9}$$

Let $C_{ik}^{(l)}$ denote the matrix from the lth summand with 1 in the (i, k) position and 0 elsewhere. (9) can now be written in the form

$$\sum_{1 \leqslant l \leqslant q} \sum_{1 \leqslant i,k \leqslant r} C_{ik}^{(l)} \Phi$$

Let $f \in \mathrm{Hom}_\Phi(\mathfrak{A}, L)$ be such that $f\phi$ is the identity on L, where ϕ is the natural homomorphism of R onto $\mathfrak{A}$. We now construct $p \in \mathrm{Hom}_\Phi(\mathfrak{A}, J)$ satisfying (5). Of fundamental importance in this construction is the fact that $\exp J = 2$. Indeed, because of this, we may turn the radical J of R into an $(\mathfrak{A}, \mathfrak{A})$-bimodule by defining

$$\bar{a}\omega = f(\bar{a})\omega \qquad \text{and} \qquad \omega\bar{a} = \omega f(\bar{a}) \qquad \text{for} \quad \omega \in J, \quad \bar{a} \in \mathfrak{A} \tag{10}$$

Set

$$p(\bar{a}) = \sum_{\substack{1 \leqslant l \leqslant q \\ 1 \leqslant i \leqslant r_l}} C_{i1}^{(l)}(f(C_{1i}^{(l)}) \cdot f(\bar{a}) - f(C_{1i}^{(l)} \cdot \bar{a})), \qquad \bar{a} \in \mathfrak{A}$$

To show p satisfies (5) it is sufficient to verify that (5) holds in the case

$$\bar{a}_1 = C_{ht}^{(s)}, \qquad \bar{a}_2 = \bar{a}\phi$$

since p is linear,

$$\begin{aligned} & f(C_{ht}^{(s)})\, p(\bar{a}) - p(C_{ht}^{(s)} \cdot \bar{a}) + p(C_{ht}^{(s)})\, f(\bar{a}) \\ &= f(C_{ht}^{(s)}) \cdot \sum_{i,l} f(C_{i1}^{(l)})[f(C_{1i}^{(l)}) \cdot f(\bar{a}) - f(C_{1i}^{(l)}\bar{a})] \\ &\qquad - \sum_{i,l} f(C_{i1}^{(l)}) \cdot [f(C_{1i}^{(l)})\, f(C_{ht}^{(s)} \cdot \bar{a}) - f(C_{1i}^{(l)} C_{ht}^{(s)} \cdot \bar{a})] \\ &= \sum_{i,l} f(C_{i1}^{(l)}) \cdot [f(C_{1i}^{(l)}) \cdot f(C_{ht}^{(s)}) - f(C_{1i}^{(l)} C_{ht}^{(s)})] \cdot f(\bar{a}) \end{aligned}$$

where the expressions in the square brackets are in the radical J. Since Definition (10) turns J into a left $\mathfrak{A}$-module and since the mutually orthogonal

$C_{ik}^{(l)}$ are contained in $\mathfrak{A}$, the right-hand side of the above equation is equal to

$$f(C_{h1}^{(s)})[f(C_{1t}^{(s)}) \cdot f(\bar{a}) - f(C_{1t}^{(s)} \cdot \bar{a})]$$

$$- \sum_{i,l} f(C_{i1}^{(l)}) f(C_{1i}^{(l)})[f(C_{ht}^{(s)}\bar{a}) - f(C_{ht}^{(s)}) f(\bar{a})]$$

$$+ f(C_{h1}^{(s)}) \cdot [f(C_{1t}^{(s)} \cdot \bar{a}) - f(C_{1t}^{(s)}) \cdot f(\bar{a})]$$

Here the first and third lines cancel. In the second line the expression in square brackets is the element $-h(C_{ht}^{(s)}, \bar{a})$ contained in the left $\mathfrak{A}$-module J. The second line is therefore equal to

$$f\left(\sum_{i,l} C_{i1}^{(l)}\right) \cdot h(C_{ht}^{(s)}, \bar{a}) = f(1) \cdot h(C_{ht}^{(s)}, \bar{a}) = h(C_{ht}^{(s)}, \bar{a})$$

and hence

$$f(C_{ht}^{(s)}) \cdot p(\bar{a}) - p(C_{ht}^{(s)}\bar{a}) + p(C_{ht}^{(s)}) \cdot f(\bar{a}) = h(C_{ht}^{(s)}, \bar{a})$$

In particular, since, by Theorem VI.1.12, the Φ-algebra R/J is separable if Φ is the center of the simple algebras in $R/J = \bar{\mathfrak{A}}_1 \oplus \cdots \oplus \bar{\mathfrak{A}}_q$, Theorem 1 is valid in this case. ▮

Malcev has shown that the semisimple subring S in the direct sum $R = S \oplus J$ is uniquely determined up to an inner automorphism of R (cf. Curtis and Reiner [1962], Theorem 72.19).

Rings with Identity

CHAPTER VII

The theory of semisimple artinian rings was brought to a satisfactory conclusion in the preceding chapters. In this and the next two chapters we shall present some aspects of the theory of rings with nonzero radical. The theory developed will not be as complete as in the semisimple case and to avoid complications we shall assume that rings have an identity.

If R has the descending chain condition on left ideals, ${}_RR$ is a direct sum of finitely many indecomposable left ideals, i.e. left ideals which cannot be expressed as a nontrivial direct sum of two left ideals. In the first section we shall investigate this decomposition fully. We shall see that, here too, idempotents will play the most important part. The main results obtained will be for the class of semiperfect rings with identity, i.e. the class of rings R such that $R/J(R)$ is artinian and each idempotent of $R/J(R)$ can be "lifted." Every artinian ring is semiperfect but not conversely. For example, the ring of power series in one indeterminate over a division ring is semiperfect but not artinian. If in addition $R/J(R)$ is a simple ring, R can be represented as a matrix ring over a local ring T, i.e. a ring T such that $T/J(T)$ is a division ring. Theorem VII.1.10 deals with the structure of a general semiperfect ring and later in Theorem VII.1.19 we give a proof of the essential uniqueness of the decomposition obtained in Theorem 10.

The indecomposable left ideals are the prototypes of the so-called projective modules. These modules will be discussed in Section 2. As an application we shall develop the theory of Asano orders in a quotient ring. This is a generalization to noncommutative rings of the classical theory of orders of the algebraic integers in an algebraic number field.

The lattice of left ideals of a semisimple artinian ring is complemented. This gives rise to the concept of an injective module, i.e. a module which is a direct summand of any containing module. These modules will be dealt with in Section 3 and it will be shown that each module is embeddable in a smallest injective. The theory developed will be applied to self-injective rings R, i.e. rings R such that ${}_RR$ is an injective module. If in addition R has the descending chain condition on both left and right ideals, R is a quasi-Frobenius ring.

1. Semiperfect rings

Although a nonsemisimple artinian ring R with 1 is not completely reducible (Theorem IV.1.4), ${}_RR$ is still the direct sum of a finite number of *indecomposable* left ideals.

Definition 1. 1. The direct summands of the R-module M are called *components* of M.

2. An R-module $M \neq (0)$ is *indecomposable* if the only components of M are (0) and M.

3. A *decomposition* of M *into components* M_i is a representation of M as the direct sum of the M_i.

The existence of a decomposition ${}_RR = L_1 \oplus \cdots \oplus L_r$ is a consequence of the minimal condition on left ideals in R: Let $\mathfrak{F}$ be the family of left ideals of R which are not expressible as a direct sum of a finite number of indecomposable left ideals. Assume by way of contradiction that $\mathfrak{F} \neq \varnothing$. Then by the minimal condition on left ideals there exists L_0 minimal in $\mathfrak{F}$. L_0 is decomposable, say $L_0 = L_0' + L''$. Since L_0' and L'' are properly contained in L_0, L_0' and L'' are not members of $\mathfrak{F}$. Hence each is a finite direct sum of indecomposable modules showing that $L_0 \in \mathfrak{F}$, a contradiction. Therefore $\mathfrak{F} = \varnothing$.

If the ring has 1 we can establish a connection between the components of ${}_RR$ and the representations of 1 as the sum of a finite number of orthogonal idempotents.

Let the ring R, considered as a left R-module, be expressed as a direct sum of left ideals L_i, $i = 1, \ldots, r$,

$$ {}_RR = L_1 \oplus \cdots \oplus L_r \tag{1} $$

The identity 1 of R is then expressible in the form,

$$ 1 = e_1 + \cdots + e_r \qquad \text{where} \quad e_i \in L_i \tag{2} $$

The e_i are idempotent and mutually orthogonal, i.e.

$$e_ie_j = \delta_{ij}e_i, \qquad \text{where} \quad \delta_{ij} \quad \text{is the Kronecker delta} \tag{3}$$

The e_i generate the L_i,

$$L_i = Re_i, \qquad i = 1, ..., r \tag{4}$$

Proof. Since $e_i \in L_i$, $Re_i \subseteq L_i$. Conversely, if $l_i \in L_i$

$$l_i = l_i1 = l_ie_1 + \cdots + l_ie_i + \cdots + l_ie_r \tag{5}$$

By the directness of the sum in (1),

$$l_ie_k = \delta_{ik}l_ie_i \in Re_i \tag{6}$$

Hence by (5), $l_i \in Re_i$ and (4) is proved. Substituting e_i for l_i in (6) we get the orthogonality,

$$e_ie_k = 0 \qquad \text{if} \quad i \neq k$$

The idempotence of the e_i is proved by observing that

$$e_1 + \cdots + e_r = 1 = e_1{}^2 + \cdots + e_r{}^2$$

and again using the fact that the sum in (1) is direct.

Conversely, if $1 = e_1 + \cdots + e_r$, where the e_i are mutually orthogonal idempotents, then

$${}_RR = Re_1 + \cdots + Re_r$$

Moreover, this sum is direct since, if

$$a = x_1e_1 + \cdots + x_re_r \tag{7}$$

then $ae_k = x_ke_k$. Hence the Re_i are components of ${}_RR$. Finally, if L is any component of ${}_RR$, i.e. ${}_RR = L \oplus L'$ for some left ideal L', then by what we have shown above,

$$L = Re \qquad \text{and} \qquad L' = R(1 - e)$$

for some idempotent e.

We collect these results in the following:

Theorem 1. Let R be a ring with identity.

1. Each decomposition of ${}_RR$ into a direct sum of finitely many left

ideals gives rise to a representation of the identity 1 of R as a sum of orthogonal idempotents and conversely:

$$1 = e_1 + \cdots + e_r \Leftrightarrow {}_RR = Re_1 \oplus \cdots \oplus Re_r$$

2. The left ideal Re is an indecomposable component of ${}_RR$ if and only if e cannot be written as the sum of two nonzero orthogonal idempotents. Such idempotents are called *primitive*.

3. If R is artinian, ${}_RR$ may be decomposed as a direct sum of indecomposable left ideals. ▌

The primitivity of an idempotent of a ring R may be equivalently expressed as follows:

Definition 2. An idempotent e of a semigroup is primitive if for any idempotent f, the equalities

$$f = ef = fe$$

imply $e = f$.

Theorem 2. An idempotent $e \neq 0$ in a ring R is primitive in the ring-theoretic sense if and only if it is primitive in the multiplicative semigroup S of R.

Proof. Let $e = g + h$, with $g \neq 0$, $h \neq 0$ orthogonal idempotents. Then $ge = g^2 + gh = g$ and $eg = g^2 + hg = g$. Therefore $ge = eg = g$ but $e \neq g$ showing that e is not primitive in the multiplicative semigroup of R. Conversely, if there exists $g \in R$ such that $0 \neq g^2 = g \neq e$ and $g = ge = eg$, then g and $e - g$ are nonzero idempotents whose sum is e. Therefore e is not primitive in the ring theoretic sense. ▌

Let $J = J(R)$ be the radical of the ring R with 1. Suppose that R/J is artinian and hence completely reducible. The indecomposable left ideals in R/J are the minimal left ideals. Denote the coset $u + J$ by $\bar{u}$. Then $\bar{R} = R/J = \bar{R}\bar{e}_1 + \cdots + \bar{R}\bar{e}_r$ where the $\bar{e}_i$ are primitive orthogonal idempotents whose sum is $\bar{1}$. If we can "lift" this set of idempotents $\{\bar{e}_1, ..., \bar{e}_r\}$ to a set $\{e_1, ..., e_r\}$ of orthogonal idempotents in R with $\bar{e}_i = e_i + J$ and such that $e_1 + \cdots + e_r = 1$, then we can deduce that the e_i are also primitive; for if $e_i = g + h$, g and h nonzero orthogonal idempotents, then $\bar{e} = \bar{g} + \bar{h}$. But neither $\bar{g}$ nor $\bar{h}$ can be zero since the radical can contain no idempotent. In fact, if $0 \neq g = g^2 \in J$ and $g + y - yg = 0$, then $g^2 + yg - yg^2 = 0$ and hence $g = 0$, a contradiction. In the following we shall show that if for each idempotent $\bar{u} \in R/J$ we can find an idempotent $e \in R$ such that $\bar{e} = \bar{u}$, then we can lift any countable family of mutually orthogonal idempotents of $\bar{R}$ to a family of orthogonal idempotents of R. This discussion motivates the following definition:

Definition 3. A ring R with identity 1 and radical J is semiperfect if
1. R/J is artinian and
2. each idempotent $\bar{u} \in \bar{R}$ can be lifted to an idempotent $e \in R$.

Theorem 3. Let $M \subseteq J = J(R)$ be an ideal of R and suppose idempotents can be lifted from R/M. Let $g = g^2 \in R$ and suppose $u^2 \equiv u (\text{mod } M)$. If $0 \equiv ug \equiv gu (\text{mod } M)$, there exists an idempotent $e \in R$ such that $e \equiv u (\text{mod } M)$ and $eg = ge = 0$.

Proof. By hypothesis there exists $f \in R$ such that $f^2 = f \equiv u (\text{mod } M)$. Therefore $fg \equiv 0 \text{ mod } M$. Since $1 \in R$ and $fg \in J$, $1 - fg$ is a unit. Let $f' = (1 - fg)^{-1} f(1 - fg)$. f' is also idempotent and since

$$f(1 - fg)g = fg - f^2 g^2 = 0, \qquad f'g = 0.$$

Furthermore, $f' \equiv (1 - fg)f' = f(1 - fg) \equiv f (\text{mod } M)$. Let $e = (1 - g)f'$. Then $ge = eg = 0$ and $e^2 = e(1 - g)f' = ef' = (1 - g)f' = e$. ∎

The proof that every countable family $\{\bar{u}_1, \bar{u}_2, ...\}$ of mutually orthogonal nonzero idempotents can be lifted follows easily:

Theorem 4. Let $M \subseteq J$, R a ring with 1, M an ideal. Assume idempotents modulo M can be lifted. Then, given any countable family $\{\bar{u}_1, \bar{u}_2, ...\}$ of orthogonal nonzero idempotents in R/M, there exists an orthogonal family $\{e_1, e_2, ...\}$ of idempotents in R such that $e_i \equiv u_i \text{ mod } M$, $i = 1, 2, ...$.

Proof. By induction. Assume that the idempotents $\bar{u}_1, ..., \bar{u}_{r-1}$ have been lifted to the mutually orthogonal idempotents $e_1, ..., e_{r-1}$. Let

$$g = e_1 + \cdots + e_{r-1}$$

Then

$$u_r g = \sum_{i=1}^{r-1} u_r e_i \equiv \sum_{i=1}^{r-1} u_r u_i \equiv 0 \ (\text{mod } M)$$

and similarly $gu_r \equiv 0 (\text{mod } M)$. By Theorem 3, there exists $e_r \in R$ such that $e_r^2 = e_r \equiv u_r (\text{mod } M)$ and $e_r g = ge_r = 0$. But

$$e_r e_i = e_r g e_i = 0 = e_i e_r \quad ∎$$

In particular Theorem 4 implies that in a semiperfect ring R any finite family of orthogonal primitive idempotents $\{\bar{u}_1, ..., \bar{u}_r\}$ in R/J such that $\bar{u}_1 + \cdots + \bar{u}_r = \bar{1}$ can be lifted to orthogonal idempotents $e_1, ..., e_r$ in R. We can show under these circumstances that $e_1 + \cdots + e_r = 1$. To this end set $e = e_1 + \cdots + e_r$. Then

$$(1 - e)^2 = 1 - 2e + e^2 = 1 - e \equiv 1 - (u_1 + \cdots + u_r) \equiv 0 \bmod J$$

Hence $1 - e$ is an idempotent in J showing that $1 - e = 0$. We have therefore proved the following:

Theorem 5. A semiperfect ring R contains a family of orthogonal primitive idempotents whose sum is 1. ${}_RR$ is then expressible as the direct sum of finitely many indecomposable left ideals generated by these idempotents. ∎

Theorem 6. Every artinian ring with identity is semiperfect. This is a corollary to the following:

Theorem 7. Let M be a nil ideal in the ring R with 1. Let $\bar{x} = x + M$ be idempotent:

$$x^2 \equiv x \bmod M \tag{8}$$

Then there exists an idempotent $e \in R$ which is expressible as a linear combination of the elements x, x^2, x^3, ..., with coefficients in $\mathbb{Z}$ and such that $e \equiv x \bmod M$.

Proof (due to Herstein). Since $x^2 - x \in M$ and M is nil, there exists a natural number n such that

$$(x^2 - x)^n = 0 \tag{9}$$

The various powers of x commute with each other and we may therefore apply the binomial theorem to get

$$0 = x^n(1 - x)^n - x^n \sum_{j=0}^{n} (-1)^j \binom{n}{j} x^j = x^n - x^{n+1}g \tag{10}$$

where

$$g = \sum_{j=1}^{n} (-1)^{j-1} \binom{n}{j} x^{j-1}$$

Let

$$e = x^n g^n \tag{11}$$

From (10) we obtain the following equalities:

$$e^2 = x^{2n}g^{2n} = x^{n-1}x^{n+1}gg^{2n-1} = x^{n-1}x^n g^{2n-1}$$
$$= x^{2n-1}g^{2n-1} = \cdots = x^n g^n = e$$

The idempotent e is congruent to x mod M as may be seen from the following:

$$xg \equiv \sum_{j=1}^{n} (-1)^{j-1} \binom{n}{j} x \bmod M$$

since $x^s \equiv x \bmod M$ for all $s > 0$. But

$$0 = (1-1)^n = \sum_{j=0}^{n} (-1)^j \binom{n}{j}$$
$$= 1 - \sum_{j=1}^{n} (-1)^j \binom{n}{j}$$

Therefore

$$\sum_{j=1}^{n} (-1)^{j-1} \binom{n}{j} = 1$$

Hence $xg \equiv x \bmod M$ and $x^n g^n \equiv x^n \bmod M$ from which it follows that

$$e \equiv x \bmod M \qquad \text{since} \quad x^2 \equiv x \bmod M \quad \blacksquare$$

Theorem 8. Let R be a ring with identity. Let e be an idempotent in R. Then

$$J(eRe) = eRe \cap J(R) = eJ(R)e$$

Proof. Let $a \in J(eRe)$. Then $a = eae$ and $exa \in J(eRe)$ for all $x \in R$. Let y be a quasi-inverse of exa in eRe so that $(e - y)(e - exa) = e$. It follows that $yxa - y = exa$. Therefore

$$(1 - y)(1 - xa) = 1 - y - xa + yxa = 1 - (1 - e)\,xa$$

Since $(1 + [1 - e]\,xa)(1 - [1 - e]\,xa) = 1$ and $a(1 - e) = 0$, $1 - xa$ has $(1 + [1 - e]\,xa)(1 - y)$ as a left inverse. Hence for all $x \in R$, $1 - xa$ has a left inverse proving that $a \in J(R)$. Therefore $a = eae \in eJ(R)e$. Conversely, let $a \in eJ(R)e$ and let $x \in eRe$. Since $eJ(R)e \subseteq J(R)$, $1 - xa$ has a left inverse $1 - y \in R$. Hence

$$(e - eye)(e - xa) = e(1 - y)\,e(e - xa) = e \cdot 1 = e \quad \blacksquare$$

Theorem 9. Let $e \neq 0$ be an idempotent in the semiperfect ring R. Then eRe is also semiperfect.

Proof. By the isomorphism theorem and the previous result

$$\frac{eRe}{eJe} \approx \frac{eRe + J}{J}$$

Let $L/eJe \subset L'/eJe$ be left ideals in eRe/eJe. Consider the left ideals $(RL + J)/J$ and $(RL' + J)/J$ in R/J. Suppose if possible that $RL + J = RL' + J$. Then $e(RL + J)e = e(RL' + J)e$. Hence $L + eJe = L' + eJe$ since $eRLe = L$ and $eRL'e = L'$. Therefore $L = L'$, a contradiction proving that $(RL + J)/J \subset (RL' + J)/J$. Hence eRe/eJe is artinian. Now let $u \in Re$ and $u^2 - u \in eJe \subseteq J$. Since R is semiperfect and $u(1 - e) = (1 - e)u = 0$, there exists by Theorem 3 an idempotent $f \in R$ with $f - u \in eJe$ and $f(1 - e) = (1 - e)f = 0$. Thus $f = fe = ef = f^2$ and it follows that $f \in eRe$. ∎

Since the factor ring R/J of a semiperfect ring R is artinian semisimple

$$R/J = \bar{R}_1 \oplus \cdots \oplus \bar{R}_q \qquad \text{where the } \bar{R}_i \text{ are simple rings}$$

Let $\bar{1} = \bar{g}_1 + \cdots + \bar{g}_q$, $\bar{g}_i \in \bar{R}_i$. The set $\{\bar{g}_1, \bar{g}_2, ..., \bar{g}_q\}$ of orthogonal idempotents of R/J may be lifted to a set of orthogonal idempotents $\{g_1, ..., g_q\}$ in R. Furthermore, $g = g_1 + \cdots + g_q$ is the identity of R since $1 - g$ is an idempotent contained in J. By Theorem 9, the rings g_iRg_i are semiperfect and $g_iRg_i/g_iJg_i \approx \bar{R}_i$ where $\bar{R}_i$ is simple. We therefore obtain a decomposition of R as follows:

$$R = 1R1 = \sum_{1 \leqslant i \leqslant q} g_iRg_i + \sum_{i \neq j} g_iRg_j$$

The first sum on the right-hand side is a direct sum of rings g_iRg_i since $g_ig_j = \delta_{ij}g_i$. The second sum is a subgroup M of the additive group of the radical since $g_iRg_j \equiv g_ig_jR \equiv 0 \bmod J$. In general M is not an ideal because, for example, $g_1Rg_2 \cdot g_2Rg_1 \subseteq g_1Rg_1$. We have proved the following important theorem:

Theorem 10. Let R be a semiperfect ring. Then considered as an additive group, $R = S \oplus M$, where M is a subgroup of the additive group of the radical of R and S is a subring of R with the following properties: S, considered as a ring, is a finite direct sum of semiperfect rings $R_1, ..., R_q$ whose factor rings R_i/J_i, $i = 1, ..., q$ are simple. Furthermore, $R_i = g_iRg_i$, $J_i = g_iJ(R)g_i$, $M = \sum_{i \neq j} g_iRg_j$ and $\{g_1, ..., g_q\}$ is a family of orthogonal idempotents with sum 1, lifted from the set of identities of the rings $\bar{R}_i$ in the decomposition of R/J as a direct sum of the simple rings $\bar{R}_i$. ∎

The structure of the rings R_i above will be further investigated in Theorem 21. We shall also show that the subring S and the subgroup of $\{J, +\}$ are uniquely determined up to an inner automorphism of R.

To this end we first prove that if e is an idempotent in R, the ring eRe is antiisomorphic to the endomorphism ring of the R-module Re. More generally we have:

Theorem 11. Let e be an idempotent of R and let $f \in R$. Then for each $a \in eRf$, the mapping

$$\phi_a\colon xe \to xea \qquad \text{for all} \quad x \in R$$

is an element of $\mathrm{Hom}_R(Re, Rf)$ and

$$\eta\colon a \to \phi_a \qquad \text{for all} \quad a \in eRf$$

is an isomorphism of the additive group of eRf onto the additive group of $\mathrm{Hom}_R(Re, Rf)$. If $e = f$, η is a ring antiisomorphism.

Proof. Let $\phi \in \mathrm{Hom}_R(Re, Rf)$. ϕ is uniquely determined by the image of e under ϕ. Since $\phi(e) \in Rf$, $\phi(e^2) = e\phi(e) = eaf$ for some $a \in R$. η clearly preserves addition and therefore eRf and $\mathrm{Hom}_R(Re, Rf)$ are isomorphic as groups. The second assertion follows from

$$\begin{aligned}[\eta(ab)]\, xe &= \phi_{ab}(xe) = xeab = (xea)b \\ &= \phi_b\phi_a(xe) = [\eta(b)\, \eta(a)]\, xe \quad \blacksquare\end{aligned}$$

In a semiperfect ring the primitivity of an idempotent e is equivalent to eRe being a local ring:

Definition 4. A ring with identity is *local* if R/J is a division ring.

Theorem 12. Let e be a nonzero idempotent in the ring R with identity. Then the following two assertions are equivalent:

1. e is primitive
2. Re is an indecomposable left ideal

Furthermore, Condition 3 implies Assertion 1:

3. eRe is a local ring

If R is semiperfect, Assertions 1, 2, and 3 are equivalent.

Proof. The equivalence of 1 and 2 was proved in Theorem 1. Now assume $Re = L_1 \oplus L_2$ and let π_i be the projection of $L_1 \oplus L_2$ onto L_i. Then π_1 and π_2 are nonzero orthogonal idempotents in $\mathrm{Hom}_R(Re, Re)$

whose sum is the identity of the ring. By Theorem 11, eRe contains two nonzero idempotents e_1 and e_2 such that $e_1 + e_2 = e$. Since the radical eJe of eRe contains no idempotent, $e_1 + eJe$ and $e_2 + eJe$ are nonzero orthogonal idempotents in eRe/eJe. Hence this ring is not a division ring. Conversely, if R is semiperfect and if eRe/eJe is not a division ring, it contains two orthogonal idempotents with sum $\bar{e}$ since eRe is semiperfect by Theorem 9. These idempotents may be lifted to orthogonal idempotents e_1 and e_2 with sum e. ∎

Theorem 13. Let e be an idempotent in the semiperfect ring R with radical J. Let $\nu: x \to \bar{x} = x + J$ be the natural homomorphism of R to R/J. Then Re is an indecomposable left ideal in R if and only if $\bar{R}\bar{e}$ is a minimal left ideal in $\bar{R} = R/J$.

Proof. $\bar{e}\bar{R}\bar{e} = \overline{eRe} = (eRe + J)/J \approx eRe/eJe$. By Theorem 12, Re (resp. $\bar{R}\bar{e}$) is indecomposable if and only if eRe (resp. $\bar{e}\bar{R}\bar{e}$) is local. Since $\bar{e}\bar{R}\bar{e} \approx eRe/eJe$ and the latter ring is semisimple artinian, so is $\bar{e}\bar{R}\bar{e}$. Hence eRe is local if and only if $\bar{e}\bar{R}\bar{e}$ is. The indecomposable left ideals in $\bar{R}$ are minimal left ideals since the lattice $V_{\bar{R}}({}_{\bar{R}}\bar{R})$ is complemented, by Theorem IV. 4.1. Therefore Re is indecomposable if and only if $\bar{R}\bar{e}$ is minimal. ∎

Theorem 14. Let e and f be idempotents in the ring R. The R-modules Re and Rf are isomorphic if and only if there exist u, v in R such that $uv = e$ and $vu = f$.

Proof. Suppose $Re \approx Rf$ and let $xe \to xeu'f$, $x \in R$, be the isomorphism (cf. Theorem 11). Let $yf \to yfv'e$, $y \in R$ be its inverse. Then $e \cdot eu'ffv'e = e$ and $ffv'eeu'f = f$. Let $u = eu'f$ and $v = fv'e$. Conversely, suppose $uv = e$ and $vu = f$. Then the homomorphism $xe \to xeu = xuv \cdot u = xu \cdot vu = xuf \in Rf$ for $xe \in Re$ has the mapping $yf \to yfv$ for $yf \in Rf$ as inverse. ∎

If $Re \approx Rf$ for idempotents e and f in R it follows that $\bar{R}\bar{e} \approx \bar{R}\bar{f}$ in R/J. The converse is a consequence of the following:

Theorem 15. Let e and f be idempotents in the ring R with 1. Let J be the radical of R. Suppose $u'v' \equiv e \bmod J$ and $v'u' \equiv f \bmod J$. Then there exist elements u and v in R such that $uv = e$ and $vu = f$.

Proof. $fv' \equiv v'u'v' \equiv v'e$ and $eu' \equiv u'v'u' \equiv u'f \bmod J$ imply $eu'fv' \equiv u'fv' \equiv u'v'e \equiv e^2 \equiv e \bmod J$. Therefore $x = e - eu'fv' \in J$ and since $x = ex$, $eu'fv' = e(1 - x)$. Now $x \in J$ and hence there exists $y \in J$ such that $(1 - x)(1 - y) = 1$. Let $v = fv'(1 - y)$ and $u = eu'f$. Then

$uv = eu'fv'(1 - y) = e(1 - x)(1 - y) = e$ and hence $(vu)^2 = vuvu = veu = vu$. This implies $(f - vu)^2 = f - fvu - vuf + vu = f - vu$. But since $vu = fv'(1 - y)\, eu'f \equiv fv'eu'f \equiv fv'u'f \equiv f^3 \equiv f \pmod J$, $f - vu \in J$. Hence $f = vu$. ∎

Theorem 16. Let e and f be idempotents in the ring R with identity and let J be the radical. The left ideals Re and Rf are isomorphic R-modules if and only if $\bar{R}\bar{e}$ and $\bar{R}\bar{f}$ are isomorphic R/J-modules.

Proof. Theorems 14 and 15. ∎

By Theorem 13 the indecomposable left ideals in a semiperfect ring R are precisely those whose homomorphic images in R/J under the natural homomorphism are minimal left ideals of the artinian ring R/J. By Theorem IV.3.3, these are isomorphic to the irreducible R-modules and therefore as a simple consequence of Theorem 16 we have:

Theorem 17. Let R be semiperfect. Then there is a one-one correspondence between the isomorphism classes of irreducible R-modules and the isomorphism classes of indecomposable left ideals. ∎

In Theorems 11, 14, and 15 we did not assume that R is semiperfect and in Theorem 12 we only needed the assumption to prove the implication $1 \Rightarrow 3$. We may therefore apply these results in the proof of the following theorem:

Theorem 18. Let $\{e_1, ..., e_m\}$ and $\{f_1, ..., f_n\}$ be two orthogonal families of local idempotents in the ring R. Let $\sum_{i=1}^{m} e_i = \sum_{j=1}^{n} f_j = 1$. Then $m = n$ and there is an inner automorphism of R mapping the set $\{e_1, ..., e_m\}$ onto the set $\{f_1, ..., f_m\}$.

Proof. **1.** Let $x \to \bar{x} = x + J$ be the natural homomorphism of R onto R/J. We first show that $\bar{R}\bar{e}_i$ is a minimal left ideal in $\bar{R} = R/J$ (cf. Theorem 13). Set $e = e_i$ and assume $m > 1$. Since $\bar{R} = \bar{R}(\bar{1} - \bar{e}) + \bar{R}\bar{e}$, it follows that $\bar{R}(\bar{1} - \bar{e}) \neq \bar{R}$. Applying Zorn's lemma we obtain a left ideal $\bar{L} \subsetneq \bar{R}$ maximal subject to containing $\bar{R}(\bar{1} - \bar{e})$. Clearly $\bar{R} = \bar{L} + \bar{R}\bar{e}$. This sum is also direct: For suppose $(\bar{L} \cap \bar{R}\bar{e}) \neq (\bar{0})$. By semisimplicity of $\bar{R}$, $(\bar{L} \cap \bar{R}\bar{e})^2 \neq (\bar{0})$. Let x in R be such that $\bar{x}\bar{e} \in \bar{L}$ and $\bar{R}\bar{e} \cdot \bar{x}\bar{e} \neq \bar{0}$. The element $\bar{e}\bar{x}\bar{e} \neq 0$ has an inverse $\bar{y}$ in the division ring $\bar{e}\bar{R}\bar{e}$. Hence $\bar{y} \cdot \bar{e}\bar{x}\bar{e} = \bar{e} \in \bar{L}$. Therefore $\bar{R} = \bar{L}$, a contradiction. As the complement of a maximal left ideal in $\bar{R}$, $\bar{R}\bar{e}$ is minimal.

2. By hypothesis and what we have just shown in Part 1,

$$\bar{R}\bar{e}_1 + \cdots + \bar{R}\bar{e}_m = \bar{R}\bar{f}_1 + \cdots + \bar{R}\bar{f}_n$$

are two decompositions of ${}_{\bar{R}}\bar{R}$ as direct sums of minimal (and hence irreducible) R-modules. By the Jordan–Hölder Theorem (Theorem V.4.2), $m = n$ and there is a permutation π of the numbers $1, 2, \ldots, n$ such that $\bar{R}\bar{e}_i \approx \bar{R}\bar{f}_{\pi i}$, $i = 1, 2, \ldots, n$. Theorem 14 guarantees the existence of elements $\bar{u}_i$, $\bar{v}_i$ in $\bar{R}$ with $\bar{u}_i\bar{v}_i = \bar{e}_i$ and $\bar{v}_i\bar{u}_i = \bar{f}_{\pi i}$. By Theorem 15 we may assume that $u_iv_i = e_i$ and $v_iu_i = f_{\pi i}$, $i = 1, \ldots, n$. It now follows that $v_ie_iu_i = v_iu_iv_ie_i = f_{\pi i}$ and $u_if_{\pi i}v_i = e_i$. Set

$$u = \sum_i e_iu_if_{\pi i} \qquad \text{and} \qquad v = \sum_i f_{\pi i}v_ie_i$$

Then $uv = \sum_i e_i = 1 = f_{\pi i} = vu$ and $ve_j = f_{\pi j}v_je_j = f_{\pi j}v$. The element v is therefore a unit in the ring R and it induces the inner automorphism $x \to vxv^{-1}$ which maps e_j onto $f_{\pi j}$, $j = 1, 2, \ldots, r$. ▮

As a first application of this theorem we prove:

Theorem 19. Let R be a semiperfect ring and let $R = S \oplus M$ where S and M are as in Theorem 10. Then this representation of R is unique up to an inner automorphism of R.

Proof. Each idempotent g_i in Theorem 10 is the sum of an orthogonal family of primitive (and hence local) idempotents g_{ij}. The g_{ij} are mutually orthogonal for all i, j and $\sum_{i,j} g_{ij} = \sum_i g_i = 1$. Let $\{g_1, \ldots, g_n\}$ and $\{h_1, \ldots, h_n\}$ be two sets of idempotents which give rise to a decomposition of R as in Theorem 10. By Theorem 18 there exists an inner automorphism $x \to vxv^{-1}$ mapping $\{g_{ij}\}$ onto $\{h_{ij}\}$. This mapping takes the summands g_{ij} and g_{ih} of g_i onto the summands h_{rs} and h_{rt} of h_r since $g_p g_{ij} = \delta_{pi} g_{ij}$. This shows the existence of a permutation π of the numbers $1, 2, \ldots, n$ such that $vg_iv^{-1} = h_{\pi i}$. ▮

A further application of Theorem 8 is given in the following characterization of semiperfect rings.

Theorem 20 (B. Müller). A ring with 1 is semiperfect if and only if 1 is the sum of a finite number of orthogonal, local idempotents.

Proof. 1. We have already shown the necessity in Theorem 5.

2. Conversely, let $1 = e_1 + \cdots + e_n$, e_i local orthogonal idempotents.

As was shown in the first part of the proof of Theorem 18, $\bar{R}\bar{e}_i$, $i = 1, ..., n$ is a minimal left ideal in $\bar{R}$. $\bar{R}$ therefore has a finite maximal chain of left ideals $\bar{R}\bar{e}_1 \oplus \cdots \oplus \bar{R}\bar{e}_i$, $i = 1, ..., n$. By Theorem V.4.2 (Jordan–Hölder) all such chains have the same length, or, in other words, $\bar{R}$ is artinian.

3. Let $f \equiv f^2 \bmod J$ and $\bar{f} = \bar{f}_1 + \cdots + \bar{f}_r$, $\bar{1} - \bar{f} = \bar{f}_{r+1} + \cdots + \bar{f}_n$ be a decomposition of $\bar{1}$ in $\bar{R}$ as a sum of orthogonal primitive idempotents in the artinian ring $\bar{R} = R/J$. By Theorem 18 there is an inner automorphism $\bar{x} \to \bar{u}\bar{x}\bar{v}$ with $\bar{u}\bar{v} = \bar{v}\bar{u} = \bar{1}$ which maps the set of $\bar{e}_i$ onto the set of $\bar{f}_j$. By renumbering we may assume that $\bar{u}\bar{e}_i\bar{v} = \bar{f}_i$, $i = 1, 2, ..., n$ and therefore $\bar{u}(\bar{e}_1 + \cdots + \bar{e}_r)\bar{v} = \bar{f}$. Set $e = e_1 + \cdots + e_r$. If u and v are representatives of the cosets $\bar{u}$ and $\bar{v}$ respectively, $uv = 1 - \omega$ for some $\omega \in J$. Therefore $(1 - \omega')\, uv = 1$ where $(1 - \omega')(1 - \omega) = 1$. Let $u_1 = (1 - \omega')u$ and $v_1 = v$. Then $u_1 \equiv u \bmod J$ and $u_1v_1 = 1$. It follows that $v_1u_1v_1u_1 = v_1u_1 \equiv vu \equiv 1 \bmod J$ and hence $1 - v_1u_1 = (1 - v_1u_1)^2 \in J$. This implies $v_1u_1 = 1$. The mapping $x \to u_1xv_1$ is therefore an inner automorphism of R and in particular $u_1ev_1 = u_1(e_1 + \cdots + e_r)\, v_1 \equiv f \bmod J$. Hence idempotents may be lifted modulo J. ▮

The structure of the direct summands R_i of S in Theorem 10 may be described in terms of the following concept:

Definition 5. An R-module M is *free* if it is a direct sum of modules each isomorphic to ${}_RR$.

Theorem 21. Let R be a semiperfect ring, J its radical and R/J a simple ring. Then R is isomorphic to the full endomorphism ring $\mathrm{Hom}_T(M, M)$ of a finitely generated free right T-module M over the local ring $T = eRe$, where e is a suitable primitive idempotent of R.

Proof. Let $\bar{x}$ be the image of the element x of R under the natural homomorphism of R onto $\bar{R} = R/J$. Theorem 5 applied to R_R guarantees the existence of an orthogonal family of primitive idempotents $\{e_1, ..., e_n\}$ such that $e_1 + \cdots + e_n = 1$. Let $e = e_1$. By Theorem 13, $\bar{e}\bar{R}$ is a minimal right ideal of $\bar{R}$. By hypothesis, $\bar{R}$ is a simple artinian ring and therefore the $\bar{e}_i\bar{R}$, $i = 1, ..., n$ are isomorphic to each other (Theorem IV.3.3). By Theorems 14 and 15 there exist elements u_i, v_i in R such that $u_iv_i = e_i$, $v_iu_i = e$. Since $1 \in R$, R is isomorphic (as a ring) to $\mathrm{Hom}_R(R_R, R_R)$, the ring of left multiplications of R by elements of R (cf. Theorem I.2.3). We first show

$$\mathrm{Hom}_R(R_R, R_R) \approx \mathrm{Hom}_{eRe}(Re, Re) \tag{12}$$

where eRe is the centralizer of the R-module Re and Re is considered as a right eRe-module. (If $J = 0$, (12) has already been proved in Theorem IV.3.2.)

Proof of (12). Let $\phi \in \mathrm{Hom}_R(R_R, R_R)$ and set $\phi'(xe) = \phi(xe) = \phi(xe^2) = \phi(xe)e \in Re$ for $x \in R$. ϕ' is clearly an element of $\mathrm{Hom}(Re, Re)$. Furthermore, $\phi'(xe \cdot eye) = \phi(xe)\, eye = \phi'(xe)\, eye$ for all $y \in R$ and hence $\phi' \in \mathrm{Hom}_{eRe}(Re, Re)$. Let η be the mapping defined by $\eta\colon \phi \to \phi'$ for $\phi \in \mathrm{Hom}_R(R_R, R_R)$. We show η is an isomorphism onto $\mathrm{Hom}_{eRe}(Re, Re)$. Clearly η is additive and preserves multiplication. Given ϕ' as the image of ϕ, we may recover ϕ as follows: $v_i u_i = e$ implies $u_i e v_i = u_i v_i \cdot u_i v_i = e_i^2 = e_i$ for $i = 1, 2, \ldots, n$. Let $a \in R$. Then $a = e_1 a + \cdots + e_n a = \sum_{1 \leqslant i \leqslant n} u_i e v_i a$. It follows that

$$\phi a = \sum \phi(u_i e v_i a) = \sum (\phi(u_i e))\, v_i a = \sum \phi'(u_i e)\, v_i a$$

If $\phi' = \eta\phi = 0$ we therefore have $\phi = 0$ and η is one-one. Now let $\psi \in \mathrm{Hom}_{eRe}(Re, Re)$. The mapping $\phi_0\colon a = e_i a \to \sum \psi(u_i e)\, v_i e_i a$ for $a \in R$ is well-defined since $R_R = e_1 R + \cdots + e_n R$ by the orthogonality of the e_i. Since $ax = \sum e_i a x$ and $e_i a x \in e_i R$, ϕ_0 is clearly an endomorphism of R_R. To show that $\eta\phi_0 = \psi$, let $a = ae = \sum e_i a e$. Since ψ is an endomorphism of the right eRe-module Re and $\psi(u_i e) = \psi(u_i e^2) = \psi(u_i e)e$, the product $\psi(u_i e)\, v_i e_i a e = \psi(u_i e v_i e_i a e) = \psi(e_i a e)$. Therefore $(\eta\phi_0)(ae) = \phi_0(ae) = \psi(ae)$ and η is an epimorphism. Hence η is an isomorphism. We therefore have

$$R \approx \mathrm{Hom}_R(R_R, R_R) \approx \mathrm{Hom}_{eRe}\left(\oplus \sum_i e_i Re, \oplus \sum_i e_i Re\right) \tag{13}$$

We must now show that the $e_i Re$ are isomorphic as right eRe-modules to T_T, where $T = eRe$. The mappings

$$e_i xe \to v_i e_i xe, \qquad x \in R, \qquad i = 1, 2, \ldots, n \tag{14}$$

are the required isomorphisms since

$$v_i e_i xe = v_i u_i v_i xe = e v_i xe \in eRe$$

In other words, we have proved that R is isomorphic to the full ring of eRe-linear mappings of a direct sum of n copies of the ring eRe, considered as a right eRe-module.

A free right T-module M with generators $m_1, \ldots, m_n$ is the direct sum of the right T-modules $m_i T$. A T-linear mapping A of M into itself is (as in the case of vector spaces) uniquely determined by the images Am_k, $k = 1, \ldots, n$. Let $Am_k = \sum_i m_i t_{ik}$. Then $\mathrm{Hom}_T(M, M)$ is isomorphic to the ring of $n \times n$ matrices, $T_{n\times n}$, over T. ∎

We now prove:

Theorem 22. Let P be a ring with 1 and let J be its radical. Let $P_{n\times n}$ be the ring of $n \times n$ matrices over P and let C_{ik} denote the matrix with 1 in the ith row and kth column and zeros elsewhere. Then the radical

$$J(P_{n\times n}) = \sum_{i,k} J(P)\, C_{ik}$$

Proof. Let $E = C_{11} + \cdots + C_{nn}$ be the identity of $R = P_{n\times n}$. Let $Z_k = \sum_i z_{ik}C_{ik}$ where $z_{ik} \in J$, $i = 1, \ldots, n$. In particular there exists $(1 - z_{kk})^{-1} \in P$. Set $Z_k' = -Z_k(1 - z_{kk})^{-1}$. Then

$$\begin{aligned}(E - Z_k')(E - Z_k) &= E + \sum_i z_{ik}(1 - \underset{r}{z_{kk}})^{-1}C_{ik} - \sum_i z_{ik}C_{ik} \\ &= \sum_i z_{ik}(1 - z_{kk})^{-1}z_{kk}C_{ik}\end{aligned}$$

since $C_{ik} \cdot C_{jk} = \delta_{kj}C_{ik}$. Therefore $(E - Z_k')(E - Z_k) = E$ and the left ideals $\sum_i JC_{ik}$ are quasiregular for each $k = 1, \ldots, n$. Hence $\sum_{i,k} JC_{ik}$ is contained in the radical $J(R)$. Conversely, let $Z = \sum_{i,k} z_{ik}C_{ik} \in J(R)$. For each $a \in P$, $J(R)$ also contains $aE \cdot Z = \sum_{i,k} az_{ik}C_{ik}$. Then

$$\sum_h C_{hj} \sum_{i,k} az_{ik}C_{ik}C_{lh} = az_{lh}E$$

There exists $Z' = z_{ik}C_{ik} \in R$ such that

$$E = (E - Z')(E - az_{jl}E) = (E - Z')(1 - az_{jl}).$$

Therefore $1 = (1 - z_{11})(1 - az_{jl})$ for all $a \in R$ and az_j is quasiregular. Hence $J(R) \subseteq \sum_{i,k} JC_{ik}$. ∎

Now let $T_{n\times n}$ denote the ring of $n \times n$ matrices over the local ring T and let $J(T)$ be the radical of T. By Theorem 22, the radical of $T_{n\times n}$ consists of those matrices all of whose entries lie in $J(T)$. It follows that the factor ring $T_{n\times n}/J(T_{n\times n})$ is isomorphic to the ring of $n \times n$ matrices over the division ring $T/J(T)$ and is therefore a simple ring. Moreover, the identity $C_{11} + \cdots + C_{nn}$ of $T_{n\times n}$ is the sum of the orthogonal family of local idempotents C_{ii}, $i = 1, \ldots, n$. By Theorem 20, $T_{n\times n}$ is therefore a semiperfect ring with simple factor ring $T_{n\times n}/J(T_{n\times n})$. In conjunction with Theorem 21 we have the following characterization of these rings:

Theorem 23. *R is a semiperfect ring with R/J simple if and only if it is isomorphic to the ring of $n \times n$ matrices over a local ring.* ∎

As an illustration of this theorem we prove:

Theorem 24. *Let Φ be a field and $\Phi^* = \Phi[[\omega]]$ the ring of formal power* series in the indeterminate ω. Then the ring $\Phi^*_{n\times n}$ of $n \times n$ matrices is semiperfect and $\Phi^*_{n\times n}$ factored by its radical $\omega\Phi^*_{n\times n}$ is the simple ring $\Phi_{n\times n}$. However, $\Phi^*_{n\times n}$ is not artinian.

Proof. The radical of Φ^* is $\omega\Phi^*$ and $\Phi^*/\omega\Phi^* \approx \Phi$. However, Φ^* is not artinian since $\omega^m\Phi^*$, $m = 0, 1, \ldots$, is an infinite, strictly descending chain of ideals and this gives rise to the descending chain $\omega^m\Phi^*_{n\times n}$ in $\Phi^*_{n\times n}$. ▮

In Theorem 18 we proved that if we are given 2 families of orthogonal, local idempotents whose sum is 1, we can find an inner automorphism which maps one family onto the other. We now use this fact in the proof of the following theorem due to Krull and Schmidt which we shall need in the next section:

An R-module M is said to be indecomposable if M has no nontrivial direct summands. Now let $M = M_1 \oplus \cdots \oplus M_m$ and $M = N_1 \oplus \cdots \oplus N_n$ be two representations of M as direct sums of indecomposable submodules. It will be shown that $m = n$ and that for a suitable permutation π of the numbers $1, 2, \ldots, n$, $M_i \approx N_{\pi i}$ if M is both artinian and noetherian. The proof depends on the fact that such a module is local if and only if M is indecomposable.

Theorem 25. Let $\phi \in \mathrm{Hom}_R(M, M)$.

1. If M is artinian, ϕ is an epimorphism if ϕ is a monomorphism.
2. If M is noetherian, ϕ is a monomorphism if ϕ is an epimorphism.

Proof. 1. There is a natural number n such that $M \supset M\phi \supset \cdots \supset M\phi^n = M\phi^{n+1}$. Let $x \in M$. Then $x\phi^n = y\phi^{n+1}$ for some $y \in M$. But ϕ^n is a monomorphism since ϕ is and therefore $x = y\phi$.

2. Similarly $0 \subset 0\phi^{-1} \subset \cdots \subset 0\phi^{-n} = 0\phi^{-n-1}$ for some n. Let $x\phi = 0$ for some $x \in M$. ϕ^n is an epimorphism since ϕ is. Hence there exists $y \in M$ such that $y\phi^n = x$. Then $y\phi^{n+1} = x\phi = 0$ and $y \in 0\phi^{-n-1} = 0\phi^{-n}$. This implies $x = y\phi^n = 0$. ▮

Every linear transformation T of a finite-dimensional vector space gives rise to a decomposition of the space as a direct sum of the kernel of the transformation and a subspace isomorphic to the image of T. A similar situation obtains when M is artinian and noetherian

Theorem 26 (Fitting). Let M be an artinian and noetherian R-module and let $\phi \in \mathrm{Hom}_R(M, M)$. Then for some natural number n,

$$M = M\phi^n \oplus 0\phi^{-n}$$

Proof. Since M is artinian, there exists n with $M \supset M\phi \supset \cdots \supset M\phi^n = M\phi^{n+1} = \cdots = M\phi^{2n}$. Thus ϕ^n induces an epimorphism of the noetherian module $M\phi^n$. By Theorem 25 this restriction of ϕ^n to $M\phi^n$ is a monomorphism. Hence the sum $M\phi^n \oplus 0\phi^{-n}$ is direct. Moreover, if $x \in M$, $x\phi^n = y\phi^{2n}$ for some $y \in M$. Therefore $x - y\phi^n \in 0\phi^{-n}$ and $x = y\phi^n + (x - y\phi^n)$ proving that $M\phi^n \oplus 0\phi^{-n} = M$. ∎

Theorem 27. Let M be an artinian and noetherian R-module. Its endomorphism ring $\operatorname{Hom}_R(M, M)$ is local if and only if M is indecomposable. In this case each endomorphism is either an automorphism or nilpotent.

Proof. Let $\phi \in \operatorname{Hom}_R(M, M)$ and let M be indecomposable. By Theorem 26 either $M\phi^n = 0$ or $M\phi^n = M$. In the latter case ϕ^n is onto and therefore ϕ is a monomorphism by Theorem 25. In the former case, ϕ^n is nilpotent. If ϕ is not an automorphism, neither is $\alpha\phi$ for any $\alpha \in \operatorname{Hom}_R(M,M)$. Therefore every element in the left ideal of $\operatorname{Hom}_R(M, M)$ generated by ϕ is nilpotent. This implies ϕ is in the radical of the endomorphism ring. On the other hand no automorphism of M is in the radical of $\operatorname{Hom}_R(M, M)$ since the automorphisms are units in $\operatorname{Hom}_R(M, M)$. Therefore the radical consists precisely of the nilpotent elements of the endomorphism ring. It follows that each nonzero element of $\operatorname{Hom}_R(M, M)/J(\operatorname{Hom}_R(M, M))$ is a unit and therefore $\operatorname{Hom}_R(M, M)$ is local. Conversely, if M is decomposable, say $M = M_1 \oplus M_2$, then the projections π_1 and π_2 of M onto M_1 and M_2, respectively, are two orthogonal idempotents in $\operatorname{Hom}_R(M, M)$, showing that this ring is not local. ∎

Theorem 28 (Krull–Schmidt). Let M be an artinian and noetherian R-module. Let $M = M_1 \oplus \cdots \oplus M_r$ and $M = N_1 \oplus \cdots \oplus N_s$ be two decompositions of M as the direct sum of indecomposable submodules. Then $r = s$ and for some permutation π of the numbers $1, \ldots, r$, $M_i \approx M_{\pi i}$, $i = 1, 2, \ldots, r$.

Proof. Set $E = \operatorname{Hom}_R(M, M)$. Let π_i, $i = 1, \ldots, r$ be the projections of M onto M_i, η_j, $j = 1, \ldots, s$ the projections of M onto N_j. Then $\{\pi_1, \ldots, \pi_r\}$ and $\{\eta_1, \ldots, \eta_s\}$ are two orthogonal families of idempotents of E, each with sum 1, where 1 denotes the identity automorphism of M. By Theorem 27 the endomorphism rings of the M_i and N_j are local. To apply Theorem 18 we must first show that these rings are isomorphic to the $\pi_i E \pi_i$ and $\eta_j E \eta_j$, respectively. Since $\pi_i M_i = M_i$, for each $\phi \in E$, $\pi_i \phi \pi_i$ induces an endomorphism of M_i. On the other hand let $\phi_i \in \operatorname{Hom}_R(M_i, M_i)$. Set $(\sum_k m_k)\, \psi_i = m_i \phi_i$. Then $\psi_i \in \operatorname{Hom}_R(M, M)$ and $\pi_i \psi_i \pi_i = \phi_i$. Therefore the mapping $\phi_i \to \psi_i$ is an isomorphism of $\operatorname{Hom}_R(M, M)$ onto the subring

$\pi_i E \pi_i$ of E and hence $\pi_i E \pi_i$ is local. Similarly $\eta_j E \eta_j$ is local. By Theorem 18, $r = s$ and there is a unit $\zeta \in E$ (i.e. an automorphism of M) and a permutation π of 1, 2, ..., r such that $\zeta^{-1}\pi_i\zeta = \eta_{\pi i}$ for $i = 1, 2, ..., r$. Since $N_{\pi i} = M\eta_{\pi i} = M\zeta^{-1}\pi_i\zeta = M\pi_i\zeta = M_i\zeta$, M_i is mapped onto $N_{\pi i}$ under this automorphism for $i = 1, 2, ..., r$. ∎

Theorem 28 was proved under the assumption that the R-module is artinian and noetherian. In applications the following is often useful:

Theorem 29. Let R be artinian with identity and M a unital R-module. Then M is artinian if and only if it is noetherian.

Proof. Consider the following chain of submodules:

$$M \supset JM \supset \cdots \supset J^N M = 0,$$

where J is the radical of R. The factor modules $J^i M/J^{i+1}M$ are R/J-modules and hence completely reducible. If M is artinian (noetherian) so is $J^i M/J^{i+1}M$. A completely reducible module is artinian (noetherian) if and only if it is a direct sum of finitely many irreducible modules. Therefore if M is artinian (noetherian), M has a finite composition series and by Theorem V.4.2 (Jordan–Hölder), M is noetherian (artinian). ∎

Theorem 30. Let R be a ring with 1 and U a submodule of M. Then

1. M is artinian (noetherian) if and only if both U and M/U are.
2. Let M be a finitely generated R-module. Then M is artinian (noetherian) if and only if R is.

Proof. 1(a) Clearly if M is artinian (noetherian), so are U and M/U. 1(b) Let U and M/U be artinian and

$$M = M_0 \supseteq M_1 \supseteq M_2 \supseteq \cdots$$

be a descending chain of submodules of M. Then $U = M_0 \cap U \supseteq M_1 \cap U \supseteq M_2 \cap U \supseteq \cdots$ and $M/U = (M_0 + U)/U \supseteq (M_1 + U)/U \supseteq \cdots$ are descending chains of submodules of U and M/U, respectively. By hypothesis there exists N such that $M_N \cap U = M_n \cap U$ and $(M_N + U)/U = (M_n + U)/U$ for all $n \geqslant N$. Since $M_N \supseteq M_n$ and since the lattice of R-submodules is modular we have $M_n = M_n + (U \cap M_n) = M_n + (U \cap M_N) = (M_n + U) \cap M_N = (M_N + U) \cap M_N = M_N$ for all $n \geqslant N$. Similarly we may show that if U and M/U are noetherian so is M.

2. Let M be finitely generated, say $M = Rm_1 + Rm_2 + \cdots + Rm_n$ and let R be artinian (noetherian). If $n = 1$, M is cyclic and therefore isomorphic

to ${}_R R/L$, where $L = \{x \in R \mid xm_1 = 0\}$. Since ${}_R R$ is artinian (noetherian) so is every factor module. Assume inductively that the theorem holds for modules which can be generated by $n - 1$ or fewer elements. Then Rm_1 is artinian (noetherian) as we have just shown and

$$M/Rm_1 \approx (Rm_2 + \cdots + Rm_n)/Rm_1 \cap (Rm_2 + \cdots + Rm_n)$$

is artinian (noetherian) by induction hypothesis. Hence M is artinian (noetherian) by the first assertion of the theorem. ∎

2. Projective modules and Asano orders

In the last section the indecomposable direct summands of ${}_R R$ and in Theorem 1.21 the free R-modules played the most important roles. All these modules have the property expressed in Definition 1(1) below.

Definition 1. 1. Let $\cdots \to A \xrightarrow{f} B \xrightarrow{g} C \xrightarrow{h} D \to \cdots$ be a sequence of modules and homomorphisms. The sequence is said to be *exact at B* if image $f =$ kernel g. The sequence is said to be *exact* if it is exact at each module.

2. Let R be a ring with 1 and M an R-module. M is called a *projective R-module* if the left diagram can be embedded in the right diagram

$$\begin{array}{ccc} & & M \\ & & \downarrow \phi \\ A & \xrightarrow{\pi} & B \to 0 \end{array} \quad \text{exact} \qquad\qquad \begin{array}{ccc} & & M \\ & \swarrow \bar{\phi} & \downarrow \phi \\ A & \xrightarrow{\pi} & B \to 0 \end{array}$$

in such a way that $\pi\bar{\phi} = \phi$. Roughly speaking M is projective if homomorphisms of M can be lifted from factor modules B to modules A.

Theorem 1. The R-module ${}_R R$ is projective.

Proof. Let $1\phi = b \in B$ and $b = a\pi$ for some $a \in A$. Then $x\phi = xb$ and $(xa)\pi = xb$ for $x \in R$. Set $x\psi = xa$. Then $\psi \in \mathrm{Hom}_R(R, A)$ and $x\psi\pi = (xa)\pi = xb = x(1\phi) = x\phi$. ∎

Theorem 2. Let M be the direct sum of the R-modules M_i, $i \in I$. M is projective if and only if the M_i are.

Proof 1. Assume each M_i is projective and consider the diagram

$$\begin{array}{ccc} & & M \\ & & \downarrow \phi \\ A & \xrightarrow{\pi} & M \to 0 \end{array}$$

where the row is exact. Let $\phi\,|_{M_i} = \phi_i$ be the restriction of ϕ to M_i. By projectivity of M_i there exists $\psi_i \in \mathrm{Hom}_R(M_i, A)$ with $\psi_i\pi = \phi_i$. Let $\psi = \sum_i \psi_i \in \mathrm{Hom}_R(M, A)$. Then $\psi\pi = \phi$ and M is projective.

2. Conversely, if M is projective and we have the diagram

$$\begin{array}{ccc} & M_\lambda & \\ & \downarrow{\scriptstyle \phi_\lambda} & \\ A & \xrightarrow[\pi]{} B & \to 0 \end{array}$$

with exact row, extend ϕ_λ to $\phi \in \mathrm{Hom}_R(M, B)$ by defining $\phi = \delta_{i\lambda}\phi_\lambda$. Let $\psi\pi = \phi$ for some $\psi \in \mathrm{Hom}_R(M, A)$. Then the restriction

$$\psi_\lambda = \psi\,|_{M_\lambda} \in \mathrm{Hom}_R(M_\lambda, A)$$

makes the diagram commutative, i.e. $\pi\psi_\lambda = \phi_\lambda$. ∎

The following is an immediate consequence of Theorems 1 and 2:

Theorem 3. Every free R-module and every indecomposable left ideal in a ring with 1 is projective. ∎

Theorem 4. Every R-module is the homomorphic image of a free (and hence projective) R-module P.

Proof. Let $\{b_i \mid i \in I\}$ be a set of generators of the R-module B. Let R_i, $i \in I$ be a family of R-modules each equal to ${}_RR$ and let $P = \oplus \sum R_i$. The mapping $\sum_i x_i \to \sum x_i b_i$ is an epimorphism of the free R-module P onto B. ∎

Theorem 5. An R-module M is projective if and only if for every exact sequence $A \to M \to 0$ there exists a submodule U of A such that $A = U \oplus M'$, where $M' \approx M$.

Proof. 1. Sufficiency: Let P be the module in Theorem 4. Then M is isomorphic to a direct summand of the projective module P and therefore by Theorem 2, M is projective.

2. Necessity: Take $\phi = 1_M$, the identity of M in Definition 1. Then there exists a monomorphism ψ from M to A such that $\psi\pi = 1_M$. Moreover, $\pi\psi\pi\psi = \pi\psi$ and the mapping $\pi\psi$ from A onto $M\psi$ is an idempotent endomorphism of A. Hence $A = M\pi\psi \oplus A(1_A - \pi\psi)$ since $a = a\pi\psi + a(1_A - \pi\psi)$ for all $a \in A$ ∎

Theorems 2, 4, and 5 imply the first assertion of the following:

Theorem 6. 1. An R-module is projective if and only if it is isomorphic to a direct summand of a free module.

2. Let R be a semiperfect ring and let M be a finitely generated R-module. M is projective if and only if M is isomorphic to a direct sum of finitely many R-modules each isomorphic to some indecomposable left ideal of R.

Proof. All has been proved except for the necessity of the second assertion. $\bar{M} = M/JM$ is a finitely generated $\bar{R} = R/J$-module and is therefore a direct sum $\bar{M} = \sum \oplus U_i$ of irreducible $\bar{R}$-modules U_i since $\bar{R}$ is semisimple artinian. These are images of indecomposable left ideals Re_i in R under epimorphisms ϕ_i. The epimorphism ϕ whose restriction to Re_i is ϕ_i therefore maps the direct sum $P = \sum \oplus Re_i$ onto $\bar{M}$. Since P is projective by the first assertion of the theorem, there exists $\psi \in \mathrm{Hom}_R(P, M)$ with $\psi\pi = \phi$, where π: $M \to \bar{M}$. Since both ϕ and $\psi\pi$ are epimorphisms, we have $M = P\psi + JM$. By the following lemma, JM is *small* in M, i.e., $M = JM + N$ implies $M = N$. This proves $P\psi = M$. Since M is projective, ker ψ is a direct summand of P by Theorem 5. Moreover, $\phi = \psi\pi$ and hence ker $\psi \subseteq$ ker ϕ. By the definition of ϕ, ker $\phi = \sum \oplus Je_i = JP$ and is therefore small in P. Hence ker ψ is small in P and, as a direct summand of P, is the zero module. In other words, ψ is an isomorphism of P onto M.

Lemma. (Nakayama). Let R be a ring with identity and J its radical. Let M be a finitely generated R-module. Then JM is *small* in M, i.e., if $M = JM + N$ for some submodule N of M, then $M = N$.

Proof. From $M = Rm_1 + \cdots + Rm_s$ and $M = JM + N$ it follows that $m_s = w_1m_1 + \cdots + w_sm_s + v_s$ where $w_i \in J$ and $v_s \in N$. Therefore $(1 - w_s)m_s = w_1m_1 + \cdots + w_{s-1}m_{s-1} + v_s$. Multiplying this equation on the left by the inverse $(1 - w_s')$ of $(1 - w_s)$, we get

$$m_s = w_1'm_1 + \cdots + w_{s-1}'m_{s-1} + v_s'$$

Thus $M = JM_1 + \cdots + Jm_{s-1} + N$. Applying the same process to m_{s-1} as we applied to m_s and continuing in this fashion, we eventually arrive at $M = N$. ∎

A vector space M over a division ring R is a free R-module and, given a basis $\{b_i \mid i \in I\}$ each $m \in M$ is expressible in the form $M = \sum(m\psi_i)\, b_i$, where $\psi_i \in \mathrm{Hom}_R(M, {}_RR)$ maps the vector m onto its ith coordinate relative to the basis $\{b_i \mid i \in I\}$. The following is a generalization of this situation:

Theorem 7. Let R be a ring with 1 and M and R-module generated by

$\{b_i \mid i \in I\}$. Then M is projective if and only if there exists a set $\{\psi_i \mid \psi_i \in \mathrm{Hom}_R(M, {}_RR), i \in I\}$ such that

$$m = \sum_i (m\psi_i)\, b_i \qquad \text{for all} \quad m \in M \tag{1}$$

where $m\psi_i = 0$ for all but a finite number of ψ_i.

Proof. Let $P = \oplus \sum_i Re_i$ be the free R-module with basis $\{e_i \mid i \in I\}$ and π an epimorphism of P onto M taking e_i to b_i (Theorem 4). M is projective if and only if it is isomorphic to a direct summand of P. As in Theorem 5, this is the case if and only if there exists a homomorphism ψ of M to P such that $\psi\pi$ is the identity automorphism 1_M on M. If ψ exists, each image $m\psi$ is uniquely expressible in the form

$$m\psi = \sum (m\psi_i)\, e_i \tag{2}$$

where $m\psi_i$ is an element of R depending on m and i. The mapping $\psi_i\colon m \to m\psi_i$ is an R-module homomorphism of M to ${}_RR$. Applying π to both sides of (2) we get (1) since $\psi\pi = 1_M$ and $e_i\pi = b_i$. Conversely, if the condition is satisfied, then we may define an R-module homomorphism ψ of M to P by $m\psi = \sum_i (m\psi)\, e_i$ for all $m \in M$. Applying π to both sides of (1) we get $m\psi\pi = \sum_i (m\psi_i)\, b_i = m$ for all $m \in M$ since $e_i\pi = b_i$. ∎

We may now base the theory of noncommutative Asano orders (due to Robson) on this theorem. We start with the following:

Definition 2. An element a of a ring R is *regular* if it is neither a right nor a left zero divisor. A ring Q with 1 is a *quotient ring* if each regular element has an inverse. The subring R of Q is an *order* (more precisely, a left order) in the quotient ring Q if each element of Q is expressible in the form $a^{-1}b$, where $a, b \in R$ and a is regular. An additive subgroup I of Q is a *left R-ideal* provided (i) $RI \subseteq I$, (ii) I contains a regular element, and (iii) there exists a regular element $b \in R$ such that $Ib \subseteq R$. We define right R-ideals and (two-sided) R-ideals in Q analogously. An R-ideal is *integral* if it is contained in R.

Definition 3. The order R with 1 in the quotient ring Q is an Asano order if the R-ideals in Q form a group under multiplication.

In this case the group is abelian, the prime ideals in R are the maximal ideals of R and each ideal in R is uniquely expressible as a product of powers of prime ideals (cf. Theorem 11). The integers are an example of an order in the quotient ring $\mathbb{Q}$ of rationals. The $\mathbb{Z}$-ideals in $\mathbb{Q}$ are of the form $\mathbb{Z}a/b$ where $a \neq b$. The aim of the following discussion is to obtain a characterization of the Asano orders R in Q. It will be shown that they are precisely the "maximal orders" in Q with the property that each integral ideal in R is projective as a left R-module. The expression "maximal order" is precisely defined as follows:

Definition 4. An order S in the quotient ring Q is *equivalent* to the order R in Q if there exist regular elements c, d, e, f in Q such that $cRd \subseteq S$ and $eSf \subseteq R$. An order R in Q is *maximal* if it is maximal (with respect to set theoretic inclusion) among all orders equivalent to R.

Theorem 8. 1. Let R be an order in the quotient ring Q. Then for each pair (a, b) of elements in R with a regular, there exists a pair (a_1, b_1) in R with a_1 regular, such that $a_1 b = b_1 a$ is a common left multiple of a and b.

Note: In Chapter XI, we shall show that this condition (due to Ore) is a sufficient condition for the existence of a quotient ring Q containing R as an order.

Proof. The product ba^{-1} is in Q. Hence there exist $a_1, b_1 \in R$ such that $ba^{-1} = a_1^{-1} b_1$. ∎

Addition and multiplication in Q can be reduced to the corresponding operations in R once we have established the existence of common denominators.

Theorem 9. Let $a_1, a_2, \ldots, a_n$ be regular elements in the order R of the quotient ring Q. Then there exist regular elements $a, b_1, \ldots, b_n$ in R such that $a_i^{-1} = a^{-1} b_i, i = 1, \ldots, n$.

Proof. By induction on n: When $n = 1$, take $a = a_1^2$, $b_1 = a_1$. Assume inductively that regular elements $a_0, b_1', \ldots, b_{n-1}'$ can be found so that $a_i^{-1} = a_0^{-1} b_i'$, $i = 1, \ldots, n-1$. By Ore's condition (Theorem 8) there exist elements b and b_n in R with b_n regular such that $b_n a_n = b a_0$. Solving for b we get $b = b_n a_n a_0^{-1}$ and as a product of three units in Q, b is also regular. It follows that $a = b_n a_n = b a_0$ is regular and $a_0^{-1} = a^{-1} b$, $a_n^{-1} = a^{-1} b_n$. Finally, set $b_i = b b_i'$, $i = 1, 2, \ldots, n-1$. Then $a_i^{-1} = a_0^{-1} b_i' = a^{-1} b b_i' = a^{-1} b_i$. ∎

Let I be an R-ideal of Q. It may happen that I is also a left ideal with respect to an order R' properly containing R and contained in Q. To understand this we need the following:

Definition 5. Let R be an order in the quotient ring Q and I an R-ideal in Q. Then the set $O_l(I) = \{q \in Q \mid qI \subseteq I\}$ is the *left order of* I and the set $O_r(I) = \{q \in Q \mid Iq \subseteq I\}$ is the *right order of* I. Finally, we define $I^* = \{q \in Q \mid IqI \subseteq I\}$ as the *inverse* of I.

Since R is contained in both $O_l(I)$ and $O_r(I)$, each is an order in Q. Moreover, they are equivalent to R: In fact, let x be a regular element of I and b be a regular element in R with $Ib \subseteq R$. (x and b exist by definition). Then for each $q \in O_l(I)$, $1 \cdot q \cdot xb \subseteq Ib \subseteq R$ and $1 \cdot R \cdot 1 \subseteq O_l(I)$. Similarly we may show that $O_r(I)$ is equivalent to R.

Since

$$I^* = \{q \in Q \mid Iq \subseteq O_l(I)\} = \{q \in Q \mid qI \subseteq O_r(I)\} \tag{3}$$

and since $O_l(I)$ and $O_r(I)$ are subrings of Q, I^* is a subgroup of the additive group of Q for which

$$O_r(I) \cdot I^* \subseteq I^* \qquad \text{and} \qquad I^* \cdot O_l(I) \subseteq I^*$$

Let b be a regular element with $Ib \subseteq R \subseteq O_l(I)$. Then b is in I^*. Let x be a regular element in I. Then $I^*x \subseteq O_r(I)$ and $xI^* \subseteq O_l(I)$. Hence I^* is a left $O_r(I)$-ideal and a right $O_l(I)$-ideal.

It follows from (3) that $I^*I \subseteq O_r(I)$ for arbitrary I. The following theorem characterizes those R-ideals I for which $I^*I = O_r(I)$.

Theorem 10. Let I be an R-ideal in the order R of the quotient ring Q. Then $I^*I = O_r(I)$ if and only if I is a projective left $O_l(I)$-module. In this case I is a finitely generated left $O_l(I)$-ideal. If, in addition, R is a maximal order in the quotient ring Q then $O_r(I) = O_l(I) = R$.

Proof. 1. Since $1 \in O_r(I)$ and $I^*I = O_r(I)$, there exist finite sets $\{q_i \in I^* \mid i \in J\}$ and $\{b_i \in I \mid i \in J\}$ such that $\sum q_i b_i = 1$. Therefore, for each $a \in I$, $\sum aq_i b_i = a$. Set $a\psi_i = aq_i$ for $a \in I$ and $i \in J$. By (3), the ψ_i form a set of homomorphisms of the left $O_l(I)$-module I to the left $O_l(I)$-module $O_l(I)$ such that $\sum (a\psi_i)\, b_i = a$ for all $a \in I$. It follows from the criterion established in Theorem 7 that I is projective as a left $O_l(I)$-module and is generated by finitely many b_i.

2. Conversely, let I be a projective $O_l(I)$-module. Again, by Theorem 7, let $\{b_i \mid i \in J\}$ be a set of generators of I considered as a $O_l(I)$-module and let $\{\psi_i \mid i \in J\}$ be a set of $O_l(I)$-module homomorphisms of I to $O_l(I)$ with the property that $\sum_i (a\psi_i)\, b_i = a$ for all $a \in I$. Since $1 \in Q$, the endomorphism ring of ${}_QQ$ consists of the set of right multiplications by elements q of Q. The mapping $x \to xq$ for all $x \in Q$ induces an $O_l(I)$-module homomorphism of I to $O_l(I)$ if and only if $Iq \subseteq O_l(I)$; in other words, if and only if $q \in I^*$. Since I contains a regular element, it follows that distinct elements of I^* give rise to distinct elements of $\mathrm{Hom}_{O_l(I)}(I, O_l(I))$. Moreover, this monomorphism of the additive group of I^* to the additive group of $\mathrm{Hom}_{O_l(I)}(I, O_l(I))$ is an epimorphism as can be seen from the following: Since I contains a

regular element of Q (and hence a unit of Q), $QI = Q$. Furthermore, $O_l(I)$ is an order in Q and therefore each element of Q is of the form $c^{-1}x$, where c, $x \in O_l(I)$ and c is regular. Now $Q = QI = \{\sum_{i=1}^{n} c^{-1}x_i a_i \mid a_i \in I, c_i, x_i \in O_l(I)\}$. The existence of common denominators proved in Theorem 9 gives $Q = \{c^{-1}a \mid c \in O_l(I), c \text{ regular}, a \in I\}$. Now let θ be an $O_l(I)$-module homomorphism of I to $O_l(I)$. Then

$$\theta^*: c^{-1}a \to c^{-1}(a\theta) \qquad \text{for} \quad c \in O_l(I), \quad a \in I$$

is a well-defined mapping of Q into itself.

Proof. Suppose $c^{-1}a = c_1^{-1}a_1$. Then $a = cc_1^{-1}a_1$. Since $O_l(I)$ is a (left) order in Q, there exist elements d and d_1 in $O_l(I)$ such that $cc_1^{-1} = d^{-1}d_1$. Then $a = d^{-1}d_1a_1$ and hence $da = d_1a_1$. It follows that $d(a\theta) = (da)\theta = (d_1a_1)\theta = d_1(a_1\theta)$ and therefore $a\theta = d^{-1}d_1(a_1\theta) = cc_1^{-1}(a_1\theta)$ proving that $c^{-1}(a\theta) = c_1^{-1}(a_1\theta)$. Hence θ^* is well-defined. That θ^* is additive again follows from the existence of common denominators. Moreover, θ^* is a Q-module homomorphism. Proof: $c_1^{-1}a_1 \cdot c^{-1}a = c_1^{-1} \cdot d^{-1}b \cdot a$, where $b \in I$, $d \in O_l(I)$ with $a_1c^{-1} = d^{-1}b$. Then $(c_1^{-1}a_1 \cdot c^{-1}a)\,\theta^* = (dc_1)^{-1}((ba)\,\theta) = c_1^{-1}d^{-1}b(a\theta) = c_1^{-1}a_1c^{-1}(a\theta) = c_1^{-1}a_1((c^{-1}a)\,\theta^*)$. The mapping $\theta \in \mathrm{Hom}_{O_l(I)}(I, O_l(I))$ is the restriction of $\theta^* \in \mathrm{Hom}_Q({}_QQ, {}_QQ)$ to I since $(1^{-1}a)\,\theta^* = a\theta$ for all $a \in I$. It follows that to each $\theta \in \mathrm{Hom}_{O_l(I)}(I, O_l(I))$, there exists $q \in I^*$ such that $a\theta = aq$ for all $a \in I$. Let $q_i \in I^*$, $i \in J$ correspond to the ψ_i, $i \in J$ occurring in the equation

$$a = \sum_i (a\psi_i)\, b_i \qquad \text{for all} \quad a \in I$$

For a given $a \in I$, $aq_i = a\psi_i = 0$ for all but a finite number of i. Since I contains a regular element this implies that $q_i = 0$ for almost all i and hence almost all ψ_i are the zero map. It follows that since

$$a = \sum (a\psi_i)\, b_i = \sum_i aq_ib_i = a \sum_i q_ib_i \qquad \text{for all} \quad a \in I$$

I is generated as an $O_l(I)$-module by the finitely many b_i for which $q_i \neq 0$. Moreover, if we choose a regular in I, we get $1 = \sum q_ib_i \in I^*I$ and hence $I^*I \subseteq O_r(I) \subseteq I^*IO_r(I) = I^*I$. ∎

Theorem 11. Let R be a maximal order in the quotient ring Q. Let R contain the identity of Q and let each integral R-ideal in Q be projective as an R-module. Then the lattice of integral R-ideals in Q is noetherian and we have the following:

1. To each integral R-ideal I in Q there corresponds an R-ideal I^{-1} in Q such that $I^{-1}I = II^{-1} = R$.

2. Each integral R-ideal in Q is a product of maximal ideals in R. In particular, each prime ideal is maximal.

3. The R-ideals in Q form an abelian group under multiplication and therefore R is an Asano order in Q.

Furthermore, each R-ideal I in Q is expressible as a product of powers of maximal ideals in R:

$$I = \mathcal{M}_1^{k_1}\mathcal{M}_2^{k_2}\cdots\mathcal{M}_r^{k_r}, \qquad k_i \in \mathbb{Z}$$

This product is unique up to order.

Proof. 1. By Theorem 10, $O_l(I) = O_r(I) = R$ for each R-ideal I in Q since R is a maximal order in Q. The lattice of R-ideals in Q is noetherian since each R-ideal is finitely generated, by Theorem 10. Hence, *a fortiori*, the lattice of integral R-ideals is noetherian. In Theorem 10 we have already shown $I^*I = R$. We now prove that if I is a maximal ideal in R, then $II^* = R$. Assume $R = I^*$. Then $R = I^*I = RI = I$, contradicting the maximality of I. Hence $R \neq I^*$. Since $1 \in I^*$, $R \subset I^*$ and hence $I \subseteq II^* \subseteq R$. Assume $I = II^*$. Then $R = I^*I = RI^* = I^*$, a contradiction. Now let I be an arbitrary integral R-ideal. There exists a maximal R-ideal $\mathcal{M}_1$ containing I. If $I = I\mathcal{M}_1{}^* \subseteq \mathcal{M}_1\mathcal{M}_1{}^* = R$ then $\mathcal{M}_1{}^* = R\mathcal{M}_1{}^* = I^*I\mathcal{M}_1{}^* = I^*I = R$, a contradiction. Therefore $I \subset I\mathcal{M}_1{}^*$. If $I\mathcal{M}_1{}^* \neq R$, $I\mathcal{M}_1{}^*$ is contained in a maximal R-ideal $\mathcal{M}_2$. Again, as before, we have $I\mathcal{M}_1{}^*\mathcal{M}_2{}^* \subseteq R$ and $I \subset I\mathcal{M}_1{}^* \subset I\mathcal{M}_1{}^*\mathcal{M}_2{}^* \subseteq R$. After a finite number of steps we get

$$I\mathcal{M}_1{}^*\mathcal{M}_2{}^* \cdots \mathcal{M}_n{}^* = R$$

Set $I^{-1} = \mathcal{M}_1{}^* \cdots \mathcal{M}_n{}^*$. Then $II^{-1} = I^{-1}I = R$ and the first assertion is proved. Multiplying (4) on the right by $\mathcal{M}_n \cdots \mathcal{M}_1$ we get $I = \mathcal{M}_n \cdots \mathcal{M}_1$. This proves the second assertion since if P is prime, $P = \mathcal{M}_1 \cdots \mathcal{M}_n$ implies that P must be one of the $\mathcal{M}_i$.

2. If $\mathcal{M}_1$ and $\mathcal{M}_2$ are maximal ideals, $\mathcal{M}_1\mathcal{M}_2 = \mathcal{M}_1 \cap \mathcal{M}_2$. Proof: $\mathcal{M}_1 \cap \mathcal{M}_2 \subset \mathcal{M}_1$ if $\mathcal{M}_1 \not\subseteq \mathcal{M}_2$. Therefore, as we have already shown in 1, $\mathcal{M}_1 \cap \mathcal{M}_2 = \mathcal{M}_1\mathcal{G}$ for some integral R-ideal $\mathcal{G}$. This implies $\mathcal{M}_1\mathcal{G} \subseteq \mathcal{M}_2$ and therefore $\mathcal{G} \subseteq \mathcal{M}_2$ since $\mathcal{M}_1 \not\subseteq \mathcal{M}_2$ and $\mathcal{M}_2$ is prime because it is maximal. Hence $\mathcal{M}_1 \cap \mathcal{M}_2 = \mathcal{M}_1\mathcal{G} \subseteq \mathcal{M}_1\mathcal{M}_2 \subseteq \mathcal{M}_1 \cap \mathcal{M}_2$. In other words: multiplication of two maximal ideals is commutative and therefore by Part 1, multiplication of arbitrary integral ideals is commutative.

3. Each integral ideal $\mathcal{G}$ possesses an inverse R-ideal $\mathcal{G}^{-1}$ in Q. Let I be an R-ideal in Q. Let $J = \{x \in R \mid Ix \subseteq R\}$. Then J is a subgroup of the additive group of R for which $RJ = J = JR$. By the definition of an R-ideal, R contains a regular element a such that $Ia \subseteq R$. Clearly a lies in J and therefore J is an integral R-ideal as well as the product IJ. By what we proved

in Part 1, $I = IR = IJJ = IJ \cdot J^{-1}$ and hence $I(J \cdot (IJ)^{-1}) = (IJ)(IJ)^{-1} = R$. Therefore each R-ideal in Q has an inverse in the multiplicative semigroup of R-ideals in Q with R as identity. We have therefore proved the first part of the third assertion of the theorem. Finally, if $IJ = \mathcal{M}_1\mathcal{M}_2 \cdots \mathcal{M}_n$ and $J = \mathcal{M}_{n+1} \cdots M_r$, then

$$I = IJ \cdot J^{-1} = \mathcal{M}_1 \cdots \mathcal{M}_n \mathcal{M}_{n+1}^{-1} \cdots \mathcal{M}_r^{-1}$$

It follows that the representation of I as a product of maximal ideals is uniquely determined up to order since the $\mathcal{M}_i$ are prime. ∎

We have just proved that if R is a maximal order, fulfilling the other hypotheses in Theorem 11, in a quotient ring Q it is an Asano order. The converse is also true:

Theorem 12. Let R be an Asano order in the quotient ring Q. Then R is a maximal order and each integral R-ideal in Q is a projective R-module.

Proof. 1. We first show that R is a maximal order in Q. Suppose not and let T be equivalent to R with $R \subset T$. By Definition 1, there exist regular elements $a^{-1}b$ and $c^{-1}d$ in Q with $a^{-1}bTc^{-1}d \subseteq R$ where a, b, c, and d are regular elements of R. It follows that

$$bTd = bTcc^{-1}d \subseteq bTc^{-1}d \subseteq aR \subseteq R$$

Hence $U = R + TdR$ is a subring of T containing R and therefore U is an order in Q. Moreover, $bU = bR + bTdR \subseteq R$. If $R \neq U$, set $S = U$. If $R = U$, then $Td \subseteq TdR \subseteq U$. Set $S = T$ in this case. In either case S is an order in Q which is equivalent to R. Furthermore, $R \subset S$ and $bS \subseteq R$ (or $Sd \subseteq R$) for a suitable element b (resp. d) in R. Suppose $bS \subseteq R$. Then RbS is an integral R-ideal. By hypothesis, there exists an R-ideal C in Q which is the inverse of RbS. Therefore $C \cdot RbS = R$. It follows that $S = RS = C \cdot RbS \cdot S = CRbS = R$, a contradiction.

2. Now let I be an R-ideal in Q. As we have just shown in Part 1, R is a maximal order in Q and therefore by (3), $I^*I \subseteq O_r(I) = R$. By hypothesis, I has an inverse R-ideal I^{-1} in Q. Now $II^{-1} = R$ implies $II^{-1}I = I$ from which it follows that $I^{-1} \subseteq I^*$. Therefore, $R = I^{-1}I \subseteq I^*I \subseteq R$ and hence $I^*I = R = O_r(I)$. By Theorem 10 this proves that I is projective as an R-module. ∎

3. Injective modules and self-injective rings

The following definition is the "dual" of the definition of a projective module given in Section 2:

Definition 1. Let R be a ring with identity and let B be an arbitrary R-module. Then an R-module M is *injective* if homomorphisms of submodules of B into M can be lifted to homomorphisms of B into M. In terms of diagrams, M is injective if the diagram

$$\begin{array}{ccccc} 0 & \rightarrow & A & \xrightarrow{i} & B \\ & & \downarrow{\scriptstyle\phi} & & \\ & & M & & \end{array}$$

where i is the embedding of A into B, can be embedded in the commutative diagram

$$\begin{array}{ccccc} 0 & \rightarrow & A & \xrightarrow{\quad i \quad} & B \\ & & \downarrow{\scriptstyle\phi} & \swarrow{\scriptstyle\psi} & \\ & & M & & \end{array}$$

If M is injective, take $L = A$, L a left ideal of R, and take $R = B$. Then $x\psi = (x \cdot 1)\psi = x(1\psi)$ for all $x \in R$. Hence there exists $m_0 \in M$ (namely, 1ψ) such that $l\phi = lm_0$ for all $l \in L$. The converse of this remark is also true and hence we have the following characterization of injectivity:

Theorem 1. The R-module M is injective if and only if for each left ideal L in R and each R-module homomorphism ϕ of L into M, there exists $m_0 \in M$ such that $l\phi = lm_0$ for all $l \in L$.

Proof. Suppose the condition is satisfied. Let B be an R-module and let $A \subseteq B$ be a submodule of B. Let $\phi \in \mathrm{Hom}_R(A, M)$. Consider the family $\mathcal{F}$ of pairs (A', ϕ'), where A' is a submodule of B containing A and $\phi' \in \mathrm{Hom}_R(A', M)$ is an extension of ϕ to A'. Define a relation $\subseteq$ on $\mathcal{F}$ by $(A', \phi') \subseteq (A'', \phi'')$ if $A' \subseteq A''$ and ϕ'' is an extension of ϕ'. Clearly $\subseteq$ partially orders $\mathcal{F}$. Moreover, if $\{(A_\lambda, \phi_\lambda)\}_{\lambda\in\Lambda}$ is a chain in $\mathcal{F}$ and ϕ is the homomorphism defined on $\bigcup_{\lambda\in\Lambda} A_\lambda$ by

$$\bar{\phi}: a \rightarrow a\phi_\mu \qquad \text{if} \quad a \in A_\mu, \quad \mu \in \Lambda$$

the pair $(\bigcup_{\lambda\in\Lambda} A_\lambda, \bar{\phi}) \in \mathcal{F}$ is an upper bound for the chain. By Zorn's lemma, there exists a maximal element (B_0, ψ_0) in $\mathcal{F}$. We must now show that $B_0 = B$. Suppose not and let $b_1 \in B \backslash B_0$. Then the set $\{x \in R \mid xb_1 \in B_0\}$ is a left ideal L of R and the mapping $x \rightarrow (xb_1)\,\psi_0$ for $x \in L$ is an R-module homomorphism of L to M. By hypothesis there exists $m_0 \in M$ such that $(xb_1)\,\psi_0 = xm_0$. Set $\psi_1: b_0 + yb_1 \rightarrow b_0\psi_0 + ym_0$ for $b_0 \in B_0$ and $y \in R$. We show ψ_1 is well defined: Suppose $b_0 + yb_1 = b_0' + y'b_1$. Then $b_0 - b_0' = (y' - y)\, b$, and therefore $y' - y \in L$. It follows that $(b_0 - b_0')\,\psi_0 = ((y' - y)\, b_1)\,\psi_0 = (y' - y)\, m_0$ and hence $b_0\psi_0 + ym_0 = b_0'\psi_0 + y'm_0$, showing that ψ_1 is well defined. Clearly $\psi_1 \in \mathrm{Hom}_R(B_0 + Rb_1, M)$ extends ψ_0 and $B_0 + Rb_1 \supset B_0$, contradicting maximality of (B_0, ψ_0). Therefore $B_0 = B$. ▌

The following is analogous to Theorem VII.2.2:

Theorem 2. The complete direct sum M of modules M_i, $i \in I$ is injective if and only if each M_i is injective.

Proof. Suppose each M_i is injective and let $\phi \in \mathrm{Hom}_R(A, \sum_c M_i)$ where A is a submodule of a given module B. Let π_i be the projection of M onto M_i. Now $\phi\pi_i \in \mathrm{Hom}_R(A, M_i)$ and by injectivity of M_i there exists $\psi_i \in \mathrm{Hom}_R(B, M_i)$ such that $\psi_i|_A = \phi\pi_i$. Then $\psi = \sum \oplus \psi_i \in \mathrm{Hom}_R(B, M)$ is an extension of ϕ to B and hence M is injective. Conversely, suppose M is injective and let $\phi_i \in \mathrm{Hom}_R(A, M_i)$. Denote the embedding of M_i in M by μ_i. Then $\phi = \phi_i\mu_i \in \mathrm{Hom}_R(A, M)$. By hypothesis there exists $\psi \in \mathrm{Hom}_R(B, M)$ with $\psi|_A = \phi = \phi_i\mu_i$ and $\psi\pi_i$ is an extension of $\phi_i = \phi_i\mu_i\pi_i$ to B. ▌

Injective $\mathbb{Z}$-modules (i.e. abelian groups considered as modules over the integers) are characterized by the following:

Theorem 3. An Abelian group M is injective as a $\mathbb{Z}$-module if and only if it is *divisible*, i.e. for each $m \in M$ and each natural number N, there exists $m_0 \in M$ such that $Nm_0 = m$.

Proof. Let $L = \mathbb{Z}l_0 \neq 0$ and let ϕ be an additive map of L to M. Let $l_0\phi = m \in M$. If M is divisible there exists $m_0 \in M$ such that $m = l_0 m_0$. Clearly $l\phi = lm_0$ for all $l \in L$ and by Theorem 1, M is injective. Conversely, suppose M is $\mathbb{Z}$-injective. The mapping ϕ from $\mathbb{Z}l_0$ to M defined by $\phi: zl_0 \to zm$ is additive for arbitrary $m \in M$, $l_0 \in \mathbb{Z}$. If M is injective there exists $\psi: \mathbb{Z} \to M$ which is an extension of ϕ. Let $1\psi = m_0$. Then $lm = l_0(1\psi) = (l_0\psi) = l_0\phi = m$. Hence M is divisible. ▌

The rational numbers under addition form an abelian group $\mathbb{Q}$ containing as a subgroup the additive group of integers $\mathbb{Z}$. The group $\mathbb{Q}/\mathbb{Z}$ is divisible and hence $\mathbb{Z}$-injective. Using this fact we may construct more injective modules. We begin with the following:

Definition 2. Let M be a right R-module. Then the additive group

$\operatorname{Hom}_{\mathbb{Z}}(M, \mathbb{Q}/\mathbb{Z})$ can be made into an R-module M^* by defining for $a \in R$ and $\chi \in M^*$

$$m(a\chi) = (ma)\chi \qquad \text{for all} \quad m \in M \tag{1}$$

The R-module M^* is called the *character module* of M.

The R-module property of M^* under (1) follows from

$$m((ab)\,\chi) = (mab)\,\chi = ((ma)\,b)\,\chi = (ma)(b\chi) = m(a(b\chi)) \qquad \text{for all} \quad m \in M \tag{2}$$

Theorem 4. If F is a free, right R-module, its character module F^* is an injective left R-module.

Proof. Let L be a left ideal in R and let $\phi \in \operatorname{Hom}_R(L, F^*)$. By Theorem 1, it is sufficient to show that ϕ can be extended to ${}_RR$. Let $\{f_i \mid i \in I\}$ be an R-basis of the free right R-module F. Then $FL = \{f_{i_1}l_1 + \cdots + f_{i_N}l_N \mid l_i \in L\}$ is a subgroup of the additive group of F. In the representation of an element of FL as a sum of f_il_i, the l_i are uniquely determined since $\{f_i \mid i \in I\}$ is an R-basis of F. Define a mapping ψ from FL to $\mathbb{Q}/\mathbb{Z}$ by

$$\left(\sum f_il_i\right)\psi = \sum_i f_i(l_i\phi) \tag{3}$$

Clearly $\psi \in \operatorname{Hom}_{\mathbb{Z}}(FL, \mathbb{Q}/\mathbb{Z})$. Since by Theorem 3, $\mathbb{Q}/\mathbb{Z}$ is injective as a $\mathbb{Z}$-module, there exists $\chi \in \operatorname{Hom}_Z(F, \mathbb{Q}/\mathbb{Z}) = F^*$ which extends ψ. By the definition of F^* the mapping

$$\eta\colon a \to a\psi \qquad \text{for} \quad a \in R$$

is an R-module homomorphism of ${}_RR$ to F^* and hence an element of $\operatorname{Hom}_R(R, F^*)$. By (3) the mapping η is such that

$$f(l\eta) = f(l\chi) = (fl)\,\chi = (fl)\,\psi = f(l\phi)$$

Therefore η extends ϕ to the whole of ${}_RR$. ∎

In Theorem 7 we shall show that an R-module is injective if and only if it is a direct summand of the character module of some free right R-module. To this end we first prove:

Theorem 5 (Baer). Every right R-module M is isomorphic to a submodule of the character module of some free module. Hence every right R-module is a submodule of an injective module.

Proof. 1. We first prove if $0 \neq m_0 \in M$, there exists $\chi \in M^*$ such that $m_0\chi \neq 0$. Suppose first that $zm_0 \neq 0$ for all $0 \neq z \in \mathbb{Z}$. Set $(zm_0)\,\phi = (z/2)\,\pi$ where π is the natural homomorphism of $(\mathbb{Q}, +)$ onto $\mathbb{Q}/\mathbb{Z}$. Since $\phi \in \operatorname{Hom}_{\mathbb{Z}}(\mathbb{Z}m_0, \mathbb{Q}/\mathbb{Z})$, ϕ can be extended to $\chi \in \operatorname{Hom}(M, \mathbb{Q}/\mathbb{Z}) = M^*$ and $m_0\chi = m_0\phi = \frac{1}{2}\pi \neq 0$. Next suppose $zm_0 = 0$ for some $0 \neq z \in \mathbb{Z}$. Let z_0 be the natural number for which $z_0 m_0 = 0$. Define ϕ by

$$\phi\colon (zm_0) \to (z/z_0)\,\pi$$

Clearly $\phi \in \operatorname{Hom}_{\mathbb{Z}}(\mathbb{Z}m_0, \mathbb{Q}/\mathbb{Z})$ and again it can be extended to $\chi \in M^*$. Moreover, $m_0\chi = (1/z_0)\,\pi \neq 0$.

2. M is isomorphically embedded as a submodule of M^{**}. Proof: For each $m \in M$ define the mapping $\hat{m}$ by

$$\hat{m}\colon \chi \to m\chi \qquad \text{for} \quad \chi \in M^*$$

Clearly $\hat{m} \in \operatorname{Hom}(M^*, \mathbb{Q}/\mathbb{Z})$ and the mapping $m \to \hat{m}$ is additive. $\hat{m} = \hat{0}$ implies $m\chi = 0$ for all $\chi \in M^*$ and hence by Part 1, $m = 0$. Moreover,

$$\widehat{ma}\chi = (ma)\chi = m(a\chi) = \hat{m}(a\chi) = (\hat{m}a)\chi$$

for all $\chi \in M^*$.

Therefore $\widehat{ma} = \hat{m}a$ and $m \to \hat{m}$ is an R-module monomorphism of M into M^{**}.

3. Let ϕ be an R-module homomorphism of the right module U to the right module W and let ϕ^* be the mapping of W^* to U^* defined by

$$\chi\phi^*\colon u \to (\phi u)\,\chi \qquad \text{for} \quad u \in U$$

ϕ^* is an R-module homomorphism since for $u \in U$ and $a \in R$ we have $u((a\chi)\,\phi^*) = (\phi u)(a\chi) = (\phi(ua))\,\chi = u(a(\chi\phi^*))$. If ϕ is an epimorphism, ϕ^* is a monomorphism. Proof: Let $\chi \in W^*$ and suppose $\chi\phi^* = 0$. Then $(\phi U)\,\chi = 0$ and hence $W\chi = 0$ proving that $\chi = 0$.

4. As we showed in Part 2, M is isomorphic to a submodule of $M^{**} = (M^*)^*$. But M^* is the homomorphic image of a free module ${}_RF$. By Part 3 it follows that $(M^*)^*$ is isomorphic to a submodule of F^*. ▮

Theorem 6. The R-module M is injective if and only if M is a direct summand of any containing module.

Proof. (Cf. Proof of Theorem 2.5.) 1. Suppose M is injective and let $M \subseteq A$. By injectivity of M the identity map 1_M on M extends to a homomorphism ψ of A to M. Now $A(\psi^2 - \psi) = A\psi(\psi - 1_M) = 0$ and hence ψ is idempotent. Therefore

$$A = A\psi \oplus A(1_A - \psi)$$

is a representation of A as a direct sum of the submodules $A\psi = M$ and $A(1_A - \psi)$.

2. If M satisfies the condition of the theorem then in particular M is a direct summand of the module F^* of the previous theorem. By Theorem 2, M itself is injective. ∎

As an immediate corollary to Theorem 6 we have:

Theorem 7. The R-module M is injective if and only if it is isomorphic to a direct summand of the character module of a free right R-module. ∎

This theorem is analogous to the first assertion of Theorem 2.6 which states that an R-module is projective if and only if it is isomorphic to a direct summand of a free module.

If R is an algebra over the field Φ and if M is a right algebra R-module we can substitute Φ for $\mathbb{Q}/\mathbb{Z}$ in Definition 2. The elements of $\mathrm{Hom}_\Phi(M, \Phi)$ are called *linear functionals* on M. They form an R-module, again denoted by M^*, if we define for $a \in R$ and $\chi \in \mathrm{Hom}_\Phi(M, \Phi)$

$$m(a\chi) = (ma)\chi \qquad \text{for all} \quad m \in M$$

Except for minor alterations in the terminology, Theorem 4 remains unchanged and the same proof carries through since the field Φ is an injective Φ-module. The same is true of Theorems 5 and 6. In the context of Φ-algebras, Theorem 7 becomes:

Theorem 7′. Let R be an algebra with identity over the field Φ. The R-(algebra)-module M is injective if and only if it is isomorphic to a direct summand of the dual module F^* of a free R-(algebra)-module F. ∎

By Theorem 5 every R-module can be embedded in an injective R-module. Our next objective is to prove that an R-module can be embedded in a minimal injective and that any two minimal injectives are isomorphic. For this purpose we introduce the following concept:

Definition 3. The R-module U is an *essential extension* of the submodule M if every nonzero submodule of U intersects M nontrivially.

The following theorem shows that every injective extension I of M isomorphically contains every essential extension U of M.

Theorem 8. Let I be an injective extension of the R-module M and let U be an essential extension of M. Then there exists a monomorphism ϕ of U into I whose restriction to M is the identity 1_M.

Proof. From the exactness of the sequence $0 \to M \to U$ and the injectivity of I follows the existence of $\phi \in \mathrm{Hom}_R(U, I)$ with $\phi\,|_M = 1_M$. This implies that $\ker \phi \cap M = (0)$. Since U is an essential extension of M it follows that $\ker \phi = (0)$. Hence ϕ is a monomorphism. ∎

Theorem 9. The R-module M is injective if and only if it has no *proper* essential extension, i.e. no essential extension properly containing M.

Proof. 1. If M is injective and U is an extension of M, then M is a direct summand of U. Hence $M = U$ if we assume U is essential over M.

2. Suppose M has no proper essential extensions and let I be an extension of M. Let M' be a submodule of I, maximal subject to $M \cap M' = (0.)$ (M' exists by Zorn's lemma.) Now $I/M' \subseteq (M + M')/M' \approx M$. Moreover, if $M' \subseteq U \subseteq I$ and $(U/M') \cap (M + M')/M' = (\bar{0})$, then $U \cap (M + M') \subseteq M'$ and hence $U \cap M \subseteq M' \cap M = (0)$. By the maximality of M' this implies $U = M'$ proving that I/M' is an essential extension of $(M + M')/M'$. By hypothesis this implies that $I = M \oplus M'$. If we choose I to be an injective extension of M, M will be a direct summand of I and hence, by Theorem 2, M is injective. ∎

Now let M be a submodule of the module I. The set theoretic union of an ascending chain of essential extensions of M in I is itself an essential extension of M. By Zorn's lemma there exists a submodule U of I maximal subject to containing M essentially. If U' is an essential extension of U contained in I then any nonzero submodules S of U' intersects U nontrivially. Since U is an essential extension of M and since

$$S \cap U \neq (0), \qquad S \cap U \cap M = S \cap M \neq (0)$$

Hence any essential extension of U in I is also an essential extension of M. Therefore U has no proper essential extensions *in I*. Moreover, if I is injective, we can show that U has no proper essential extensions *in any* containing module: Let U' be any essential extension of U. By Theorem 8 applied to the essential extension U' of U, there exists a monomorphism ϕ of U' into I such that $U \subseteq U'\phi$. Clearly $U'\phi$ is an essential extension of U in I. But U is maximal essential in I and therefore $U = U'\phi$. Since U has no proper essential extension it is an injective R-module by Theorem 9. We have therefore shown that if I is an injective extension of M, each maximal essential extension U of M in I is an injective R-module. On the other hand, by Theorem 9, no proper submodule $\breve{U}$ of U containing M is injective since U is a proper essential extension of $\breve{U}$. Therefore U is a minimal injective extension of M and is uniquely determined (up to isomorphism) by M: In fact, if U' is any essential extension of M, then by Theorem 8 there exists a monomorphism ϕ of U' into U. If U' is also a maximal essential extension of M, so is $U'\phi$. Hence $U'\phi = U$.

Before collecting the results proved in the above discussion we make the following:

Definition 4. The R-module I is an *injective hull* of its submodule M if

1. I is injective
2. $M \subseteq L \subseteq I$, L injective implies $L = I$.

Theorem 10. 1. Every R-module M has an injective hull.

2. Every injective extension of M contains an injective hull of M.

3. An R-module U containing M is an injective hull of M if and only if it is a maximal essential extension of M.

4. If I and I' are injective hulls of M then there exists an isomorphism of I onto I' restricting to the identity on M. ∎

Since the injective hulls of a given module M are isomorphic we speak of "the" injective hull $H(M)$ of M.

This last theorem can be used as the basis of an investigation of those rings R with are injective considered as R-modules over themselves. Let us assume that $\breve{U}$ is an indecomposable component in such a ring and suppose $\breve{U}$ contains at least one minimal left ideal M of R. By hypothesis ${}_RR$ is injective and since $\breve{U}$ is a direct summand, $\breve{U}$ itself is also injective. By Theorem 10, $\breve{U}$ contains an injective hull $H(M)$ of M. By Theorem 6, $H(M)$ is a direct summand of $\breve{U}$ and hence $H(M) = \breve{U}$ since $\breve{U}$ was assumed to be indecomposable. Again by Theorem 10, $\breve{U}$ is an essential extension of M. If $\breve{U}$ contained another minimal left ideal M', then $M' \cap M = (0)$ contradicting the fact that $\breve{U}$ contains M essentially. Therefore $\breve{U}$ contains exactly one minimal left ideal. Now if $\breve{U}$ and $\breve{U}'$ are two isomorphic indecomposable components of ${}_RR$ each containing minimal left ideals M and M', respectively, then M and M' are uniquely determined by $\breve{U}$ and $\breve{U}'$, respectively, and since $\breve{U} \approx \breve{U}'$, $M \approx M'$. Conversely if $M \approx M'$ it follows that $\breve{U} \approx \breve{U}'$ since the injective hulls $\bar{U} = H(M)$ and $\breve{U}' = H(M')$ are uniquely determined up to isomorphism by Theorem 10. Finally, let M be a minimal left ideal in R and let $H(M)$ be an injective hull of M contained in ${}_RR$. Then $H(M) = S \oplus T$, $S \neq (0)$ implies $S \cap M \neq (0)$ since $H(M)$ contains M essentially and hence $M \subseteq S$ by minimality of M. Since S is a direct summand of an injective it is injective (Theorem 2). Therefore $T = (0)$ since $H(M)$ is minimal injective and hence $H(M)$ is indecomposable.

If the ring R is artinian with identity and ${}_RR$ is injective, then there is a one-one correspondence between the isomorphism classes of indecomposable left ideals and the isomorphism classes of minimal left ideals. On the other hand, we showed in Theorem 1.17 that there is a one-one correspondence between the isomorphism classes of irreducible R-modules and the isomorphism classes of indecomposable submodules of ${}_RR$. We have therefore proved the following:

Theorem 11. Let R be artinian with identity and let ${}_RR$ be injective. Then there is a one-one correspondence between the classes of irreducible R-modules and the classes of indecomposable left ideals in R. Moreover, each indecomposable left ideal contains exactly one minimal left ideal. ∎

Theorem 12. Let R be artinian with identity. Then the following are equivalent:

1. ${}_RR$ is an injective R-module
2. A finitely generated R-module M is injective if and only if it is projective.

Definition 5. A ring R is *self-injective* if ${}_RR$ is an injective R-module.

Proof of Theorem 12. 1. Assume Part 2. By Theorem 2.1 ${}_RR$ is projective and therefore ${}_RR$ is injective.

2. Conversely, suppose ${}_RR$ is injective and let M be a finitely generated projective R-module. By Theorem 2.6, M is a direct sum of a finite number of submodules, each isomorphic to some indecomposable left ideal of R. Since each of these is a direct summand of the injective module ${}_RR$, each is injective and therefore by Theorem 2 M is injective.

3. Let ${}_RR$ be injective and let M be a finitely generated injective module. Since R is artinian and M is finitely generated the lattice $V_R(M)$ of submodules of M satisfies the minimal condition by Theorem 1.30. The module M is therefore the direct sum of a finite number of indecomposable submodules. Let U be one of these. As a direct summand of M, U is also injective. Let W be a minimal submodule of U. Then $H(W)$, the injective hull of W, is contained in U. As above, $U = H(W)$. Since W is an irreducible submodule of M, by Theorem 11 it is isomorphic to a minimal left ideal L in R whose injective hull $H(L)$ is an indecomposable left ideal of R. Since $W \approx L$ it follows from Theorem 10 that $H(W) \approx H(L)$. Now $H(L)$ is an indecomposable left ideal and hence is a direct summand of ${}_RR$. Therefore $H(L)$ is projective proving that $U = H(W)$ is also projective. Hence M is projective. ∎

The lattice $V({}_RR)$ of left ideals L of R may be mapped into the lattice of right ideals $V(R_R)$ by means of the right annihilator mapping r defined by

$$L \to r(L) = \{x \in R \mid Lx = 0\}$$

Similarly the lattice $V(R_R)$ of right ideals W is mapped into $V({}_RR)$ by the left annihilator l defined by

$$W \to l(W) = \{x \in R \mid xW = 0\}$$

The following hold:

$$r(L_1 + L_2) = r(L_1) \cap r(L_2) \quad \text{and} \quad l(W_1 + W_2) = l(W_1) \cap l(W_2) \tag{4}$$

$$r(L_1 + r(L_2) \subseteq r(L_1 \cap L_2) \quad \text{and} \quad l(W_1) + l(W_2) \subseteq l(W_1 \cap W_2) \tag{5}$$

In (5) the inclusions could be strict. It is an open question whether the mappings are lattice isomorphisms between $V({}_RR)$ and $V(R_R)$.

Let R be a ring with identity. In the following it will be shown that the inclusions in (5) may be replaced by equalities if R is a self-injective left artinian and right noetherian ring.

We first prove that for principal right ideals aR of R we have

$$r(l(aR)) = aR \tag{6}$$

Proof. $x \in l(aR)$ implies $xa = 0$. Hence $a \in r(x)$ and therefore $aR \subseteq r(l(aR))$ always holds. Conversely, let $b \in r(l(aR))$. Then $xb = 0$ for all $x \in l(aR)$. Since $l(aR) \subseteq l(bR)$, the mapping

$$\phi\colon xa \to xb \qquad \text{for} \quad x \in R$$

is an element of $\mathrm{Hom}_R(Ra, Rb)$. Since ${}_RR$ is injective by Theorem 1 there exists $c \in R$ such that $xb = xa\phi = xac$ for all $x \in R$. Therefore $b = ac \in aR$ and (6) is proved.

To sharpen the first inclusion in (5) to an equality let $b \in r(L_1 \cap L_2)$. Define $\phi_i \in \mathrm{Hom}_R(L_i, {}_RR)$, $i = 1, 2$ as follows:

$$l_1\phi_1 = l_1 \quad \text{for } l_1 \in L_1 \qquad \text{and} \qquad l_2\phi_2 = l_2(1 - b) \quad \text{for } l_2 \in L_2$$

The mapping $(l_1 + l_2)\,\phi = l_1\phi_1 + l_2\phi_2$ is well defined: $l_1 + l_2 = l_1' + l_2'$ implies $l_1 - l_1' = -l_2 + l_2' \in L_1 \cap L_2$. But $b \in r(L_1 \cap L_2)$ and therefore $l_2 b = l_2' b$ showing that $(l_1 + l_2)\,\phi = (l_1' + l_2')\,\phi$. Since ${}_RR$ is injective, by Theorem 1 there exists $c \in R$ such that $(l_1 + l_2)\,\phi = (l_1 + l_2)\,c$. This implies

$$l_1 + l_2(1 - b) = (l_1 + l_2)\,\phi = (l_1 + l_2)\,c$$

and therefore $l_1(1 - c) + l_2(1 - b - c) = 0$ for all $l_1 \in L_1$, $l_2 \in L_2$. It follows that $1 - c \in r(L_1)$ and $1 - b - c \in r(L_2)$. Therefore

$$b = (1 - c) - (1 - b - c) \in r(L_1) + r(L_2)$$

This shows

$$r(L_1 \cap L_2) = r(L_1) + r(L_2) \qquad \text{for} \quad L_1, L_2 \in V({}_RR) \tag{7}$$

and the mapping $L \to r(L)$ is a lattice homomorphism of $V({}_RR)$ into $V(R_R)$.

If in addition R is right noetherian, i.e. if $V(R_R)$ satisfies the maximal condition, the mapping $L \to r(L)$ is also an epimorphism. Proof: Each right ideal W in R is the sum of a finite number of principal right ideals: $W = a_1R + \cdots + a_sR$. By (4), (7), and (6) it follows that

$$\begin{aligned} r(l(W)) &= r(l(a_1R + \cdots + a_sR)) = r(l(a_1R) \cap \cdots \cap l(a_sR)) \\ &= r(l(a_1R)) + \cdots + r(l(a_sR)) = a_1R + \cdots + a_sR = W \end{aligned}$$

Hence $W \in V(R_R)$ is the image of the left ideal $l(W)$ under the mapping $L \to r(L)$. Finally this mapping is a monomorphism if R is also artinian. In this case we apply Theorem 11 as follows: Assume $L \subset l(r(L))$ for some left ideal L. Then there exists a left ideal $L_0 \subseteq l(r(L))$ such that L_0 contains L and L_0/L is an irreducible R-module. By Theorem 11, L_0/L is isomorphic to a minimal left ideal in R. Therefore there exists $\phi \in \mathrm{Hom}_R(L_0, {}_RR)$ with $L\phi = 0$ and L_0 minimal in $V({}_RR)$. Since ${}_RR$ is injective there exists $c \in R$ such that $l_0\phi = l_0c$ for all $l_0 \in L_0$. In particular, $Lc = 0$ and therefore $c \in r(L)$. But $L_0 \subseteq l(r(L))$ and hence $L_0c = 0$, a contradiction, since $L_0c = L_0$. Therefore $l(r(L)) = L$ for all $L \in V({}_RR)$ and the mapping $L \to r(L)$ is one-one. We gather these results in the following:

Theorem 13. Let R be a self-injective left artinian and right noetherian ring. Then the mapping $L \to r(L)$, L a left ideal, is a lattice antiisomorphism of the lattice $V({}_RR)$ of left ideals in R onto the lattice $V(R_R)$ of right ideals in R. The inverse of this mapping is the mapping $W \to l(W)$, $W \in V(R_R)$. In particular we have $l(r(L)) = L$ for all $L \in V({}_RR)$ and $r(l(W)) = W$ for all $W \in V(R_R)$. ▌

Definition 6. A left and right artinian ring with identity whose left and right ideals satisfy

$$l(r(L)) = L \qquad \text{and} \qquad r(l(W)) = W \tag{8}$$

respectively is called a quasi-Frobenius ring.

Theorem 13 shows that a self-injective left and right artinian ring with identity is a quasi-Frobenius ring. The converse is also true [cf. Curtis and Reiner (1962), Theorem 58.6]. In the next chapter, as a corollary to Theorem VIII.2.11, we shall prove the converse in the special case when R is an algebra of finite dimension over a field.

Frobenius Algebras

CHAPTER VIII

In Chapter VII we defined a quasi-Frobenius ring R as one which is both left and right artinian and satisfies the annihilator conditions

$$l(r(L)) = L \qquad \text{and} \qquad r(l(W)) = W \tag{1}$$

for left ideals L and right ideals W of R. If R is a finite-dimensional Φ-algebra, we call R a Frobenius-algebra provided, in addition to the annihilator conditions in (1), the following equalities hold for the Φ-dimension of left and right ideals:

$$\dim_\Phi(L) + \dim_\Phi r(L) = \dim_\Phi(R) = \dim_\Phi(W) + \dim_\Phi l(W) \tag{2}$$

As examples of Frobenius algebras we have the group algebras $\Phi[G]$, G a finite group and Φ a field. By a result of Maschke (Theorem VIII.5.3), $\Phi[G]$ is semisimple if and only if the characteristic of Φ does not divide the order of the group G. In general $\Phi[G]$ is only a Frobenius algebra.

Instead of working with the one-sided conditions (1) and (2) in this chapter, we shall find it more convenient to introduce an equivalent more symmetric definition of a Frobenius algebra which uses the concept of duality. The equivalence of this definition with that given above will not be proved until the end of Section 2 in Theorem 12 (due to Nakayama).

1. The dual of an algebra module

The minimal condition on left and right ideals of a ring R is satisfied in the particular case that R is a finite-dimensional algebra over its ground field Φ since each one-sided ideal is a finite-dimensional Φ-subspace of the Φ-space R. If in addition, R has an identity, by Theorem VII.1.12 the left ideal Re generated by the idempotent e is indecomposable if and only if the right ideal eR is. This motivates the further investigation of the relationship between the indecomposable left and right ideals of such algebras.

In order to compare a right R-module M with some left R-module we try turning M into a left R-module by defining

$$a \circ m = ma \qquad \text{for all} \quad m \in M \quad \text{and} \quad a \in R \tag{1}$$

Naturally the additive structure of M remains unaltered but in general M is not a left R-module since it follows from (1) that

$$(a \cdot b) \circ m = m(a \cdot b) = (ma)\, b = b \circ (a \circ m) \tag{2}$$

and in a noncommutative ring $ab \neq ba$ for all a and b. However, M with the operation of elements of R defined as in (1), is a left R'-module where R' is the ring with the same additive structure as R but with multiplication defined by $a * b = ba$.

In the case R is an algebra over a field Φ and M a right algebra R-module we can resolve the difficulty encountered in (2) by constructing from the given right module M the so-called dual module M^*.

Definition 1. Let R be a Φ-algebra and M a right algebra R-module. The elements of the Φ-space $\mathrm{Hom}_\Phi(M, \Phi)$ are called *Φ-linear forms on M*. They form a left R-module M^* if for $\phi \in \mathrm{Hom}_\Phi(M, \Phi)$ and $a \in R$ we define $a\phi$ by

$$(a\phi)\, m = \phi(ma) \qquad \text{for all} \quad m \in M \tag{3}$$

M^* is called the *dual of M*. Analogously the dual of a left R-module M is the right R-module M^*, where

$$(\phi a)\, m = \phi(am) \qquad \text{for all} \quad \phi \in M^*, \quad a \in R, \quad \text{and} \quad m \in M$$

Since

$$((ab)\, \phi)\, m = \phi(m(ab)) = \phi((ma)\, b) = (b\phi)(ma) = a((b\phi)\, m) = (a(b\phi))\, m \tag{4}$$

it follows that M^* is indeed a left R-module.

If the Φ-dimension of the right R-module M is finite, M^* and M are isomorphic as Φ-spaces. To prove this we need only show that

$$\dim_\Phi M^* = \dim_\Phi M \tag{5}$$

Let $\{m_1, ..., m_n\}$ be a Φ-basis of the Φ-space M. Relative to this basis define the Φ-linear forms ϕ_i on M by

$$\phi_i m_k = \delta_{ik}, \qquad i, k = 1, ..., n \tag{6}$$

and extend linearly.

Let $\phi \in M^*$. Then the linear form $\sum_{1 \leqslant i \leqslant n} \phi_i(\phi m_i)$ coincides with ϕ on $\{m_1, ..., m_n\}$ since $\sum_{1 \leqslant i \leqslant n} [\phi_i(\phi m_i)](m_j) = \phi_j(m_j)\, \phi(m_j) = \phi(m_j)$ by (6). But a linear form is uniquely determined by its action on a basis. It follows that

$$\phi = \sum_{1 \leqslant i \leqslant n} \phi_i(\phi m_i) \tag{7}$$

The linear independence of $\{\phi_1, ..., \phi_n\}$ is also an easy consequence of (6). $\{\phi_1, ..., \phi_n\}$ is called the basis *dual* to $\{m_1, ..., m_n\}$.

The first assertion of the following theorem points to the fact that when $\dim_\Phi M < \infty$, nothing is "lost" by the transition from M to M^*.

Theorem 1. Hypotheses: Let R be a finite-dimensional algebra over the field Φ and let M be a right R-module with $\dim_\Phi M < \infty$. Conclusions:

1. $M^{**} \approx M$ as right R-modules.
2. If M is the direct sum $M_1 \oplus M_2$ of the right R-modules M_1 and M_2, M^* is isomorphic (as a left R-module) to the direct sum $M_1{}^* \oplus M_2{}^*$ of the dual modules $M_i{}^*$, $i = 1, 2$.
3. For a given submodule U of M we define $U^\perp$ by

$$U^\perp = \{\phi \mid \phi \in M^*, \phi u = 0 \text{ for all } u \in U\}$$

$U^\perp$ is an R-submodule of the left R-module M^* and the module U^*, the dual of U, is isomorphic as a left R-module to the factor module $M^*/U^\perp$.

4. The mapping τ: $U \to U^\perp$ is a lattice antiisomorphism of the lattice of R-submodules of M onto the lattice of R-submodules of M^*.

Proof. 1. For each $m \in M$ denote by μ_m the Φ-linear form on M^* which maps the element $\phi \in M^*$ to $\phi m \in \Phi$:

$$\mu_m\colon \phi \to \phi m \tag{8}$$

The mapping

$$\eta\colon m \to \mu_m \qquad \text{for} \quad m \in M \tag{9}$$

is a Φ-linear map of M to the dual M^{**} of the left R-module M^*. Since

$$(\mu_m a)\, \phi = \mu_m(a\phi) = (a\phi)\, m = \phi(ma) = \mu_{ma}\phi \qquad \text{for} \quad \phi \in M^*$$

the mapping in (9) is a homomorphism of the right R-module M to the right R-module M^{**}. If $\mu_m = 0$, $m = 0$ for all $\phi \in M^*$. If $m \neq 0$ we may complete $m = m_1$ to a Φ-basis $\{m_1, ..., m_n\}$ of M and we may construct as in (6) a Φ-linear form ϕ_1 with $\phi_1 m = \phi_1 m_1 = 1$. Hence the mapping in (9) is an R-module monomorphism of M into M^{**} which is also a Φ-linear mapping of the Φ-space M into the Φ-space M^{**}. Applying (5) we have

$$\dim_\Phi \eta M = \dim_\Phi M = \dim_\Phi M^* = \dim_\Phi M^{**}$$

showing that η is an epimorphism.

2. For $\phi \in M^*$ define

$$\eta\colon \phi \to (\phi\,|_{M_1}, \phi\,|_{M_2})$$

where $\phi\,|_{M_i}$ is the restriction of ϕ to M_i. Clearly $\phi\,|_{M_i} \in \mathrm{Hom}_\Phi(M_i, \Phi)$, $i = 1, 2$. Moreover, since M_i is a submodule, $a\phi\,|_{M_i} \in \mathrm{Hom}_\Phi(M_i, \Phi)$ for all $a \in R$. Hence η is an R-module homomorphism of M^* to $M_1^* \oplus M_2^*$. If $\eta(\phi) = 0$, then $\phi\,|_{M_i} = 0$, $i = 1, 2$ and hence $\phi = 0$. Therefore η is a monomorphism. Finally, given $(\phi_1, \phi_2) \in M_1^* \oplus M_2^*$, define ϕ by

$$\phi m = \phi_1 m_1 + \phi_2 m_2 \tag{10}$$

where $m = m_1 + m_2$, $m_i \in M_i$. Clearly $\phi \in M^*$ and $\eta(\phi) = (\phi_1, \phi_2)$ proving that η is an epimorphism.

3. If $\phi \in U^\perp$, $u \in U$, and $a \in R$, then $(a\phi)\,u = \phi(ua) = 0$. Therefore $U^\perp$ is an R-submodule of M^*. For $\phi \in M^*$ define

$$\lambda\colon \phi \to \phi\,|_U$$

Clearly $\phi\,|_U \in U^*$ and $\ker \lambda = U^\perp$. Furthermore, given $\psi \in U^*$ it can be extended to $\phi \in M^*$ since U is a direct summand of M. Thus λ is an epimorphism of the left R-module M^* onto the left R-module U^*. Therefore $M^*/U^\perp \approx U^*$ as left R-modules.

4. The implication

$$U_1 \subseteq U_2 \Rightarrow U_1^\perp \supseteq U_2^\perp \tag{11}$$

where U_1 and U_2 are submodules of M is an immediate consequence of the definition of $U^\perp$. To prove the mapping τ is one-one we define for a submodule L of M^* the R-submodule $L^\perp$ of M by

$$L^\perp = \{m \mid m \in M,\ \phi m = 0 \text{ for all } \phi \in L\} \tag{12}$$

and show that

$$U^{\perp\perp} = U \tag{13}$$

as follows: Clearly $U \subseteq U^{\perp\perp}$. Let $m \in U$. Choose a basis $\{m_1, ..., m_r\}$ of U and extend to a basis $\{m_1, ..., m_r, m = m_{r+1}, ..., m_n\}$ of M. Let

$\{\phi_1, ..., \phi_r, \phi_{r+1}, ..., \phi_n\}$ be the dual basis as defined in (6). Then $\phi_{r+1} \in U$ but $\phi_{r+1}m_{r+1} = 1 \neq 0$. Hence $m \notin U^{\perp\perp}$. Therefore $U = U^{\perp\perp}$ and the mapping $U \to U^{\perp}$ is one-one. In a similar fashion we show that $L^{\perp\perp} = L$ for all submodules L of M^*, proving that $U \to U^{\perp}$ is onto. ▌

2. Characterization of Frobenius algebras

The concept of the dual of a module allows us to compare the left R-module ${}_RR$ with the dual $R_R{}^*$ of the right R-module R_R in the case R is a finite dimensional algebra with 1 over the field Φ.

Definition 1. Let R be a finite-dimensional algebra with identity over the field Φ. Then R is a *Frobenius algebra* if the left R-modules ${}_RR$ and $R_R{}^*$ are isomorphic. We recall that $R_R{}^*$ is the abelian group $\mathrm{Hom}_R(R_R, \Phi)$ made into a left R-module by defining $a\phi$ by

$$(a\phi)\, x = \phi(x \cdot a) \qquad \text{for all } a, x \in R \quad \text{and} \quad \phi \in R_R{}^* \tag{1}$$

Let R be a Frobenius algebra over Φ and let η be an R-module isomorphism of ${}_RR$ onto $R_R{}^*$. By (1),

$$(\eta(a \cdot b))\, x = (a\eta(b))\, x = (\eta b)(x \cdot a) \qquad \text{for all} \quad a, b, x \in R \tag{2}$$

This implies that f defined by

$$f(x, y) = (\eta(y))\, x \qquad \text{for} \quad x, y \in R \tag{3}$$

is a nondegenerate associative Φ-bilinear form on R in the sense of the following:

Definition 2. 1. A bilinear form on the Φ-space M is a mapping f from the cartesian product $M \times M = \{(m_1, m_2) \mid m_i \in M, i = 1, 2\}$ to Φ which is Φ-linear in each of the variables. More specifically

$$f(\alpha_1 a_1 + \alpha_2 a_2, a_3) = \alpha_1 f(a_1, a_3) + \alpha_2 f(a_2, a_3)$$
$$f(a_1, \alpha_2 a_2 + \alpha_3 a_3) = \alpha_2 f(a_1, a_2) + \alpha_3 f(a_1, a_3)$$

for all $a_1, a_2, a_3 \in M$ and $\alpha_1, \alpha_2, \alpha_3 \in \Phi$.

2. F is nondegenerate if we can deduce that $a = 0$ from either $f(a, M) = 0$ or $f(M, a) = 0$.

3. If R is a Φ-algebra, the bilinear form f on R is associative provided

$$f(ab, c) = f(a, bc) \qquad \text{for all} \quad a, b, c \in R$$

The bilinear form on the Frobenius algebra R defined in (3) is nondegenerate since on the one hand $f(R, y) = 0$ implies that $\eta(y)$ is the zero mapping of $R_R{}^*$ and thus $y = 0$ since η is an isomorphism; on the other hand, if $f(x, R) = 0$, x is annihilated by every linear form in $R_R{}^*$ and hence $x = 0$. Associativity is a consequence of (2) since

$$f(x \cdot y, z) = (\eta z)(x \cdot y) = (y(\eta z))\, x = (\eta(y \cdot z))\, x = f(x, y \cdot z)$$

Conversely a given nondegenerate associative Φ-bilinear form f on a finite dimensional Φ-algebra R with 1 gives rise to an isomorphism of ${}_RR$ onto $R_R{}^*$ as follows: For each $y \in {}_RR$ define $\eta(y) \in R_R{}^*$ by

$$(\eta(y))\, x = f(x, y) \qquad \text{for all} \quad x \in R_R \tag{4}$$

Since $(\eta(a \cdot y))\, x = f(x, a \cdot y) = f(x \cdot a, y) = (\eta y)(x \cdot a) = (a(\eta y))\, x$, η is an R-homomorphism of ${}_RR$ to $R_R{}^*$. Since f is nondegenerate, $(\eta y)\, x = 0$ for all $x \in R$ implies that $y = 0$. Hence η is one-one. By VIII.1.(5) R_R and $R_R{}^*$ have the same Φ-dimension and therefore η is an R-isomorphism of ${}_RR$ onto $R_R{}^*$. This discussion proves

Theorem 1. The finite dimensional Φ-algebra R with 1 is a Frobenius algebra over Φ if and only if there exists a nondegenerate associative Φ-bilinear form on R. ▮

We obtain a second characterization of Frobenius algebras in

Theorem 2 (Brauer and Nesbitt). The finite dimensional Φ-algebra R with identity is a Frobenius algebra if and only if there is a linear form $\phi \in \mathrm{Hom}_\Phi(R, \Phi)$, the kernel of which contains no left or right ideal different from (0).

Proof. 1. Let R be a Frobenius algebra and let f be a bilinear form on R guaranteed by Theorem 1. Define $\phi \in R_R{}^*$ by

$$x = f(x, 1) \qquad \text{for all} \quad x \in R \tag{5_1}$$

If the right ideal xR is contained in $\ker \phi$, then

$$\phi(xR) = f(xR, 1) = f(x, R) = 0$$

and hence $x = 0$. Similarly, if $Rx \subseteq \ker \phi$

$$\phi(Rx) = f(Rx, 1) = f(R, x) = 0$$

and hence $x = 0$.

2. Conversely, let $\phi \in R_R{}^*$ satisfy the condition of Theorem 2. Then

$$f(x, y) = \phi(x \cdot y) \tag{5_2}$$

is nondegenerate since $f(x, R) = 0$ implies $xR \subseteq \ker \phi$ and hence $x = 0$. Similarly if $f(Rx) = 0$, $Rx \subseteq \ker \phi$ and hence $x = 0$. The associativity of f is a consequence of the associativity $(x \cdot y) \cdot z = x \cdot (y \cdot z)$ in R. ∎

A much deeper characterization of Frobenius algebras was discovered by Nakayama. To arrive at this characterization we first derive some consequences of the existence of a nondegenerate associative bilinear form on a Frobenius algebra R (cf. Theorem 3).

Thus let f be a nondegenerate bilinear form on a finite-dimensional Φ-space R with values in Φ. Then for each $y \in R$ the mapping

$$\phi_y\colon x \to f(x, y), \qquad x \in R \tag{6}$$

is a linear form on R with values in Φ. Moreover

$$\chi\colon y \to \phi_y\,, \qquad y \in R \tag{7}$$

is a one-one Φ-linear mapping of the Φ-space R onto the dual space $R^* = \mathrm{Hom}_\Phi(R, \Phi)$. The Φ-linearity of χ is clear. If ϕ_y is the zero linear form, then by (6), $f(R, y) = 0$ and hence $y = 0$, proving that χ is one-one. Since $\dim_\Phi R = \dim_\Phi R^*$, χ is onto. Now let U be a Φ-subspace of R and let

$$U^\perp = \{a \mid a \in R, f(a, U) = 0\}$$

In terms of the notation introduced in (6)

$$U^\perp = \{a \mid \phi_u a = 0 \text{ for all } u \in U\}$$

In other words, $U^\perp = \bigcap_{u \in U} \ker \phi_u$. Since χ is one-one and linear, $\dim_\Phi\{\phi_u \mid u \in U\} = \dim_\Phi U$. On the other hand

$$\dim_\Phi U^\perp = \dim_\Phi R - \dim_\Phi\{\phi_u \mid u \in U\}$$

because $U^\perp$ is given by the linear equations $\phi u = 0$, $u \in U$. Hence

$$\dim_\Phi U + \dim_\Phi U^\perp = \dim_\Phi R \tag{8}$$

Now let the Φ-space R be a finite-dimensional Φ-algebra with identity and let f be a nondegenerate associative bilinear form on R with values in Φ. If W is a right ideal of R and a an element of R such that $f(a, W) = 0$, then

$$0 = f(a, WR) = f(aW, R)$$

Since f is nondegenerate, this implies $aW = 0$. Hence $W^{\perp}$ is contained in $l(W)$, the left annihilator of W. Conversely, if $a \in l(W)$, $f(a, W) = f(1, aW) = 0$. Therefore $l(W) = W^{\perp}$ and by (8)

$$\dim_{\Phi} W = \dim_{\Phi} l(W) = \dim_{\Phi} R \qquad (*)$$

Analogously we may show that if L is a left ideal,

$$r(L) = L^{\perp} \qquad \text{and} \qquad \dim_{\Phi} L + \dim_{\Phi} r(L) = \dim_{\Phi} R \qquad (**)$$

where

$$L^{\perp} = \{a \mid a \in R, f(L, a) = 0\}$$

and $r(L)$ is the right annihilator of L. Since $L \subseteq l(r(L))$ and $W \subseteq r(l(W))$, the equalities in the next theorem follow from $(*)$ and $(**)$.

Theorem 3. If R is a Frobenius algebra and L and W are left and right ideals, respectively, then

$$l(r(L)) = L, \qquad r(l(W)) = W \qquad (9)$$

and

$$\dim_{\Phi} L + \dim_{\Phi} r(L) = \dim_{\Phi} R = \dim_{\Phi} W + \dim_{\Phi} l(W) \quad \blacksquare \qquad (10)$$

The conditions in (9) and (10) are sufficient for R to be a Frobenius algebra over Φ (cf. Theorem 12); in fact (9) alone has important consequences for the structure of R.

Definition 3. Let Φ be a field and R a finite-dimensional Φ-algebra with 1. Then R is a *quasi-Frobenius algebra* if all left ideals L and right ideals W satisfy the annihilator conditions

$$l(r(L)) = L \qquad \text{and} \qquad r(l(W)) = W \qquad (9)$$

Note that in Definition VII.3.4 we defined a quasi-Frobenius ring to be a left and right artinian ring with 1, satisfying the annihilator conditions (9). In particular, therefore, a quasi-Frobenius algebra is a quasi-Frobenius ring.

Theorem 4. In a quasi-Frobenius algebra:

1. W is a maximal right ideal $\Leftrightarrow$ $l(W)$ is a minimal left ideal.
2. L is a maximal left ideal $\Leftrightarrow$ $r(L)$ is a minimal right ideal.

Proof. Obviously $W_1 \subseteq W_2$ if and only if $l(W_1) \supseteq l(W_2)$. Now let W be a maximal right ideal and L' a left ideal such that $(0) \subseteq L' \subseteq l(W)$. Then

$R = r(0) \supseteq r(L') \supseteq r(l(W)) = W$. Since W is maximal, $r(L') = R$ or $r(L') = W$. Hence $L' = l(R)$ or $L' = l(W)$. Since $1 \in R$, $l(R) = (0)$. Therefore either $L' = l(W)$ or $L' = (0)$. Conversely, if $l(W)$ is a minimal left ideal and $W \subseteq W' \subseteq R$ for some right ideal W', then $l(W) \supseteq l(W') \supseteq l(R) = (0)$. Thus $l(W') = l(W)$ or $l(W') = (0)$. In the former case $W' = W$ and in the latter, $W' = R$ by (9). The second assertion of the theorem follows similarly. ∎

Later on we shall prove the following two assertions: (1) Each indecomposable left ideal of a quasi-Frobenius algebra contains exactly one minimal left ideal and each indecomposable right ideal contains a minimal right ideal. (2) The sum of all minimal left ideals of a quasi-Frobenius algebra is the same as the sum of all minimal right ideals. In connection with the second assertion we make the following:

Definition 4. Let R be a ring with minimal left ideals. The sum of all minimal left ideals of R is called the *left socle* of R and is denoted by S_L. The *right socle* of R is defined similarly and is denoted by S_R.

Theorem 5. The left socle S_L of an artinian ring with identity is the right annihilator $r(J)$ of the radical J of R. Similarly $S_R = l(J)$.

Proof. A minimal left ideal L is an irreducible R-module. This implies $JL = 0$ and hence $JS_L = 0$. Therefore $S_L \subseteq r(J)$. $r(J)$ can be turned into a left R/J-module in the obvious way. Since R/J is semisimple artinian, $r(J)$ is completely reducible by Theorem IV.1.4. Therefore $r(J) \subseteq S_L$. Similarly we may show that $S_R = l(J)$. ∎

Theorem 6. In a quasi-Frobenius algebra R the left socle S_L is the same as the right socle S_R. $S = S_L = S_R$ is called the *socle* of R.

Proof. By Theorem 4 the left annihilator $l(W)$ of a minimal right ideal W is a maximal left ideal since W is expressible as $W = r(l(W))$. Therefore the irreducible left R-module $R/l(W)$ is annihilated by J, i.e. $JR \subseteq l(W)$ and thus $J \subseteq l(W)$. This implies $S_L = r(J) \supseteq r(l(W)) = W$. Hence each minimal right ideal of R is contained in S_L and therefore $S_R \subseteq S_L$. Similarly $S_R \supseteq S_L$. ∎

In Theorem VII.1.1 we proved that the indecomposable left and right ideals of an artinian ring R with identity are generated by primitive idempotents; in particular therefore, this is the case for the minimal left and right ideals of a semisimple artinian ring.

Theorem 7. Let e be a primitive idempotent of the quasi-Frobenius algebra R with socle S. Then Se is a minimal left ideal and eS is a minimal right ideal of R.

Proof. Since e is an idempotent and $r(J)$ an ideal,

$$er(J) = eR \cap r(J)$$

The left annihilator of this right ideal contains $1 - e$ and the radical J. Hence

$$l(er(J)) = l(eR \cap r(J)) \supseteq R(1 - e) + J \tag{10}$$

Since e is primitive so is $\bar{e} = e + J$; otherwise a decomposition of $\bar{e}$ into a sum of orthogonal idempotents can be lifted to a decomposition of e into a sum of orthogonal idempotents by Theorem VII.1.10. Thus $\bar{R}\bar{e}$ is a minimal left ideal of $\bar{R}$ and consequently $\bar{R}(\bar{1} - \bar{e})$ is a maximal left ideal of $\bar{R}$. Hence $R(1 - e) + J$ is a maximal left ideal of R. It follows that

$$l(er(J)) = R(1 - e) + J \tag{11}$$

By Theorem 4, $r(R(1 - e) + J)$ is a minimal right ideal and therefore

$$eS = er(J) = r(l(er(J))) = r(R(1 - e) + J)$$

proving that eS is a minimal right ideal of R. Similarly, Se is a minimal left ideal of R. ∎

If e is a primitive idempotent of the quasi-Frobenius algebra R, Re is an indecomposable left ideal by Theorem VII.1.1 and moreover, each indecomposable left ideal is obtained in this manner. We may prove as in Theorem 5 above that the sum of the minimal left ideals of R contained in Re is the intersection of Re with $r(J)$. This intersection,

$$Re \cap r(J) = r(J) \cdot e = Se$$

is the minimal left ideal Se by Theorem 7. Hence:

Theorem 8. Every indecomposable left ideal Re of a quasi-Frobenius algebra R with socle S contains precisely one minimal left ideal, namely Se. Similarly eR contains the unique minimal right ideal eS. ∎

We now show that the structure of a given indecomposable left ideal Re is determined by the structure of the unique minimal left ideal contained in Re.

Theorem 9. In a quasi-Frobenius algebra two indecomposable left ideals are isomorphic if and only if the uniquely determined minimal left ideals contained in each are isomorphic. If e and e' are primitive idempotents generating the indecomposable left ideals this may be expressed as follows:

$$eS \approx e'S \Leftrightarrow eR \approx e'R \Leftrightarrow \bar{e}\bar{R} \approx \bar{e}'\bar{R}$$
$$Se \approx Se' \Leftrightarrow Re \approx Re' \Leftrightarrow \bar{R}\bar{e} \approx \bar{R}e'$$

Here $\bar{R} = R/J$.

Proof. By Theorem VII.1.9 Re and Re' are isomorphic if and only if $\bar{R}\bar{e}$ and $\bar{R}\bar{e}'$ are. Suppose η is an R-module isomorphism of Re onto Re'. Then $\eta e = ze'$ for some $z \in R$ and

$$\eta(xe) = x\eta e = xze' \qquad \text{for all} \quad x \in R$$

In particular

$$\eta(se) = sze' \qquad \text{for all} \quad s \in S$$

Since $Se \subseteq Re$, $\eta(Se) = Sze' \subseteq Se'$. But Se' is minimal. Hence $\eta(Se) = Se'$. Conversely, suppose Se and Se' are isomorphic as R-modules. To show that this implies that Re and Re' are isomorphic, it is sufficient by Theorem VII.1.9 to show $\bar{R}\bar{e} \approx \bar{R}\bar{e}'$. To this end we first note that the irreducible left R-module Se is a *faithful* $\bar{R}_l$-module for exactly one of the simple direct summands in

$$\bar{R} = R/J = \bar{R}_1 \oplus \cdots \oplus \bar{R}_q \tag{12}$$

and is annihilated by the others (cf. Theorem IV.2.3). Therefore

$$gSe \neq 0 \tag{13}$$

for precisely those primitive idempotents g for which $\bar{g} = g + J \in \bar{R}_l$. Choose one such g and consider gS. Again gS is a faithful right $\bar{R}_k$-module for some k and is annihilated by the other $\bar{R}_i$. Hence $gSf \neq 0$ for precisely those primitive idempotents f for which $\bar{f} = f + J \in \bar{R}_k$. But $gSs \neq 0$ and therefore $\bar{e} \in \bar{R}_k$. By hypothesis, Se and Se' are isomorphic left R-modules. Thus $gSe' \neq 0$. Hence $\bar{e}' \in \bar{R}_k$ and by Theorem IV.2.3, $\bar{R}\bar{e} \approx \bar{R}\bar{e}'$. ∎

With the same notation as above, the following three statements are equivalent:

α. $gSe \neq 0$
β. $gS \approx \bar{e}\bar{R}$
γ. $Se \approx \bar{R}\bar{g}$

Proof. $\alpha \Rightarrow \beta$. As above, gS is a faithful irreducible $\bar{R}_k$-module where $\bar{e} \in \bar{R}_k$. Since $\bar{R}_k$ is simple, gS is isomorphic to any minimal right ideal of $\bar{R}_k$. Since e is primitive, $\bar{e}\bar{R}$ is such a minimal right ideal. Therefore $gS \approx \bar{e}\bar{R}$. $\beta \Rightarrow \alpha$ is clear. The equivalence of α and γ follows similarly.

As a consequence of Theorem 9 there exists therefore a permutation π of $\bar{R}_1, ..., \bar{R}_q$ with the properties stated in the following:

Theorem 10. Hypotheses: Let R be a quasi-Frobenius algebra over Φ with radical J and socle S. Let

$$\bar{R} = R/J = \bar{R}_1 + \cdots + \bar{R}_q \tag{12'}$$

be the decomposition of the semisimple algebra $\bar{R}$ as a direct sum of the simple algebras $\bar{R}_i$.

Conclusions: 1. There is a permutation

$$\pi = \begin{pmatrix} 1 & 2 & \cdots & q \\ \pi(1) & \pi(2) & \cdots & \pi(q) \end{pmatrix} \tag{13'}$$

of the numbers 1, 2, ..., q such that if the image $\bar{W}$ of the indecomposable right ideal W under the natural homomorphism of R to R/J lies in $\bar{R}_k$, then the unique minimal right ideal contained in W is isomorphic to any one of the minimal right ideals of $\bar{R}_{\pi(k)}$.

2. The permutation π may be described in terms of primitive idempotents g and e in R by means of the following equivalence

$$gSe \neq 0 \Leftrightarrow \{\bar{g} \in \bar{R}_k, \bar{e} \in \bar{R}_{\pi(k)}\} \quad \blacksquare \tag{14}$$

If we pass from the indecomposable right ideal gR of the Frobenius algebra R over Φ to its dual $(gR)^*$, then by Theorem VIII.1.1, the left R-module $(gR)^*$ contains the submodule

$$(gS)^\perp = \{\phi \mid \phi \in (gR)^*, \ \phi(gs) = 0 \text{ for all } s \in S\} \tag{15}$$

Moreover, $(gS)^\perp$ is the unique maximal submodule of $(gR)^*$ since gS is the unique minimal right ideal contained in gR. If $\bar{g} = g + J$ is in $\bar{R}_k$, the irreducible left R-module

$$(gR)^*/(gS)^\perp \approx \bar{R}\bar{e} \tag{16}$$

for each primitive idempotent e for which $\bar{e} = e + J \in \bar{R}_{\pi(k)}$.

Proof. If $\phi \in (gR)^*$, $e\phi$ is the Φ-linear form which maps $gx \in gR$ to $\phi(gxe)$ (Definition VIII.1.1). Thus by (15), $e\phi$ is contained in $(gS)^\perp$ if and only if $\phi(gSe) = 0$. By hypothesis and the second assertion of the previous theorem, the Φ-subspace gSe of gS is not the zero subspace. Therefore there exists a Φ-linear form $\phi_1 \in (gR)^*$ such that

$$\phi_1(gSe) \neq 0 \tag{17}$$

This proves that $\bar{e}((gR)^*/(gS)^\perp) \neq 0$ and hence, by Theorem IV.2.3, $(gR)^*/(gS)^\perp \approx \bar{R}\bar{e}$ since $(gR)^*/(gS)^\perp$ is a faithful $\bar{R}_{\pi(k)}$-module. The mapping

$$\eta\colon ye - (ye)\,\phi_1\,, \qquad y \in R \tag{18}$$

is an R-module homomorphism of the indecomposable left ideal Re to $(gR)^*$. Furthermore, η is an epimorphism since $(gS)^\perp$ is the only maximal R-submodule of $(gR)^*$ and $e\eta = e\phi_1 \notin (gS)^\perp$ implies $(Re)\,\eta = (gR)^*$. The unique minimal submodule of Re is Se and the image of Se under η is nonzero since, by (17), $((Se)\,\phi_1)\,g = \phi_1(gSe) \neq 0$. Hence η is a monomorphism.

We have therefore proved the first assertion of the following:

Theorem 11. Hypothesis: Let R be a quasi-Frobenius algebra over Φ with radical J and socle S. Let π be the permutation of Theorem 10.

1. Conclusion: Each indecomposable left ideal L of R for which $\bar{L} \subseteq \bar{R}_{\pi(k)}$ is isomorphic (as an R-module) to the dual module W^* of an indecomposable right ideal W of R for which $\bar{W} \subseteq \bar{R}_k$.

2. Conclusion: A quasi-Frobenius algebra is a self-injective ring (cf. Theorem VII.3.13).

Proof 2. By Theorem VII.3.2 we must show that each indecomposable left ideal Re of R is injective. By the first assertion of the theorem, Re is isomorphic to W^*, where W is an indecomposable right ideal of R. Since W is a direct summand of R_R, W^* is a direct summand of $(R_R)^*$. By Theorem VII.3.7′, $(R_R)^*$ is an injective R-module and hence so is the direct summand W^* by Theorem VII.3.2. ∎

This characterizes quasi-Frobenius algebras as finite-dimensional algebras which are self-injective, a result already mentioned at the end of Section VII.3.

By Theorem VIII.1.1, a direct sum decomposition of the quasi-Frobenius algebra R into indecomposable right ideals gives rise to a direct sum decomposition of the dual module $(R_R)^*$. In this latter decomposition the direct summands are isomorphic to indecomposable left ideals of R by Theorem 11. Thus $(R_R)^* \approx L_1' \oplus \cdots \oplus L_t'$, L_i' indecomposable. The factor ring R/J decomposes into the direct sum of simple rings $\bar{R}_k$, $k = 1, \ldots, q$. These rings $\bar{R}_k$ are in one-one correspondence with the isomorphism classes of the indecomposable left ideals in R. Now let ${}_RR = L_1 \oplus \cdots \oplus L_t$, where the L_i are indecomposable left ideals. It is important to determine for a fixed k

the number of those L_i which map under the natural homomorphism onto a minimal left ideal of $\bar{R}_k$. Each decomposition of ${}_R R$ as a direct sum of indecomposable left ideals can be obtained from a decomposition of ${}_R\bar{R}$ into minimal left ideals by lifting idempotents orthogonally. Thus ${}_R\bar{R} = \bar{L}_1 \oplus \cdots \oplus \bar{L}_t$, where the L_i are minimal left ideals. Let $f(k)$ be the number of $\bar{L}_i$ isomorphic to $\bar{L}_k \subseteq \bar{R}_k$. By Theorem IV.1.1, $f(k)$ is therefore the dimension of a faithful irreducible left $\bar{R}_k$-module over its centralizer D_k. By Theorem 10, Se is such a module, where $\bar{e} \in \bar{R}_{\pi(k)}$. Again by Theorem 10, $Se \approx \bar{R}\bar{g}$ and its centralizer D_k is isomorphic to $\bar{g}\bar{R}\bar{g}$ by Theorem II.6.4. Hence

$$f(k) = \dim_{D_k} \bar{R}\bar{g} \qquad \text{for each primitive idempotent} \quad \bar{g} \in \bar{R}_k \tag{19}$$

The dimension of D_k over the ground field Φ of the algebra R is calculated as follows: Since Se is isomorphic to $\bar{R}\bar{g}$ as a left R-module, the Φ-spaces gSe and $\bar{g}\bar{R}\bar{g}$ have the same dimension. Therefore

$$\dim_\Phi D_k = \dim_\Phi \bar{g}\bar{R}\bar{g} = \dim_\Phi gSe \tag{20}$$

Since $Se \approx \bar{R}\bar{g}$ and by (19) and (20) we obtain

$$\begin{aligned} \dim_\Phi Se &= \dim_\Phi \bar{R}\bar{g} = \dim_{D_k} \bar{R}\bar{g} \cdot \dim_\Phi D_k \\ &= f(k) \dim_\Phi D_k = f(k) \dim_\Phi gSe \end{aligned} \tag{21}$$

Analogously, since $gS \approx \bar{e}\bar{R}$,

$$\begin{aligned} \dim_\Phi gS &= \dim_\Phi \bar{e}\bar{R} = \dim_{D_{\pi(k)}} \bar{e}\bar{R} \cdot \dim_\Phi D_{\pi(k)} \\ &= f(\pi(k)) \dim_\Phi D_{\pi(k)} = f(\pi(k)) \dim_\Phi gSe \end{aligned} \tag{22}$$

If we assume now that R satisfies (10) $\dim_\Phi L + \dim_\Phi r(L) = \dim_\Phi R = \dim_\Phi W + \dim_\Phi l(W)$ we can prove

$$f(\pi(k)) = f(k) \tag{23}$$

as follows:

$$\begin{aligned} \dim_\Phi Se &= \dim_\Phi R - \dim_\Phi r(Se) && \text{by (10)} \\ &= \dim_\Phi R - \dim_\Phi r(l(J)\, e) && \text{by Theorem 5} \\ &= \dim_\Phi R - \dim_\Phi((1-e)\, R + J) && \text{by (11)} \\ &= \dim_\Phi (R/((1-e)\, R + J)) \\ &= \dim_\Phi \bar{e}\bar{R} = \dim_\Phi gS \\ &= f(\pi(k)) \dim_\Phi gSe && \text{by (22)} \end{aligned}$$

But by (21), $\dim_\Phi Se = f(k) \dim_\Phi gSe$. Therefore (23) is proved.

Now let ${}_RR = L_1 \oplus \cdots \oplus L_t$ be a decomposition of ${}_RR$ as a direct sum of indecomposable left ideals. From (23) it follows that the number $f(k)$ of L_i isomorphic to a given L_k, where $\bar{L}_k \subseteq \bar{R}_k$, is the same as the number $f(\pi(k))$ of L_j' isomorphic to L_k where $(R_R)^* \approx L_1' \oplus \cdots \oplus L_t'$. This shows that the R-modules ${}_RR$ and $(R_R)^*$ are isomorphic, proving the assertion following Theorem 3.

Theorem 12 (Nakayama). Hypothesis: Let R be a finite-dimensional algebra with identity over the field Φ. Conclusion: R is a Frobenius algebra if and only if the following hold:

$$l(r(L)) = L \qquad \text{and} \qquad r(l(W)) = W \tag{9'}$$

and

$$\dim_\Phi L + \dim_\Phi r(L) = \dim_\Phi R = \dim_\Phi W + \dim_\Phi l(W) \tag{10'}$$

for all left ideals L and right ideals W of R. ▮

3. Examples

Definition 1. Let G be a finite group and let Φ be a field. Then the *group algebra* $\Phi[G]$ consists of all functions $\alpha: G \to \Phi$. $\Phi[G]$ is endowed with the structure of a ring by defining

$$\begin{aligned} (\alpha + \beta)(g) &= \alpha(g) + \beta(g) \qquad \text{for all} \quad g \in G \\ (\alpha\beta)(g) &= \sum_{cd=g} \alpha(c)\,\beta(d) \qquad \text{for all} \quad g \in G \end{aligned} \tag{1}$$

The elements of $\Phi[G]$ may be expressed as formal sums

$$\alpha = \sum_{g \in G} g\alpha_g \qquad \text{where} \quad \alpha_g \in \Phi \tag{2}$$

Theorem 1. The group algebra $\Phi[G]$ is a Frobenius algebra.

Proof. $\Phi[G]$ is clearly an algebra of finite dimension over Φ with basis $\{g1 \mid g \in G\}$ since the associativity of multiplication in $\Phi[G]$ is a consequence of the associativity in G. By Theorem VIII.2.2 it is sufficient to prove the existence of a linear form $\phi \in R^* = \mathrm{Hom}_\Phi(R, \Phi)$, the kernel of which contains no left or right ideal different from (0). Let e denote the identity of G. The mapping

$$\phi: \sum_{g \in G} g\alpha_g = \alpha_e$$

is an element of R^*. Now suppose $Ra \subseteq \ker \phi$. Then for each $h \in G$,

$$0 = \phi(h^{-1}a) = \phi \sum_g h^{-1}g\alpha_g = \alpha_h \qquad \text{where} \quad a = \sum_g g\alpha_g$$

Thus $a = 0$. Similarly, if $aR \subseteq \ker \phi$, $a = 0$. ∎

We note that no mention of the characteristic of the field Φ was made in the previous theorem. We shall eventually show in Theorem VIII.4.3 that if the characteristic of Φ does not divide the order of G, $\Phi[G]$ is semisimple.

A Frobenius algebra is a generalization of a semisimple finite-dimensional algebra:

Theorem 2. Every semisimple finite dimensional algebra R is a Frobenius algebra over its ground field Φ.

Proof. By Theorem IV.1.4 the left R-module ${}_RR$ is completely reducible and hence the indecomposable left ideals are minimal. Since R has an identity (Theorem III.3.2), by Theorem VII.1.1 each left and each right ideal of R is generated by an idempotent e. From $Re \cdot x = 0$ it follows that $ex = 0$ and hence $x = ex + (1 - e)\, x = (1 - e)\, x$. Therefore

$$r(Re) = (1 - e)\, R \tag{3}$$

Similarly

$$l(eR) = R(1 - e) \tag{4}$$

Applying (3) to (4) we get

$$r(l(eR)) = eR$$

Similarly

$$l(r(Re)) = Re$$

R is therefore a quasi-Frobenius algebra. By Nakayama's theorem (Theorem VIII.2.12) it is sufficient to show that $\dim_\Phi l(eR) + \dim_\Phi eR = \dim_\Phi R$: From ${}_RR = Re \oplus R(1 - e)$ and (4) it follows that

$$\dim_\Phi l(eR) = \dim_\Phi R(1 - e) = \dim_\Phi R - \dim_\Phi Re \tag{5}$$

The idempotent e may be decomposed into a sum of orthogonal primitive idempotents:

$$e = f_1 + \cdots + f_s$$

Hence

$$eR = \sum_{1 \leqslant i \leqslant s} \oplus f_iR \qquad \text{and} \qquad Re = \sum_{1 \leqslant i \leqslant s} \oplus Rf_i \tag{6}$$

Suppose $f_i \in R_k$, one of the simple components in the decomposition

$$R = R_1 \oplus \cdots \oplus R_q, \qquad R_i \text{ simple}$$

Then $D_k = f_i R f_i$ is the centralizer of both $f_i R$ and $R f_i$ and

$$R_k \approx (D_k)_{n_k \times n_k} \qquad \text{where} \qquad n_k = \dim_{D_k} f_i R = \dim_{D_k} R f_i$$

(cf. Theorem II.3.5). It follows that

$$\dim_\Phi f_i R = n_k \cdot \dim_\Phi D_k = \dim_\Phi R f_i$$

and hence because of (6),

$$\dim_\Phi eR = \dim_\Phi Re$$

This implies

$$\dim_\Phi l(eR) + \dim_\Phi eR = \dim_\Phi R$$

for right ideals of R. Using (3) we obtain a similar result for left ideals of R. ∎

An important fact often used in the investigation of Frobenius algebras is the existence of pairs of dual bases:

Theorem 3. Let R be a Frobenius algebra over Φ and f a nondegenerate associative bilinear form on R with values in Φ (Theorem VIII.2.1). Then there exists a pair of Φ-bases $\{a_1, \ldots, a_N\}$, $\{b_1, \ldots, b_N\}$ such that $f(a_i, b_j) = \delta_{ij}$, $i, j = 1, \ldots, N$. Bases which are so related are said to be *dual with respect to* f.

Proof. 1. The mapping

$$b \to \phi_b \qquad \text{where} \quad \phi_b x = f(x, b) \quad \text{for all } b, x \in R \tag{7}$$

is an isomorphism of the R-module ${}_R R$ onto $(R_R)^*$ as may be seen as follows:

$$(a\phi_b)\, x = \phi_b(xa) = f(xa, b) = f(x, ab) = \phi_{ab} x$$

showing that (7) is an R-module homomorphism. Moreover it is a monomorphism since

$$\phi_b(R) = 0 \qquad \text{implies} \quad f(R, b) = 0 \quad \text{and hence} \quad b = 0$$

Finally, the Φ-linear map (7) is an epimorphism since $\dim_\Phi({}_R R) = \dim_\Phi {R_R}^*$.

2. Now let $\{a_1, \ldots, a_N\}$ be a Φ-basis of R. Define $\phi_i \in (R_R)^*$ for $i = 1, \ldots, N$

by $\phi_i: a_j \to \delta_{ij}$. Since the mapping in (7) is an isomorphism, there exist $b_1, ..., b_N \in R$ such that $\phi_{b_j} = \phi_j$. Hence

$$f(a_i, b_j) = \phi_{b_j}(a_i) = \phi_j(a_i) = \delta_{ij}, \qquad i, j = 1, ..., N$$

Clearly $b_1, ..., b_N$ are linearly independent over Φ. ∎

We have seen that the tensor product of two semisimple algebras is not necessarily semisimple. However, for Frobenius algebras we have the following:

Theorem 4. The tensor product of two Frobenius algebras A and B over Φ is a Frobenius algebra over their common ground field Φ.

Proof. Let $\phi: A \to \Phi$ and $\psi: B \to \Phi$ be linear forms such that ker ϕ (resp. ker ψ) contains no right or left ideal $\neq (0)$ of A (resp. B) (see Theorem VIII.2.2). Define $\lambda \in \mathrm{Hom}_\Phi(A \otimes_\Phi B, \Phi)$ by

$$\lambda: \sum_j x_j \otimes y_j \to \sum_j \phi(x_j)\, \psi(y_j) \qquad \text{for} \quad \sum_j x_j \otimes y_j \in A \otimes_\Phi B \tag{8}$$

That λ is well-defined follows from Theorem V.3.1 and the fact that the mapping $\mu: A \otimes B \to \Phi$ defined by $\mu(x, y) = \phi(x)\, \psi(y)$ is balanced. We must now show that ker λ contains no left or right ideals of $A \otimes_\Phi B$ different from (0). To this end, let $\{a_1, ..., a_r\}$, $\{a_1', ..., a_r'\}$ be a pair of dual bases of A with respect to the bilinear form $\phi(x_1 \cdot x_2)$ and let $\{b_1, ..., b_s\}$, $\{b_1', ..., b_s'\}$ be a pair of dual bases of B with respect to the bilinear form $\psi(y_1 y_2)$. These two pairs of bases exist by Theorem 3, since $\phi(xy)$ and $\psi(xy)$ are non-degenerate associative bilinear forms. Now suppose the left ideal

$$(A \otimes_\Phi B) \sum_j u_j \otimes v_j$$

of $A \otimes_\Phi B$ is contained in ker λ, i.e. suppose

$$\lambda\left((x \otimes y) \sum u_j \otimes v_j\right) = 0 \qquad \text{for all} \quad x \in A, \quad y \in B$$

Then

$$\sum \phi(xu_j)\, \psi(yv_j) = 0 \qquad \text{for all} \quad x \in A, \quad y \in B$$

But the element $\sum u_j \otimes v_j$ is expressible in the form

$$\sum_{1 \leqslant k \leqslant s} z_k \otimes b_k' \qquad \text{with} \quad z_k \in A \tag{9}$$

since $\{b_1', ..., b_s'\}$ is a Φ-basis of B. Thus (9) becomes

$$\sum_k \phi(xz_k)\, \psi(yb_k') = 0 \tag{10}$$

Now substituting $y = b_i$, $i = 1, ..., s$ in (10) we get

$$0 = \sum_k \phi xz_k)\, \psi(b_k b_k') = \sum_k \phi(xz_k)\, \delta_{ik} = \phi(xz_i) \qquad \text{for} \quad x \in A, \quad i = 1, ..., s$$

since the bases $\{b_1, ..., b_N\}$, $\{b_1', ..., b_N'\}$ are dual with respect to the bilinear form $\psi(y_1 y_2)$. Since A is a Frobenius algebra, $\phi(Az_i) = 0$ implies $z_i = 0$, $i = 1, ..., s$. Hence the element

$$\sum_j u_j \otimes v_j = \sum_k z_k \otimes b_k' = 0$$

Similarly by using the dual bases $\{a_1, ..., a_r\}$, $\{a_1', ..., a_r'\}$ we can prove that $\ker \lambda$ contains no right ideal of $A \otimes_\Phi B$ different from (0). ∎

In particular, Theorem 4 shows that although a semisimple algebra R with identity, finite dimensional over the ground field Φ does not necessarily remain semisimple when Φ is extended to a finite-dimensional overfield Λ, $R_\Phi = R \otimes_\Phi \Lambda$ is nevertheless still a Frobenius algebra (see the example at the beginning of Section VI.1).

4. Injective modules of a Frobenius algebra R

We make use of the existence of a pair of dual bases in a Frobenius algebra to prove that a Φ-linear mapping of an R-module M into itself may be turned into an R-module homomorphism of M into itself:

Theorem 1. Let R be a Frobenius algebra over Φ and let f be a non-degenerate associative bilinear form on R with values in Φ (see Theorem VIII.2.1). Let M be a left R-module and let $\phi \in \mathrm{Hom}_\Phi(M, M)$. If $\{a_1, ..., a_N\}$, $\{b_1, ..., b_N\}$ is a pair of dual bases with respect to f, then

$$\sum_{1 \leqslant i \leqslant N} b_i \phi a_i \in \mathrm{Hom}_R(M, M) \tag{1}$$

where for $m \in M$, $(b_i \phi a)\, m = b_i(\phi(a_i m))$. ∎

Proof. The second regular representation of R with respect to the basis $\{a_1, ..., a_N\}$ is defined by

$$a \to (\gamma_{ji}'(a))^T \in \Phi_{N \times N} \qquad \text{where} \quad a_i a = \sum_j a_j \gamma_{ji}'(a) \tag{1}$$

and where ^{T} denotes transpose and *the first with respect to* $\{b_1, ..., b_N\}$ by

$$a \to (\gamma_{ji}(a)) \in \Phi_{N\times N} \qquad \text{where} \quad ab_i = \sum b_j\gamma_{ji}(a) \tag{1'}$$

Since f is associative, it follows that

$$f(a_i\, a, b_k) = \sum_{\gamma} f(a_j\, , b_k)\, \gamma'_{ji}(a) = \gamma'_{ki}(a) = f(a_i\, , ab_k)$$

$$= \sum_j f(a_i\, , b_j)\, \gamma_{jk}(a) = \gamma_{ik}(a) \tag{2}$$

Therefore

$$(\gamma_{ik}(a)) = (\gamma'_{ik}(a))^T \tag{3}$$

This proves that *the second regular representation of R with respect to the first basis of a dual pair of bases coincides with the first regular representation of R with respect to the second basis of the dual pair.*

Now

$$a\left[\left(\sum_i b_i\phi a_i\right) m\right] = \sum_{i,j} b_j\gamma_{ij}(a)\ \phi a_i m = \sum_{i,j} b_j\gamma'_{ij}(a)\ \phi a_i m \tag{4}$$

and

$$\left(\sum_i b_i\phi a_i\right)(am) = \sum_{i,j} b_i\phi a_j\gamma'_{ji}(a)\, m = \sum_{i,j} b_j\gamma'_{ij}(a)\ \phi a_i m \tag{5}$$

proving

$$a\left[\left(\sum_i b_i\phi a_i\right) m\right] = \left(\sum_i b_i\phi a_i\right)(am)$$

In (4) and (5), besides (3), we use the Φ-linearity of ϕ and the left multiplication of an element $m \in M$ by an element $a \in R$. ▌

By Theorem VII.3.6 a module M over a ring R with identity is injective if and only if it is a direct summand of any containing module. This characterizing property may be used as the definition of an injective R-algebra module:

Definition 1. Let R be an algebra with identity over Φ. The R-module M is *injective* if M is a direct summand of any containing module, i.e. if $M \subseteq U$, there exists a submodule M' such that

$$U = M \oplus M'$$

A very useful criterion for the injectivity of a module over a Frobenius algebra is proved in:

Theorem 2 (Gaschutz–Ikeda). Let R be a Frobenius algebra over Φ and let $\{a_1, ..., a_N\}$, $\{b_1, ..., b_N\}$ be a pair of dual Φ-bases with respect to the nondegenerate associative bilinear form f. Then a left R-module M is injective if and only if there exists $\phi \in \mathrm{Hom}_\Phi(M, M)$ such that

$$\sum_{1 \leqslant i \leqslant N} b_i \phi a_i = 1_M \tag{6}$$

where 1_M is the identity map on M.

Proof. 1. Assume the condition holds. Let M be an R-submodule of U and let $\pi \in \mathrm{Hom}_\Phi(U, U)$ be the projection of U onto M:

$$\pi U \subseteq M, \quad \pi m = m \qquad \text{for all} \quad m \in M$$

Let $\phi \in \mathrm{Hom}_\Phi(M, M)$ satisfy (6). By Theorem 1 applied to the R-module U and the Φ-linear map $\phi\pi$,

$$\pi' = \sum_i b_i \phi \pi a_i \in \mathrm{Hom}_R(U, U)$$

Now $\pi' U \subseteq M$ and moreover, π' is a projection of U onto M since $\pi' m = (\sum b_i \phi \pi a_i)\, m = \sum b_i \phi \pi (a_i m) = (\sum b_i \phi a_i)\, m = m$ by (6) and the fact that $\pi m = m$ for all $m \in M$. If the identity on U is denoted by 1_U we have $a(1_U - \pi')\, U = (1_U - \pi')\, aU \subseteq (1_U - \pi')\, U$. It follows that $(1_U - \pi')\, U$ is an R-submodule and hence

$$U = \pi' U \oplus (1_U - \pi')\, U = M \oplus (1_U - \pi')\, U$$

since $\pi'(1_U - \pi') = \pi' - \pi'^2 = 0$.

2. To prove the converse we use (2) of Theorem 1. Let M be an injective R-module. We turn $R \otimes_\Phi M$ into a left R-module by defining $a(x \otimes m) = ax \otimes m$. The mapping

$$\eta \colon m \to \sum_i b_i \otimes (a_i m) \qquad \text{for all} \quad m \in M \tag{7}$$

is an embedding of M in $R \otimes_\Phi M$ as may be seen as follows: First, η is uniquely determined by the given dual pair of bases. Furthermore, η is an R-module homomorphism since by (2), applied to the dual bases $\{a_1, ..., a_N\}$, $\{b_1, ..., b_N\}$,

$$a\eta m = \sum_i ab_i \otimes a_i m = \sum_{i,j} b_j \gamma_{ji}(a) \otimes a_i m = \sum_{i,j} b_j \otimes \gamma_{ji}(a)\, a_i m$$

$$= \sum_{i,j} b_j \otimes a_i \gamma'_{ij}(a) m = \sum_j b_j \otimes a_j a m = \eta(am)$$

Now since the $a_i m$, $i = 1, ..., N$ lie in M and $\{b_1, ..., b_N\}$ is a Φ-basis of R, $\sum b_i \otimes a_i m = 0$ implies that $a_i m = 0$ for $i = 1, ..., N$. But the identity 1 of R is expressible as

$$1 = \sum_i a_i \xi_i \qquad \text{with} \quad \xi_i \in \Phi \tag{8}$$

Thus $a_i m = 0$ for $i = 1, ..., N$ implies $m = 0$. Hence η is a monomorphism and M is embedded in $R \otimes_\Phi M$. The next step consists in showing that the mapping

$$\psi: \sum_i b_i \otimes m_i \to \sum_i 1 \otimes \xi_i m_i \tag{9}$$

where the ξ_i are as in (8), is an element of $\mathrm{Hom}_\Phi(R \otimes_\Phi M, R \otimes_\Phi M)$ for which

$$\sum_i b_i \psi a_i = 1_{R \otimes M} \tag{10}$$

Once this is proved, the existence of ϕ satisfying the conclusion of the theorem is easily proved. Each element of $R \otimes_\Phi M$ is uniquely expressible in the form $\sum_i b_i \otimes m_i$ since the b_i form a Φ-basis. Therefore ψ is well defined and clearly Φ-linear. Now for all $m \in M$

$$\begin{aligned}\Big(\sum_i b_i \psi a_i\Big)(b_j \otimes m) &= \sum_i b_i \psi \sum_k b_k \gamma_{kj}(a_i) \otimes m \\ &= \sum_i b_i \psi \Big(\sum_k b_k \otimes \gamma_{kj}(a_i) m\Big) = \sum_i b_i \Big(\sum_k 1 \otimes \xi_k \gamma_{kj}(a_i) m\Big) \\ &= \sum_i b_i \otimes \sum_k \xi_k \gamma_{kj}(a_i) m\end{aligned}$$

Using (2) again, the matrix entries $\gamma_{kj}(a_i)$ are expressible in terms of the bilinear form f as

$$\gamma_{kj}(a_i) = f(a_k a_i, b_j)$$

Thus, because of (8),

$$\sum_k \xi_k \gamma_{kj}(a_i) = f\Big(\sum_k \xi_k a_k \cdot a_i, b_j\Big) = f(1a_i, b_j) = \delta_{ij}$$

This gives

$$\sum_i (b_i \psi a_i)(b_j \otimes m) = \sum_i b_i \otimes \delta_{ij} m = b_j \otimes m$$

proving (10). Now set

$$\phi = \eta^{-1} \pi \psi \eta \tag{11}$$

where $\pi \in \text{Hom}_R(R \otimes_\Phi M, R \otimes_\Phi M)$ is the projection of $R \otimes_\Phi M$ onto $\eta(M)$ and η is the embedding defined in (7). The projection π exists since we are assuming that M is injective and hence $\eta(M)$ is a direct summand of $R \otimes_\Phi M$ as an R-module. Clearly $\phi \in \text{Hom}_\Phi(M, M)$. Moreover

i. η is an R-module homomorphism and thus $\eta a_i = a_i \eta$ and $b_i \eta^{-1} = \eta^{-1} b_i$

and

ii. π is an R-module epimorphism and hence $b_i \pi = \pi b_i$.
These observations coupled with (10) prove that

$$\sum_i b_i \phi a_i m = \sum_i b_i \eta^{-1} \pi \psi \eta a_i m = \eta^{-1} \pi \sum_i b_i \psi a_i \eta m = \eta^{-1} \pi \eta \dot{m} = \eta^{-1} \eta m = m$$

Hence $\sum b_i \phi a_i = 1_M$. ▮

We have shown in Theorem VIII.3.1 that if G is a finite group and Φ a field, the group algebra $\Phi[G]$ is Frobenius. We now apply the preceding theorem to $\Phi[G]$ to prove the following result which is of fundamental importance in the theory of representations of finite groups:

Theorem 3 (Maschke). Let G be a finite group and Φ a field of characteristic p (We allow $p = 0$). The group algebra $\Phi[G]$ is semisimple if and only if p does not divide the order of G.

Proof. 1. By Theorem IV.1.4, to prove the semisimplicity of $\Phi[G]$ it is sufficient to prove that every $\Phi[G]$-module is completely reducible. This is equivalent to proving that every $\Phi[G]$-module M is injective. To apply the criterion of Theorem 2 we note that the linear form

$$\phi\colon \sum_{g \in G} g\alpha_g \to \alpha_e$$

gives rise to the bilinear form f defined by

$$f\Big(\sum_g g\xi_g, \sum_g g\eta_g\Big) = \phi\Big(\Big(\sum_g g\xi_g\Big)\Big(\sum_g g\eta_g\Big)\Big) = \sum_g \xi_g \eta_{g^{-1}}$$

$\{g_i \mid i = 1, ..., N\}$ and $\{g_i^{-1} \mid i = 1, ..., N\}$ (where the g_i run through the elements of G) are a pair of dual Φ-bases with respect to this bilinear form f. Since p does not divide $|G|$, the mapping

$$\psi\colon m \to \frac{m}{|G|} \qquad \text{for} \quad m \in M$$

is an element of $\mathrm{Hom}_\Phi(M, M)$ such that

$$\sum_i g_i \psi g_i^{-1} m = \sum_i g_i \frac{1}{|G|} g_i^{-1} m = m$$

Hence ψ satisfies condition (6) of Theorem 2 and M is an injective $\Phi[G]$-module.

2. The element $s = g_1 + \cdots + g_N$, $N = |G|$ is such that $s = g_k s = s g_k$, $k = 1, ..., N$. It follows that $s\Phi$ is an ideal in $\Phi[G]$ with the property that $(s\Phi)^2 = Ns\Phi$. If $p \mid N$, $s\Phi$ is a nilpotent ideal of $\Phi[G]$ different from (0). ∎

Since a Frobenius algebra is in particular a quasi-Frobenius algebra and hence a quasi-Frobenius ring, the following is a consequence of Theorem VII.3.9 and 2. of Theorem VIII.1.1:

Theorem 4. Let R be a quasi-Frobenius algebra. Then a finitely generated R-module M is injective if and only if it is projective. ∎

5. Behavior of semisimple algebras under separable extensions of the ground field

Let the field Λ be a separable finite-dimensional extension of the field Φ. Since Λ is semisimple, Λ is a Frobenius algebra by Theorem VIII.3.2. Therefore by Theorem VIII.2.2 there exists $\phi \in \mathrm{Hom}_\Phi(\Lambda, \Phi)$ such that $\ker \phi$ contains no left or right ideals of Λ different from (0). For ϕ we may take the trace Tr of the first regular representation of Λ in Φ. This is the linear form defined by

$$\mathrm{Tr}\,\lambda = \sum_{1 \leqslant i \leqslant n} \alpha_{ii}(\lambda), \qquad \lambda \in \Lambda \tag{1}$$

where $\lambda\beta_i = \sum_{1 \leqslant k \leqslant n} \beta_k \alpha_{ki}(\lambda)$, $\alpha_{ki}(\lambda) \in \Phi$ and $\{\beta_1, ..., \beta_n\}$ is a Φ-basis of Λ. To prove Tr satisfies the condition of Theorem VIII.2.2 we proceed as follows: By Theorem V.4.1, $\mathrm{Tr}\,\lambda \neq 0$ for some $\lambda \in \Lambda$ since Λ is separable over Φ. Since the only left or right ideals in Λ are (0) and Λ, the kernel of the linear form Tr contains no one-sided ideal different from (0). As in the proof of Theorem VIII.3.2, $f(\lambda, \mu) = \mathrm{Tr}(\lambda, \mu)$ is a nondegenerate associative bilinear form on Λ. Let $\{\beta_1, ..., \beta_n\}$, $\{\beta_1', ..., \beta_n'\}$ be a pair of dual bases of Λ with respect to f. Then the following fundamental inequality holds:

$$\sigma = \sum_{1 \leqslant i \leqslant n} \beta_i \beta_i' \neq 0 \tag{2}$$

Proof. Suppose $\sigma = 0$. Then since

$$\beta_j\beta_i = \sum_k \beta_k\alpha_{ki}(\beta_j)$$

it follows that

$$0 = \beta_j\sigma = \sum_{i,k} \beta_k\beta_i'\alpha_{ki}(\beta_j)$$

which in turn implies

$$0 = \sum_{i,k} \mathrm{Tr}(\beta_k\beta_i')\,\alpha_{ki}(\beta_j) = \sum_i \alpha_{ii}(\beta_j) = \mathrm{Tr}\,\beta_j$$

for $j = 1, \ldots, n$. Hence $\mathrm{Tr}\,\Lambda = 0$, a contradiction. Therefore $\sigma \neq 0$.

We collect these results in:

Theorem 1. Let Λ be a separable field extension of Φ with $\dim_\Phi \Lambda = n < \infty$. Regarded as a Frobenius algebra over Φ, Λ has a nondegenerate associative bilinear form f and a pair of dual bases $\{\beta_1, \ldots, \beta_n\}$, $\{\beta_1', \ldots, \beta_n'\}$ with respect to f such that $\sigma = \beta_1\beta_1' + \cdots + \beta_n\beta_n' \neq 0$. Moreover, for f we can take the bilinear form defined by $f(\lambda, \mu) = \mathrm{Tr}(\lambda \cdot \mu)$ where Tr is the trace of the first regular representation of Λ over Φ. ∎

Theorem 2. Let R be a semisimple algebra over Φ and Λ a separable field extension of Φ. Suppose $\dim_\Phi R$ and $\dim_\Phi \Lambda$ are both finite. Then the algebra $R_\Lambda = R \otimes_\Phi \Lambda$ is a semisimple algebra over Λ.

Proof. By Theorem IV.1.4 it is sufficient to show that every module M over the Λ-algebra $R \otimes_\Phi \Lambda$ is completely reducible. Let U be an $R \otimes_\Phi \Lambda$-submodule of M. Since $R \approx R \otimes 1$, M and U are both R-modules for the semisimple Φ-algebra R. Hence U is a direct summand of M as an R-module. Therefore there exists a projection π from M onto U which is an R-module homomorphism:

$$\pi \in \mathrm{Hom}_R(M, M), \qquad \pi M = U, \qquad \pi^2 = \pi \tag{3}$$

Let $\{\beta_1, \ldots, \beta_n\}$, $\{\beta_1', \ldots, \beta_n'\}$ be dual bases of Λ over Φ chosen as in Theorem 1 above. Define π' from M to M by

$$\pi' \colon m \to \sum_i \left(\pi\left(m\beta_i \cdot \frac{1}{\sigma}\right)\right)\beta_i' \qquad \text{where} \quad \sigma = \sum_i \beta_i\beta_i' \neq 0 \tag{4}$$

It is a simple matter to verify that $\pi' \in \mathrm{Hom}_\Lambda(M, M)$. (Note that we may write the elements of Λ on the right of M since M is a module over the Λ-algebra

$R \otimes_{\Phi} \Lambda$). Since π is an R-module homomorphism, $\pi' \in \mathrm{Hom}_R(M, M)$; for using the fact that M is an (R, Λ)-bimodule,

$$\pi'(am) = \sum_i \left(\pi\left(am\beta_i \frac{1}{\sigma}\right)\right)\beta_i' = \sum_i a\left(\pi\left(m\beta_i \frac{1}{\sigma}\right)\right)\beta_i' = a(\pi'm),$$

$$a \in R, \quad m \in M$$

For $a \otimes \lambda \in R \otimes_{\Phi} \Lambda$ we have therefore

$$\pi'((a \otimes \lambda)\, m) = \pi'(am\lambda) = a(\pi'm)\,\lambda = (a \otimes \lambda)(\pi'm)$$

This shows that π' is an $R \otimes_{\Phi} \Lambda$-homomorphism of M to itself. π' also maps M to U by (4) since $M\beta_i(1/\sigma) \subseteq M$, $\pi(M) = U$ and $U\beta_i' \subseteq U$. Moreover, π' is a projection of M onto U since by (3) and the definition of σ,

$$\pi'u = \sum_i \left(\pi\left(u\beta_i \cdot \frac{1}{\sigma}\right)\right)\beta_i' = \sum_i \left(u\beta_i \cdot \frac{1}{\sigma}\right)\beta_i' = u \qquad \text{for all} \quad u \in U$$

Hence $M = U + (1_M - \pi')\, M$ where the sum is a direct sum of $R \otimes_{\Phi} \Lambda$-modules. ▮

If the characteristic of Φ is 0, by Theorem V.1.8, every overfield Λ of Φ is separable over Φ in the field theoretic sense. Theorem 3 therefore implies

Theorem 4. If Φ is a field of characteristic 0, every semisimple algebra R over Φ is separable in the sense of Definition V.1.5. ▮

Distributively Representable Rings

CHAPTER IX

If R is a primitive ring and M a faithful irreducible R-module then trivially, $V_R(M)$, the lattice of submodules of M, is distributive. We isolate this property and call a ring distributively representable if it possesses a faithful module with distributive lattice of submodules.

In this chapter we shall study distributively representable rings with 1 having the following properties in common with semiperfect rings:

α. R/J is artinian where J is the radical of R

β. if $R/J = \bar{R}_1 \oplus \cdots \oplus \bar{R}_q$, $\bar{R}_i$ simple and $\bar{1} = \bar{e}_1 + \cdots + \bar{e}_q$, then $\{\bar{e}_1, ..., \bar{e}_q\}$ may be lifted to a family $\{e_1, ..., e_q\}$ of orthogonal idempotents whose sum is 1.

The general theory developed in Section 1 will be applied in Section 2 to artinian rings with distributive lattice of left ideals and in Section 3 to specific arithmetic rings, namely to rings R which distributively represent their multiplication ring T. Roughly speaking these are rings with distributive lattice $V_T(R)$ of ideals whose quotient ring by the radical is artinian.

1. Modules with distributive lattice of submodules

The following theorem is basic to the development of the subsequent theory:

Theorem 1. Let U_1 and U_2 be two minimal submodules of the R-module M.

Let the lattice $V_R(M)$ of R-submodules be distributive. Then U_1 and U_2 are not isomorphic as R-modules.

Proof. Suppose by way of contradiction that ϕ_2 is an R-module isomorphism of U_1 onto U_2. Then

$$U_3 = \{u_1 + u_1\phi_2 \mid u_1 \in U_1\}$$

is an R-submodule of M and

$$(0) \neq U_3 = U_3 \cap (U_1 + U_2) = (U_3 \cap U_1) + (U_3 \cap U_2)$$

We show that this leads to a contradiction by proving $U_3 \cap U_i = (0)$, $i = 1, 2$. Define ϕ_3: $U_1 \to U_3$ by

$$\phi_3\colon u_1 \to u_1 + u_1\phi_2, \qquad u_1 \in U_1$$

Clearly ϕ_3 is an epimorphism and ker $\phi_3 = U_1$ or (0) by minimality of U_1. If ker $\phi_3 = U_1$, then $U_2 = U_1\phi_2 \subseteq U_1$ and hence $U_2 = U_1$, a contradiction. Therefore ker $\phi_3 = (0)$ and ϕ_3 is an isomorphism. Hence U_3 is a minimal R-module. Now if $U_3 \cap U_1 \neq (0)$, $U_3 = U_1$. Since $U_2 = U_1\phi_2$ this gives $U_2 \subseteq U_1$ and hence $U_2 = U_1$, a contradiction. Thus $U_3 \cap U_1 = (0)$ and similarly $U_3 \cap U_2 = (0)$. ▮

The following example shows that this theorem does not hold if the submodules U_1 and U_2 of M are not minimal: Let $R = \mathbb{Z}$, the ring of integers and let $M = {}_{\mathbb{Z}}\mathbb{Z}$. Put $U_1 = \mathbb{Z}$ and $U_2 = 2\mathbb{Z}$. Then U_1 and U_2 are isomorphic as $\mathbb{Z}$-modules under

$$\phi\colon z \to 2z \qquad \text{for} \quad z \in \mathbb{Z}$$

The lattice $V_{\mathbb{Z}}(M)$ of submodules of M is the lattice $V(\mathbb{Z})$ of ideals of $\mathbb{Z}$ and is therefore distributive.

Since the irreducible R-modules of a simple artinian ring R are isomorphic to the minimal left ideals of R and these in turn are isomorphic to one another, Theorem 1 implies:

Theorem 2. The lattice $V_R(M)$ of R-submodules of a module M over a simple artinian ring R is distributive if and only if M is irreducible. ▮

More generally we prove:

Theorem 3. Hypotheses: Let R be a ring with identity and let R/J be simple artinian where J is the radical of R. Let M be a noetherian module such that $\bigcap_{s=1}^{\infty} J^s M = (0)$.

Conclusion: The lattice $V_R(M)$ of R-submodules of M is distributive if and only if it consists of the members of the Loewy series

$$M \supseteq JM \supseteq J^2M \supseteq \cdots \tag{1}$$

together with the zero submodule of M. In this case, every submodule of M is cyclic.

Proof. 1. Clearly if $V_R(M) = \{M, JM, J^2M, ..., (0)\}$, $V_R(M)$ is distributive.

2. Set $J^0 = R$. Then for $s = 0, 1, 2, ...$, the distributivity of $V_R(M)$ implies the distributivity of the sublattice consisting of submodules U such that $J^{s+1}M \subseteq U \subseteq J^sM$. Consequently the lattice of submodules of $J^sM/J^{s+1}M$ is distributive. But this factor module is an $\bar{R}$-module over the artinian ring $\bar{R} = R/J$. Therefore, by Theorem 2, it is irreducible if it is different from (0). Now let $U \neq (0)$ be a submodule of M. Since by hypothesis $\bigcap J^sM = (0)$, there is a maximal exponent $s \geqslant 0$ such that $U \subseteq J^sM$. Obviously we must show $U = J^sM$. As we have already shown there is no submodule of M lying strictly between J^sM and $J^{s+1}M$. Hence

$$J^sM = U + J^{s+1}M \tag{2}$$

Since M is noetherian by hypothesis, J^sM is finitely generated, say

$$J^sM = Rm_1 + \cdots + Rm_{N-1} + Rm_N$$

and hence by (2)

$$J^sM = U + Jm_1 + \cdots + Jm_{N-1} + Jm_N \tag{3}$$

In particular therefore

$$m_N = u + w_1m_1 + \cdots + w_{N-1}m_{N-1} + w_Nm_N$$

where $u \in U$ and $w_i \in J$, $i = 1, ..., N$. Therefore

$$(1 - w_N)\, m_N = u + w_1m_1 + \cdots + w_{N-1}m_{N-1} \tag{4}$$

Since $w_N \in J$, there exists $w' \in J$ such that $(1 - w')(1 - w) = 1$. Hence (4) implies $m_N \in U + Jm_1 + \cdots + Jm_{N-1}$. This shows

$$J^sM = U + Jm_1 + \cdots + Jm_{N-1} \tag{5}$$

By applying the same reasoning to (5) as we did to (3) we obtain

$$J^sM = U + Jm_1 + \cdots + Jm_{N-2}$$

Continuing in this fashion we arrive finally at $J^sM = U$. (Note: This is Nakayama's lemma which we have already proved in Chapter VII.)

3. If $J^sM \neq (0)$ let $n_s \in J^sM \backslash J^{s+1}M$. Then Rn_s is a submodule of J^sM and $Rn_s \subset J^{s+1}M$. Therefore $Rn_s = J^sM$. ▮

If in place of the hypothesis that R/J be simple we substitute the weaker hypothesis that the identity elements of the simple, direct summands of R/J be liftable to an orthogonal family $\{e_1, ..., e_q\}$ of idempotents with sum 1, it is still possible to describe the structure of the lattice $V_R(M)$ quite explicitly provided it is distributive. The above hypotheses hold in particular in a semiperfect ring (Theorem VII.1.10).

We start by recalling that $J_i = e_iJe_i$ is the radical of the ring $R_i = e_iRe_i$, $i = 1, ..., q$ (Theorem VII.1.8) and that $R_i/J_i = R_i/J \cap R_i \approx R_i + J/J = \bar{e}_i\bar{R}\bar{e}_i$, $i = 1, ..., q$ is simple artinian. The ring R itself, considered as an additive group, is the direct sum of R^* and U where R^* is the ring-direct sum of the R_i, $i = 1, ..., q$ and U is the subgroup $\sum_{i \neq k} e_iRe_k$ of the additive group of the radical J of R.

With this notation every submodule U of the R-module M is the direct sum (as a group)

$$U = U_1 \oplus \cdots \oplus U_q \tag{6}$$

of the R_i-submodules $U_i = e_iU$ of the R_i-modules $M_i = e_iM$. We shall first show that the lattice $V_{R_i}(M_i)$ of R_i-submodules of M_i is distributive provided $V_R(M)$ is:

Theorem 4. Let $e = e^2$ be an idempotent of the ring R and let M be an R-module. Then the mapping

$$\phi: U \to eU \qquad \text{for} \quad U \in V_R(M) \tag{7}$$

is an epimorphism of the lattice $V_R(M)$ of R-submodules of M onto the lattice $V_{eRe}(eM)$ of eRe-submodules of the eRe-module eM. In particular $V_{eRe}(eM)$ is distributive and noetherian if $V_R(M)$ is.

Proof. Since $eRe \cdot eU \subseteq eRU \subseteq eU$, eU is an eRe-module. The equality $e(U + W) = eU + eW$ is a consequence of the definition of the join of U and W in $V_R(M)$. Since $eU = U \cap eM$, it follows that

$$e(U \cap W) = eU \cap eW$$

Therefore ϕ is a lattice homomorphism of $V_R(M)$ to $V_{eRe}(eM)$. To prove ϕ is onto we merely observe that if P is an eRe-submodule of eM, RP belongs to $V_R(M)$ and $eRP = eReP = P$. ▮

Since $J_i{}^s M_i$ $(= (e_i J e_i)^s M)$ is contained in $J^s M$, $\bigcap_s J_i{}^s M_i = (0)$. Hence the hypotheses of Theorem 3 hold in this case and we deduce that the lattice $V_{R_i}(M_i)$ consists of the chain $\{J_i{}^s M_i \mid s = 0, 1, ...\}$ together with the zero submodule of M. On the other hand, by (6), for each $U \in V_R(M)$, $U = e_1 U + \cdots + e_q U$ and hence, since $e_i U = J_i^{s_i} M_i = J_i{}^s M$, each submodule U of M is uniquely expressible in the form

$$U = J_1^{s_1} M + \cdots + J_q^{s_q} M$$

where the sum is a direct sum of the additive groups $J_i^{s_i} M$. Denote by $(s_1, ..., s_q)$ the q-tuple of exponents occurring in the above decomposition of U and $(t_1, ..., t_q)$ the corresponding exponents in the decomposition of a submodule W. Then $(\min\{s_i, t_i\} \mid i = 1, ..., q)$ and $(\max\{s_i, t_i\} \mid i = 1, ..., q)$ are the q-tuples corresponding to $U + W$ and $U \cap W$, respectively. A word of warning: not every q-tuple $(x_1, ..., x_q)$ of nonnegative rational integers occurs as a q-tuple of exponents corresponding to some submodule of M.

Finally, we show that M is not only noetherian but also that every submodule of M is cylic: By Theorem 3 the R_i-submodule $e_i U$ of $e_i M$ is cyclic, say $e_i U = R_i u_i$. Let $u = u_1 + \cdots + u_q$. Then $R_i u_i = R_i e_i u \subseteq Ru$ for all $i = 1, ..., q$. Therefore $U = \sum_i R_i u_i \subseteq Ru \subseteq U$. Hence u generates U. We collect these results in the following:

Theorem 5. Hypotheses: Let R be a ring with identity and artinian factor ring R/J where J is the radical of R. Let the identity elements of the simple direct summands of the artinian semisimple ring R/J be liftable to an orthogonal family $\{e_1, ..., e_q\}$ of idempotents with $e_1 + \cdots + e_q = 1$. (We know then that R, considered as an additive group, is expressible as a direct sum of R^* and U:

$$R = R^* \oplus U$$

In this decomposition, U is a subgroup of the additive group of the radical J of R and R^* is the ring-direct sum of the subrings $R_i = e_i R e_i$, $i = 1, ..., q$. Moreover, the radical of R_i is $e_i J e_i$ and $R_i / e_i J e_i$ is simple artinian). Furthermore, let M be noetherian satisfying $\bigcap J^n M = (0)$ and let the lattice $V_R(M)$ of submodules of M be distributive.

Conclusions: Every R-submodule U of M is uniquely expressible as a group-direct sum $U = e_1 U + \cdots + e_q U$ of R_i-submodules $e_i U = J_i^{s_i} M$ where the s_i are either nonnegative rational integers or ∞. The lattice $V_R(M)$ is isomorphic under

$$\eta\colon J_1^{s_1} M + \cdots + J_q^{s_q} M \to (s_1, ..., s_q) \tag{8}$$

to a certain sublattice of the lattice consisting of q-tuples $(s_1, \dots, s_q)$, $s_i \in 0 \cup \mathbb{N} \cup \infty$, partially ordered by

$$(s_1, \dots, s_q) \leqslant (t_1, \dots, t_q) \Leftrightarrow s_i \geqslant t_i, \qquad i = 1, \dots, q \tag{9}$$

Each R-submodule of M is cylic. ▮

Remark: If a lattice V is distributive then by the proof of Theorem II.3.5, both distributive laws hold in V:

$$A \cap (B \cup C) = (A \cap B) \cup (A \cap C)$$

and

$$A \cup (B \cap C) = (A \cup B) \cap (A \cup C)$$

In the particular case of a distributive lattice $V_R(M)$ of submodules of an R-module M, intersection distributes over an infinite sum of submodules, i.e.

$$A \cap \sum_{\lambda \in \Lambda} B_\lambda = \sum_{\lambda \in \Lambda} (A \cap B_\lambda) \tag{$*$}$$

since, on the one hand, each module $A \cap B_\lambda$ is contained in $A \cap \sum B_\lambda$, and on the other, an element d is in $A \cap \sum_{\Lambda \in \lambda} B_\lambda$ if and only if $d \in A$ and $d = \sum_{i=1}^n b_{\lambda_i}$ for $b_{\lambda_i} \in B_{\lambda_i}$, $i = 1, \dots, n$. But for this finite set of λ_i, $A \cap \sum_{i=1}^n B_{\lambda_i} = \sum_{i=1}^n A \cap B_{\lambda_i} \subseteq \sum_{\lambda \in \Lambda} (A \cap B_\lambda)$. The dual of $(*)$

$$A + \bigcap_{\lambda \in \Lambda} B_\lambda = \bigcap_{\lambda \in \Lambda} (A + B_\lambda) \tag{D*}$$

is not valid in general in $V_R(M)$ as the following example shows: $R = M = \mathbb{Z}$, $A = (2)$ and $B_n = (3^n)$, $n \in \Lambda = \mathbb{N}$. In this case the left-hand side of (D*) is (2) while the right-hand side is $\mathbb{Z}$. The fact that under the hypotheses of Theorem 5 the lattice $V_R(M)$ is D^*-distributive is therefore worthy of note. Proof. Let $\eta(A) = (a_1, \dots, a_q)$ and $\eta B_\lambda = (b_{\lambda 1}, \dots, b_{\lambda q})$. Then

$$\eta\left(A + \bigcap B_\lambda\right) = (\dots, \min\{a_i, \max_\lambda b_{\lambda i}\}, \dots)$$

and

$$\eta \bigcap (A + B_\lambda) = (\dots, \max_\lambda \cdot \min\{a_i, b_{\lambda i}\}, \dots\cdot$$

The R_i-submodule $J_i{}^s M$ generates the R-submodule $RJ_i{}^s M$ of M. These modules are the $\cup$-irreducible elements of the lattice $V_R(M)$ in the sense of the following:

Definition 1. An element $I \neq 0$ of a lattice V is *join-irreducible* ($\cup$-irreducible) if I is not the join $A \cup B$ of two elements A and B of V each different from I.

It should be noted that whereas an R-module M is indecomposable if it is not the *direct* sum of two proper submodules A and B it is $\cup$-irreducible if it is not the sum (direct or not) of two proper submodules A and B.

Suppose $RJ_i{}^s M \neq (0)$ and $\cup$-reducible, say $RJ_i{}^s M = A + B$. Then

$$J_i{}^s M = e_i RJ_i{}^s M = e_i A + e_i B$$

where $e_i A = J_i{}^u M$ and $e_i B = J_i{}^v M$ for some u and v. This implies, since $e_i A + e_i B = J_i{}^t M$, $t = \min\{u, v\}$, that one of the modules $e_i A$ and $e_i B$ is contained in the other. Therefore one of the modules $A = Re_i A$ and $B = Re_i B$ is contained in the other, a contradiction.

On the other hand according to Theorem 5 each submodule U of M is expressible in the form

$$U = RU = RJ_1^{s_1} M + \cdots + RJ_q^{s_q} M \tag{10}$$

Therefore each element U of $V_R(M)$ is the join of a finite number of join irreducible elements $RJ_i{}^s M$. In a lattice satisfying this condition we have:

Theorem 6. The $\cup$-irreducible elements of a distributive lattice V form a partial order H contained in V. If each element of the distributive lattice V is the join of a finite number of $\cup$-irreducible elements, the lattice structure of V depends only on the partial order structure of H. ∎

To prove this we need the following:

Theorem 7. Let A be an element of the distributive lattice V and suppose

$$A = I_1 \cup \cdots \cup I_n \tag{11}$$

where the I_j are $\cup$-irreducible. (Such a representation does not always exist!) This representation is said to be *irredundant* if no I_k is contained in the join of the remaining I_j. Two such irredundant representations of an element A consist of the same I_k. Moreover if

$$B = I_1' \cup \cdots \cup I_r' \tag{12}$$

is an irredundant representation of B as the join of $\cup$-irreducible elements I_j', then $A \subseteq B$ if and only if for each I_k there exists I_k' such that $I_k \subseteq I_k'$.

Proof of Theorem 7. It is sufficient to prove the second assertion: From $I_k \subseteq B$ it follows by the distributivity of the lattice that

$$I_k = (I_k \cap I_1') \cup \cdots \cup (I_k \cap I_r')$$

The $\cup$-irreducibility of I_k implies that $I_k = I_k \cap I_l'$ for some l and hence $I_k \subseteq I_l'$. The converse is clear. ∎

Proof of Theorem 6. Since a lattice V is a special case of a partial order, V is known once the structure of V as a partial order is known. But this is known once we know the structure of H as a partial order since, using the notation of (11) and (12), $A = I_1 \cup \cdots \cup I_n \subseteq B = I_1' \cup \cdots \cup I_r'$ if and only if for each I_k there exists I_l' such that $I_k \subseteq I_l'$. ∎

This result may now be applied to the lattice $V_R(M)$ of submodules of an R-module M satisfying the conditions of Theorem 5 to obtain

Theorem 8. Under the hypotheses of Theorem 5 and with the same notation, every submodule U of M is the sum of at most q $\cup$-irreducible modules. The $\cup$-irreducible elements of the lattice $V_R(M)$ of submodules of M are the R-modules $RJ_i{}^sM$, $s = 0, 1, 2, \ldots$, provided these are not zero. The structure of the lattice $V_R(M)$ depends only on the structure of the partial order $H_R(M)$ of $\cup$-irreducible R-submodules of M. ∎

Remark: The $\cup$-irreducible elements $RJ_i{}^sM$ are uniquely determined by R and M and not, as seems to be the case, by the orthogonal family $\{e_1, \ldots, e_q\}$ of idempotents. This is better understood in the particular case when R is semiperfect since under these circumstances, by Theorem VII.1.19 there exists an inner automorphism $x \to uxu^{-1}$ which maps $\{e_1, \ldots, e_q\}$ onto $\{e_1', \ldots, e_q'\}$ and hence $RJ'^sM = R(e_i'Je_i')^s M = R(ue_iu^{-1}Jue_iu^{-1})^s M = R(e_iJe_i)^s M = RJ_i{}^sM$ since $u^{-1}M = M$ and $Ru = R$.

Definition 2. A ring R with identity and artinian factor ring R/J, having the further property that the identities of the simple direct summands of R/J are liftable to an orthogonal family $\{e_1, \ldots, e_q\}$ of idempotents in R with $e_1 + \cdots + e_q = 1$, is *distributively representable* if there exists a faithful noetherian R-module M with distributive lattice of submodules and such that $\bigcap_s J^sM = (0)$. R is then said to be *distributively represented by M*.

From $(\bigcap_s J^s)\, M \subseteq \bigcap (J^sM) = (0)$ and the faithfulness of M it follows that $\bigcap J^s = (0)$ in a distributively representable ring R. Theorems 5, 7, and 8 describe the structure of $V_R(M)$. By Theorem 5, M is a cyclic R-module, say $M = Rm_0$. Then $L = \{l \in R \mid lm_0 = 0\}$ is a left ideal of R and the mapping

$$\eta: x + L \to xm_0, \qquad x \in R \tag{13}$$

is an isomorphism of the factor module ${}_RR/L$ onto the module M. It is therefore possible to describe within R the structure of the partial order $H_R(M)$ of the $\cup$-irreducible elements $RJ_i{}^sM$ in $V_R(M)$:

Theorem 9. With the same hypothese and notation as in Theorem 5, assume that each J_i is nilpotent. Let n_i be the *exponent* of J_i, i.e., n_i is the smallest power such that $J_i^{n_i} = 0$. Then for a given pair $RJ_i{}^sM$ and $RJ_k{}^tM$ of irreducible R-submodules of the faithful R-module M,

$$RJ_i{}^sM \subseteq RJ_k{}^tM \tag{14}$$

if and only if

$$J_i^{n_i-1-s}RJ_k{}^t \neq 0 \tag{15}$$

Proof. The inclusion in (14) and the faithfulness of M imply

$$0 \neq J_i^{n_i-1}M = J_i^{n_i-1-s}RJ_i{}^sM \subseteq J_i^{n_i-s-1}RJ_k{}^t$$

and (15) is proved. Conversely, if (15) holds, then from

$$e_iRJ_k{}^tM = J_i{}^p$$

it folllows that

$$0 \neq J_i^{n_i-s+p-1}M$$

and hence $n_i - s + p - 1 < n_i$. Therefore, $p \leqslant s$ and consequently

$$J_i{}^sM \subseteq J_i{}^pM = e_iRJ_k{}^tM$$

This implies (14). ∎

If M and M' are modules distributively representing a semiperfect ring R, they need not be isomorphic. However, Theorem 9 together with Theorem VII.1.19 shows that at least their lattices of submodules are isomorphic.

Let R be an artinian ring with identity. Then R is semiperfect. Let M be a faithful noetherian R-module. By Theorem VII.1.29 it is artinian. Therefore M has finite length $l(M)$, i.e., there is a chain $M_0 = (0) \subset M_1 \subset M_l$ of $l + 1$ R-submodules of M which cannot be refined by the interpolation of further submodules. The following theorem provides a very simple criterion that M represent R distributively:

Theorem 10. Hypothesis: Let R be artinian with identity and M a faithful R-module. Let the notation be as in Theorem 5, let $n_i = \exp J_i$, $i = 1, \ldots, q$ and let $l(M)$ be the length of the module M.

Conclusions:

1. $l(M) \geqslant n_1 + \cdots + n_q$.

2. The lattice $V_R(M)$ of R-submodules of M is distributive if and only if

$$l(M) = n_1 + \cdots + n_q \quad \blacksquare \tag{16}$$

Remark: The significance of the criterion (16) is twofold: in the first place, $l(M)$ has a very simple module-theoretic meaning and, secondly, the right-hand side of (16) depends only on R and not on M.

Proof of Theorem 10. Let

$$M_0 = (0) \subset M_1 \subset \cdots \subset M_l = M \tag{17}$$

be a maximal chain of R-submodules M_i of M. Let $R/J = \bar{R}_1 \oplus \cdots \oplus \bar{R}_q$ where the $\bar{R}_i$ are simple. Each irreducible factor module M_i/M_{i-1} is a faithful $\bar{R}_{f(i)}$-module for some $\bar{R}_{f(i)}$ and is annihilated by the other simple components (cf. Theorem IV.2.3). Therefore

$$e_k(M_i/M_{i-1}) = \begin{cases} M_i/M_{i-1} & \text{if } k = f(i) \\ 0 & \text{if } k \neq f(i) \end{cases} \tag{18}$$

From (17) we derive the chain

$$e_kM_0 = (0) \subseteq e_kM_1 \subseteq \cdots \subseteq e_kM_t = e_kM \tag{19}$$

of R_k-submodules of e_kM. We show that this chain is again (except possibly for repetitions) a maximal chain of R_k-modules, i.e. if in (19) $e_kM_{i-1} \neq e_kM_i$, there is no R_k-module U' lying strictly between e_kM_{i-1} and e_kM_i: By Theorem 4 the mapping

$$\phi: U \rightarrow e_kU, \qquad U \in V(M)$$

is a lattice epimorphism of $V_R(M)$ onto $V_{e_kRe_k}(e_kM)$. Suppose there exists $U' \in V_{e_kRe_k}(e_kM)$ such that

$$e_kM_{i-1} \subset U' \subset e_kM_i$$

Then there exists $U \in V_R(M)$ such that $e_kU = U'$. It follows that since $e_k^2 = e_k$

$$e_k[(U + M_{i-1}) \cap M_i] = U'$$

and the R-module in the square brackets lies strictly between M_{i-1} and M_i, contradicting the maximality of the chain in (17). Since by (18) each module M_i/M_{i-1} is annihilated by all except one e_k, the length $l(M)$ of M as an R-module is the sum of the lengths $l_k(e_kM)$ of the R_k-modules e_kM:

$$l(M) = l_1(e_1M) + \cdots + l_q(e_qM) \tag{20}$$

The R-module M is faithful and hence so is the R_k-module e_kM. Hence in the Loewy series of the R_k-module e_kM,

$$J_k^{n_k-1}e_kM \neq (0)$$

Therefore

$$l_k(e_kM) \geqslant n_k, \qquad k = 1, \ldots, q \tag{21}$$

By (20) this implies the first assertion of the theorem. Now assume $V_R(M)$ is distributive. By Theorem 4, the lattice of R_k-submodules of e_kM is also distributive. In this case $l_k(e_kM) = n_k$ since the Loewy series of the R_k-module e_kM is a maximal chain (Theorem 3) consisting of n_k members since $J_k^{n_k-1}M \neq (0)$. By (20) it follows that $l(M) = n_1 + \cdots + n_q$. Conversely from (16), (20) and (21) it follows that

$$l_k(e_kM) = n_k, \qquad k = 1, \ldots, q$$

and by Theorem 3 the lattices $V_{R_k}(e_kM)$ consist of the members of the Loewy series of e_kM. Hence each $U = e_1U + \cdots + e_qU \in V_R(M)$ is uniquely expressible in the form

$$U = J_1^{u_1}M + \cdots + J_q^{n_q}M$$

and this implies that if

$$U = J_1^{u_1}M + \cdots + J_q^{u_q}M \qquad \text{and} \qquad V = J_1^{v_1}M + J_2^{v_2} + \cdots + J_q^{v_q}M$$

$$U + V = \sum_i J_i^{\min\{u_i, v_i\}}M \qquad \text{and} \qquad U \cap V = \sum_i J_i^{\max\{u_i, v_j\}}M$$

Consequently, $V_R(M)$ is distributive since

$$\operatorname{Max}\{u_1, \operatorname{Min}\{u_2, u_3\}\} = \operatorname{Min}\{\operatorname{Max}\{u_1, u_2\}, \operatorname{Max}\{u_1, u_3\}\} \tag{22}$$

for all integers u_1, u_2, u_3.

We now give an internal characterization based on Theorem 10 that an artinian ring with 1 be distributively representable: By Theorem 5, $M = Rm_0$ is cyclic and if L denotes the left ideal $\{l \in R \mid lm_0 = 0\}$ the mapping

$$\eta\colon x + L \to xm_0 \qquad \text{for} \quad x \in R \tag{23}$$

is an isomorphism of ${}_RR/L$ onto M. The R-module R/L is faithful if and only if $xR \subseteq L$ implies $x \in L$, i.e. if and only if the only (two-sided) ideal in L is (0). Therefore as a corollary to Theorem 10 we have:

Theorem 11. The artinian ring R with identity is distributively representable if and only if R contains a left ideal L such that

α. (0) is the only ideal of R contained in L
β. The length l of a maximal chain

$$L = L_0 \subset L_1 \subset L_2 \subset \cdots \subset L_t = R$$

of left ideals of R containing L is the sum of the $n_i = \exp J_i$ (cf. Theorem 5). ▌

Theorem 12. 1. Every primitive artinian ring is distributively representable.
2. Let R_i, $i \in I$ be a family of rings each of which possesses a faithful module with distributive lattice of submodules. Then the direct sum $\oplus \sum_i R_i$ also possesses a faithful module with distributive lattice of submodules.

Proof. 1. By definition every primitive ring possesses a faithful irreducible module M and $V_R(M)$ is trivially distributive.

2. For each $i \in I$ let M_i be a faithful R_i-module with distributive lattice of submodules. Form the direct sum M of the abelian groups M_i and define the action of $\oplus \sum R_i$ on M by

$$am = \sum_i a_i m_i \quad \text{where} \quad a = \sum a_i \in \oplus \sum R_i \quad \text{and} \quad m = \sum_i m_i \in \oplus \sum M_i \tag{24}$$

This definition makes $\oplus \sum M_i$ into a faithful distributive $\oplus \sum R_i$-module. In fact the intersection and join of two elements of $V_R(M)$ is obtained component-wise in the $V_{R_i}(M_i)$. Since for each $i \in I$, $V_{R_i}(M_i)$ is distributive by hypothesis, $V_R(M)$ is distributive. Clearly M is a faithful R-module by (24). ▌

2. Rings with distributive lattice of left ideals

The results of the preceding section may be applied to the particular case when ${}_R R$ itself represents R distributively. To this end we need the hypotheses of Theorem 1.5 which in this particular case read as follows:

Hypothesis 1: Let R be a left noetherian ring with identity and with artinian factor ring by its radical. Let the identity elements of the simple direct summands of the semisimple artinian ring R/J be liftable to an orthogonal family $\{e_1, ..., e_q\}$ of idempotents such that $e_1 + \cdots + e_q = 1$. R is then expressible as a group-direct sum

$$R = (R_1 \oplus \cdots \oplus R_q) + U$$

where the expression in brackets is the ring-direct sum of the rings $R_i = e_iRe_i$ and U is a subgroup of the additive group of the radical J of R. Moreover, the radical of R_i is e_iJe_i and R_i/e_iJe_i is simple artinian.

Hypothesis 2:

$$\bigcap_n J^n = (0)$$

By the distributivity of the lattice $V({}_RR)$ it follows from Theorem IX.1.3 that the R_i-submodules of the left R_i-module e_iR are precisely the members $J_i{}^aR$, $a = 0, 1, 2, \ldots$ of the Loewy series of e_iR, $i = 1, \ldots, q$. In particular, e_iR/J_iR is an irreducible left R_i/J_i-module. The mapping

$$\chi_i: x + J_iR \to xe_i + J_i\,, \qquad x \in e_iR$$

is an R_i-epimorphism of e_iR/J_iR onto the left R_i-module R_i/J_i. Now $\chi_i(e_i) = e_i + J_i \neq \bar{0}$. Therefore the irreducibility of e_iR/J_iR as a left R_i-module implies that χ_i is an R_i-module isomorphism. Hence R_i/J_i is irreducible as a left R_i/J_i-module. The simple artinian ring R_i/J_i therefore contains no left ideals other than $(\bar{0})$ and R_i/J_i. By Theorem II.1.1 it follows that R_i/J_i is a division ring.

The distributivity of $V({}_RR)$ therefore implies that the rings R_i are local, i.e. R_i/J_i is a division ring (cf. Definition VII.1.4).

The factor module e_iR/J_iR is generated as an R_i-module by $e_i + J_iR$. Therefore since $R_i = e_iRe_i$,

$$e_iR = R_i + J_iR \tag{1}$$

in the lattice $V(e_iR)$ of left R_i-submodules. Now on the one hand

$$e_iR = e_iRe_i + \sum_{j \neq i} e_iRe_j \tag{2}$$

and on the other

$$R_i + J_iR = e_iRe_i + \sum_{j \neq i} J_iRe_j \tag{3}$$

where the right-hand sides of (2) and (3) are group-direct sums. By (1) and since $e_je_k = \delta_{jk}e_j$,

$$e_jRe_j = J_iRe_j \qquad \text{for} \quad i \neq j$$

and hence

$$e_iRe_j = J_iRe_j = J_i^2Re_j = \cdots = J_i^{n_i}Re_j = 0$$

whenever $i \neq j$

This implies that

$$R = R_1 \oplus \cdots \oplus R_q \tag{4}$$

and thus R is a ring direct sum of the local rings R_i. The Loewy series of the left R_i-module e_iR consists of the $J_i^aR = J_i^a$, $a = 0, 1, \ldots$. These two-sided ideals of R_i are also two-sided ideals of R because of (4). By Theorem IX.1.5 every left ideal of R is therefore a direct sum

$$J_1^{a_1} + \cdots + J_q^{a_q} \tag{5}$$

of the (two-sided) ideals $J_i^{a_i}$ in R. Moreover every q-tuple $(a_1, \ldots, a_q)$, $a_i = 0, 1, 2, \ldots$ and $1 \leqslant i \leqslant q$ occurs as a q-tuple of exponents of some left ideal of R. The structure of $V_R({}_RR)$ has therefore been explicitly determined.

Definition 1. A local ring R is *uniserial* if the only left ideals of R are the various powers of its radical.

We state the results of our investigations above in terms of this definition:

Theorem 1. Let R be a ring satisfying Hypotheses 1 and 2 set out at the beginning of this section. Then the lattice $V({}_RR)$ of left ideals of R is distributive if and only if R is the ring-direct sum of a finite number of local uniserial rings. By (5) it is clear that the condition is sufficient.

This theorem is valid in particular when R is commutative. In this case Hypothesis 2 is a consequence of Hypothesis 1 as will be shown in Theorem 18 (Krull) of Chapter X. ▮

Finally we may further specialize Theorem 1 to the case when R itself is a local noetherian ring. In this case there is a connection with Dedekindian rings, i.e. with commutative Asano orders in a field which we introduced in Definition VII.2.3 and discussed further in Section VII.2:

Theorem 2. Let R be a local commutative ring. Then

1. R is noetherian with distributive lattice of ideals if and only if the lattice of ideals consists of powers of the maximal ideal of R.
2. R is noetherian with distributive lattice of ideals but not artinian if and only if R is a Dedekind domain with radical different from zero.

Proof. The radical J of R is the intersection of all modular maximal left ideals in R. In the case R is a commutative ring with 1, this is the intersection of all maximal ideals of R. This shows that J is the unique maximal ideal of R. The first assertion now follows from Theorem 1 together with Theorem XI. If R is artinian, its radical is nilpotent and conversely, since by Part 1 the powers of J are the only ideals in the commutative ring R provided R is noetherian with distributive lattice of ideals. We must now show that R has no divisors of zero. Assume $a \neq 0$ is a zero divisor. Then the right annihilator $r(a)$ is a nonzero ideal in R and hence $r(a) = J^\alpha$. Its left annihilator $l(r(a))$ contains a and is again a power of J, say J^β. Then $(0) = J^{\alpha+\beta}$ and J is nilpotent proving that R is artinian. ∎

Theorem 2 may be used to characterize the arithmetic rings in the class of commutative noetherian rings with identity 1. To obtain this characterization we apply the method of localization which is often used in commutative algebra and which we now discuss. Let M be a maximal ideal in the commutative ring R with identity 1. Then the complement $R\backslash M$ of M is multiplicatively closed since the maximality of M implies that M is prime. If $R\backslash M$ contains no divisors of zero we may define an equivalence relation, as in Theorem V.1.4, on the set of pairs $(x, c) \in R \times (R\backslash M)$ by

$$(x_1, x_1) \sim (x_2, c_2) \Leftrightarrow c_2 x_1 = c_1 x_2$$

We denote the class to which (x, c) belongs by x/c. The set of equivalence classes form a ring under the operations

$$\frac{x}{c} \cdot \frac{y}{d} = \frac{xy}{cd} \qquad \text{and} \qquad \frac{x}{c} + \frac{y}{d} = \frac{dx + cy}{cd}$$

since $c \notin M$ and $d \notin M$ imply $cd \notin M$. This ring is called the *ring of fractions of R with respect to $R\backslash M$* and is denoted by R_M. It contains the original ring R since $x \to x/1$ is an embedding of R in R_M. Moreover, $c^{-1}x = x/c$. These assertions may be easily verified. If however $R\backslash M$ contains zero divisors of R, we may get rid of them as follows: The set

$$N = \{n \in R \mid \text{there exists } c \in R\backslash M \text{ with } c \cdot n = 0\}$$

is an ideal in R since $c_1 n_1 = c_2 n_2 = 0$ implies $c_1 c_2 (n_1 - n_2) = 0$. Denote the natural homomorphism of R onto $\bar{R} = R/N$ by

$$\pi: x \to \bar{x} = x + N$$

Then $\pi(R\backslash M)$ is also multiplicatively closed and $\bar{M}$ is a maximal ideal of $\bar{R}$ since otherwise it would follow from $1 = m + n$ and $cn = 0$ that $c \in M \cap R\backslash M$. Moreover, $\bar{R}\backslash\bar{M}$ has no divisors of zero. For if $\bar{x} \in \bar{M}$ and $\bar{x}\bar{y} = \bar{0}$ then $xy \in N$ and $x \notin M$. Therefore there exists $c \in R\backslash M$ such that $cxy = 0$. Since $cx \in R\backslash M$ we conclude that $y \in N$ and hence $\bar{y} = \bar{0}$. Therefore as above we may form the ring of fractions of $\bar{R}$ with respect to $\bar{R}\backslash\bar{M}$. Denote this also by R_M. Its elements may be represented in the form

$$\frac{\bar{x}}{\bar{c}} = \frac{\pi x}{\pi c} \qquad \text{where} \quad \bar{c} = \pi c \in \bar{R}/\bar{M}$$

Furthermore, the ideals of R may be mapped onto the ideals of R_M :

Theorem 3. Let M be a maximal ideal of the commutative ring R with identity 1 and let N be the ideal consisting of those elements n of R for which there exists an element $c \in R \backslash M$ such that $cn = 0$. Let $\pi: x \to \bar{x} = x + N$ denote the natural homomorphism of R onto $\bar{R} = R/N$ and let $R' = R_M$ be the ring of fractions of $\bar{R}$ with respect to $\bar{R} \backslash \bar{M}$. Denote the lattice ordered semigroup of ideals of R (resp. R') under multiplication by $V(R)$ [resp. $V(R')$]. Then

$$\phi: A \to R'\bar{A} \qquad \text{for} \quad A \in V(R) \tag{6}$$

is an epimorphism of $V(R)$ onto $V(R')$. In particular R' is arithmetic (resp. noetherian) if R is arithmetic (resp. noetherian). R' is a local ring with $R'\bar{M}$ as unique maximal ideal. (For the definition of arithmetic ring see Section 3, Definition 1.)

Proof. 1. Since A is an ideal, the elements

$$\bar{c}_1^{-1}\bar{x}_1\bar{a}_1 + \cdots + \bar{c}_n^{-1}\bar{x}_n\bar{a}_n$$

of $R'\bar{A}$ are expressible in the form $\bar{c}^{-1}\bar{a}$ with $a \in A$ and $c = c_1c_2 \cdots c_n$. Clearly $R'(\bar{A} + \bar{B}) = R'\bar{A} + R'\bar{B}$ and $R'(\bar{A} \cdot \bar{B}) = (R'\bar{A})(R'\bar{B})$ and hence ϕ is both additive and multiplicative. Obviously $R'(\bar{A} \cap \bar{B}) \subseteq R'\bar{A} \cap R'\bar{B}$. Conversely, let $\alpha = \bar{a}/\bar{c} = \bar{b}/\bar{d}$ be an element of $R'\bar{A} \cap R'\bar{B}$. From $\alpha = \overline{da}/\overline{cd} = \overline{cb}/\overline{cd}$ with $da \in A$ and $cb \in B$, it follows that $\alpha \in R'(\bar{A} \cap \bar{B})$. Therefore $\phi(A \cap B) = \phi A \cap \phi B$ and ϕ is a lattice ordered semigroup homomorphism of $V(R)$ into $V(R')$.

2. Let $A' \in V(R')$ and set $A = \pi^{-1}(A' \cap \bar{R})$. Then $\phi A \subseteq A'$. Conversely, if $\bar{c}^{-1}\bar{x} \in A'$ where $c \in R \backslash M$, $x \in R$, then $\bar{x} = \bar{c}\bar{c}^{-1}\bar{x} \in \bar{R}$ and hence $x \in \pi^{-1}(A' \cap \bar{R}) = A$. Therefore $\bar{c}^{-1}\bar{x} \in R'A$. This proves $\phi A = A'$.

3. Since by Part 2 ϕ is an epimorphism, given ideals A', B', and C' of R', there exist ideals A, B, and C of R which map under ϕ onto A', B', and C', respectively. Then if R is arithmetic,

$$\begin{aligned} A' \cap (B' + C') &= \phi A \cap (\phi B + \phi C) = \phi(A \cap (B + C)) \\ &= \phi((A \cap B) + (A \cap C)) = (\phi A \cap \phi B) + (\phi A \cap \phi C) \\ &= (A' \cap B') + (A' \cap C') \end{aligned}$$

Hence R' is also arithmetic.

4. $M = R'\bar{M}$ is an ideal of R'. Let $\delta = \bar{c}^{-1}\bar{d} \in R' \backslash R'\bar{M}$. Then $\bar{d}$ is not in $\bar{M}$ and therefore the ideal in R' generated by δ contains the identity of R'. This proves that $R'\bar{M}$ is the maximum of all ideals of R' different from R'. ∎

In the next chapter we shall show that every ideal of a commutative noetherian ring with identity is representable as the intersection of a finite number of primary ideals. By primary we mean an ideal Q with the property that if $ab \in Q$ and $a \notin Q$ then

$$b \in r(Q) = \{x \in R \mid x^n \in Q \text{ for some } n \in \mathbb{N}\}.$$

[$r(Q)$ is called the primary radical of Q]. The mapping ϕ in Theorem 3 is not in general a monomorphism. However, its restriction to those primary ideals of R contained in M is one-one, i.e. we have:

Theorem 4. The restriction of the mapping ϕ of Theorem 3 to the primary ideals Q of the noetherian ring R contained in M is one-one.

Proof. Let $Q \subseteq M$, Q primary. Since R is noetherian some power of $r(Q)$ is contained in Q and hence in M. But M is prime. Therefore $r(Q) \subseteq M$. Now let N be the kernel of the homomorphism π of Theorem 3 and let $n \in N$. Then there exists $c \in R \backslash M$ such that $cn = 0 \in Q$. But since $r(Q) \subseteq M$, $c \notin r(Q)$. Therefore $n \in Q$ and hence $N \subseteq Q$. We must therefore show that $(\phi Q) \cap \bar{R} = \bar{Q}$. Clearly $\bar{Q} \subseteq (\phi Q) \cap \bar{R}$. Conversely, let $\bar{x} = \bar{c}^{-1}\bar{q} \in (\phi Q) \cap \bar{R}$ where $c \in R \backslash M$ and $q \in Q$. Then $xc = q \in Q$ but since $r(Q) \subseteq M$, $c \notin r(Q)$. It follows that $x \in Q$ since Q is primary by hypothesis. ▮

We are now ready to apply Theorem 2. Let Q be a primary ideal of R and set $P = r(Q)$. Then $P \neq R$ since $1 \notin Q$. Therefore there exists a maximal ideal M containing P. Again denote the ring of fractions R_M by R'. By Theorem 4

$$\phi Q \in V(R') \qquad \text{and} \qquad \bar{Q} = (\phi Q) \cap \bar{R}$$

Now let R be arithmetic and noetherian. Then by Theorem 3 the local ring R' is also arithmetic and noetherian and by Theorem 2 every ideal of R' different from zero is a power of the maximal ideal $R'\bar{M}$ of R'. In particular

$$\phi(Q) = (R'\bar{M})^\alpha \qquad \text{for a suitable exponent } \alpha$$

Since ϕ is multiplicative,

$$\phi Q = (\phi M)^\alpha = \phi(M^\alpha)$$

By Theorem 4 we may conclude from this that $Q = M^\alpha$ once we have proved that M^α is primary: $r(M^\alpha)$ consists of those elements x of R for which $x^n \in M^\alpha$ for a suitable positive exponent n. Since M is prime, this implies $r(M^\alpha) = M$. Hence in a normal representation (cf. Definition X.7)

$$M^\alpha = Q_1 \cap \cdots \cap Q_r$$

of M^α, the primary radicals $r(Q_i)$ of the primary ideals Q_i are each equal to M since M is maximal and $r(Q_i) \subseteq M$. But by Theorem X.8 the intersection of primary ideals with the same radical is again a primary ideal. Hence M^α is primary and by our remarks above $M^\alpha = Q$.

Theorem 5. Let R be a commutative, noetherian, and arithmetic ring with identity. Let Q be a primary ideal of R. Then $Q = M^\alpha$ where M is some maximal ideal of R and α a suitable exponent. ▮

By Theorems X.11 and X.15 each ideal $A \neq R$ in a commutative noetherian ring with identity has a representation as the intersection of a finite number of primary ideals:

$$A = Q_1 \cap \cdots \cap Q_r, \qquad Q_i \text{ primary}$$

If R is, in addition, arithmetic, each ideal Q_i is the power of some maximal ideal in R, in other words we have

$$A = M_1^{\alpha_1} \cap M_2^{\alpha_2} \cap \cdots \cap M_r^{\alpha_r}, \qquad M_i \text{ maximal} \tag{7}$$

In this case we may even assert that A is the product of these ideals because of the following:

Theorem 6. Let R be a commutative ring with identity. A pair of ideals (U, W) of R is said to be *coprime* if $U + W = R$. If (U, W) is a coprime pair then $U \cap W = U \cdot W$. Moreover, if (U, W_1) and (U, W_2) are coprime pairs, then so is (U, W_1W_2).

Proof. 1.

$$\begin{aligned} U \cap W &= (U + W) \cdot (U \cap W) = U(U \cap W) + W(U \cap W) \\ &\subseteq UW + WU = UW \subseteq U \cap W \end{aligned}$$

2. From $U + W_i = R$, $i = 1, 2$, it follows that

$$R = (U + W_1)(U + W_2) \subseteq U + W_1W_2 \subseteq R \quad ▮$$

Since the M_i in the representation (7) are maximal ideals they are pairwise coprime. Thus by the second assertion of Theorem 6, the powers of these maximal ideals are also pairwise coprime. Hence by induction on r we may prove that A has a representation as a product of powers of maximal ideals provided $A \neq (0)$ and $A \neq R$:

$$A = M_1^{\alpha_1} M_2^{\alpha_2} \cdots M_r^{\alpha_r} \tag{8}$$

The representation (8) is uniquely determined by the ideal A.

Proof. Let

$$M_1^{\alpha_1} \cdots M_r^{\alpha_r} = U_1^{\beta_1} \cdots U_s^{\beta_s} \neq (0), \qquad M_i, U_j \text{ maximal} \tag{9}$$

Since all M_i and U_j are prime, the same maximal ideals occur on both sides of (9). By renumbering we may assume $M_i = U_i$, $i = 1, ..., r = s$. Now the intersection of $R \backslash M_i$ and M_j is nonempty if $i \neq j$. Therefore under the mapping $\phi = \phi_i$ of Theorem 3 for $M = M_i$, the image $\phi M_j = R_i' = R_M$. for $i \neq j$. Hence applying ϕ_i to both sides of (9) we get

$$(R_i'\overline{M}_i)^{\alpha_i} = \phi_i\left(\prod M_k^{\alpha_k}\right) = \phi_i\left(\prod U_k^{\beta_k}\right) = (R_i'\overline{M}_i)^{\beta_i}$$

since ϕ_i is multiplicative. Since the ideal in (9) is different from (0), it follows that $\alpha_i = \beta_i$.

This proves the implication, Part 1 $\Rightarrow$ Part 2 of the following:

Theorem 7. Let R be a commutative noetherian ring with identity. Then the following are equivalent:

1. R is arithmetic
2. Every ideal different from (0) is uniquely representable as the product of maximal ideals of R.

Proof. Assume Part 2. Then $V(R)$ is distributive since

$$\prod M_i^{\alpha_i} \cap \prod M_i^{\beta_i} = \prod M_i^{\max\{\alpha_i, \beta_i\}} \text{ and } \prod M_i^{\alpha_i} + \prod M_i^{\beta_i} = \prod M_i^{\min\{\alpha_i, \beta_i\}} \quad \blacksquare$$

As a corollary to this we have the following special case:

Theorem 8. A noetherian integral domain is arithmetic if and only if it is a Dedekind domain. ∎

3. Arithmetic rings

Let us first recall once again the definition of an arithmetic ring:

Definition 1. (Fuchs) A ring R is said to be *arithmetic* if its lattice $V(R)$ of (two-sided) ideals is distributive.

In Theorem IX.2.7 of the last section we characterized commutative noetherian and arithmetic rings with identity. The results obtained in Section 1 on distributively representable rings may also be applied to non-commutative arithmetic rings provided these possess a multiplication ring T

with the properties set down in Section 1. Section 3 is independent of Section 2 and in particular we shall not need the results of commutative algebra which were used in Section 2.

We first characterize arithmetic rings as those which satisfy the conclusion of the Chinese remainder theorem:

Theorem 1 (Chinese remainder theorem). Hypothesis: Let R be the ring of integers.

Conclusion: Let $\{A_1, \ldots, A_r\}$ be a finite set of ideals of R and let

$$x \equiv x_i (\mathrm{mod}\ A_i), \qquad i = 1, \ldots, r, \quad x_i \in R \tag{1}$$

be a system of congruences for $x \in R$. The system (1) has a solution if and only if

$$x_i \equiv x_j \ \mathrm{mod}(A_i + A_j), \qquad i \neq j \ \blacksquare \tag{2}$$

The necessity of condition (2) for an arbitrary ring follows from $x_i - x_j = (x_i - x) + (x - x_j) \in A_i + A_j$. That the condition is also sufficient for arithmetic rings is a consequence of the following theorem (cf. Zariski and Samuel [1958], Theorem V.18):

Theorem 2. A (not necessarily commutative) ring R is arithmetic if and only if the conclusion of the Chinese remainder theorem holds.

Proof. 1. We first recall that we showed the equivalence of III.3(a) and III.3(11), in the proof of Theorem III.3.5 i.e. the equivalence of

$$\mathrm{D}_1. \quad A \cap (B \cup C) = (A \cap B) \cup (A \cap C)$$

and

$$\mathrm{D}_2. \quad A \cup (B \cap C) = (A \cup B) \cap (A \cup C)$$

in an arbitrary lattice.

2. For $r = 2$, i.e. for the case of two congruences, Theorem 1 is valid in any ring since the existence of $a_i \in A_i$ with $x_1 - x_2 = a_1 - a_2$ follows from $x_1 \equiv x_2 \ \mathrm{mod}(A_1 + A_2)$ and hence $x = x_1 - a_1 = x_2 - a_2$ solves the two congruences. We have already shown the necessity of the condition of Theorem 1 for arbitrary rings.

3. To prove that in an arithmetic ring condition (2) is sufficient to ensure the existence of a solution of (1) we use induction on r. Assume inductively that there exists a solution x' of the $r - 1$ congruences,

$$x \equiv x_i (\mathrm{mod}\ A_i), \qquad i = 1, \ldots, r - 1 \tag{3}$$

The system (1) has a solution if and only if the system

$$\begin{aligned} x &\equiv x^i \bmod A_1 \cap \cdots \cap A_{r-1} \\ x &\equiv x_r \bmod A_r \end{aligned} \tag{$*$}$$

has a solution. Now $x' \equiv x_i \bmod A_i$, $i = 1, \ldots, r-1$ by induction and $x_i \equiv x_r \bmod(A_i + A_r)$ by hypothesis. Therefore $x' \equiv x_i \bmod(A_i + A_r) \equiv x_r \bmod(A_i + A_r)$ and hence

$$x' \equiv x_r \bmod(A_i + A_r), \qquad i = 1, \ldots, r-1$$

It follows that

$$x' \equiv x_r \bmod(A_1 + A_r) \cap (A_2 + A_r) \cap \cdots \cap (A_{r-1} + A_r) \tag{4}$$

By (D_2), $(A_1 + A_r) \cap \cdots \cap (A_{r-1} + A_r) = (A_1 \cap \cdots \cap A_{r-1}) + A_r$. Hence $x' \equiv x_r \bmod((A_1 \cap \cdots \cap A_{r-1}) + A_r)$ and by Part 2 above the system (*) has a solution.

4. Assume now that the Chinese remainder theorem is valid in R. We show that $V(R)$ is distributive as follows:

Clearly the left-hand side of (D_2) is contained in the right-hand side. Conversely, let $d \in (A + B) \cap (A + C)$. Then

$$d = a_1 + b = a_2 + c \qquad \text{where} \quad a_1, a_2 \in A, \quad b \in B, \quad c \in C$$

We try to express d in the form $d = x + y$ with $x \in A$ and $y \in B \cap C$, i.e. we try to find a solution to the congruences

$$x \equiv 0 \bmod A, \qquad x \equiv d \bmod B, \qquad x \equiv d \bmod C \tag{5}$$

The conditions in (2) are satisfied since $0 \equiv d \bmod(A + B), 0 \equiv d \bmod(A + C)$ and $d \equiv d \bmod(B + C)$. Let x be a solution of (5) and set $y = d - x$. Then $y \in B \cap C$ and $d = x + y \in A + (B \cap C)$. ▮

To apply the theory of distributively representable rings to arithmetic rings R with identity we observe the key fact that the multiplication algebra T generated by the left multiplications

$$\mathfrak{L}_a\colon x \to ax, \qquad x \in R$$

and the right multiplications

$$\mathfrak{R}_a\colon x \to xa, \qquad x \in R$$

is distributively represented by the faithful T-module R since the ideals of R are precisely the T-submodules of R.

In order to apply Theorem IX.1.5 on distributively representable rings to the T-module R we must place enough restrictions on R so as to ensure that T satisfies the hypotheses of the theorem. Now we obtain a decomposition

$$T = (T_1 \oplus T_2 \cdots \oplus T_n) \oplus U$$

from a corresponding decomposition of R provided the factor ring of R by its radical J is artinian and the identity elements of the simple direct summands $\bar{R}_i$, $i = 1, ..., q$ of R/J are liftable to an orthogonal family of idempotents $\{e_1, ..., e_q\}$ of R with sum equal to 1. Moreover we must impose on R a chain condition strong enough to ensure that T mod its radical $J(T)$ is artinian. To this end we assume that R is a finitely generated Z-module where Z is a local noetherian ring contained in the center of R. [We shall then be able to show by Theorem VII.1.30 that $T/J(T)$ is a finite-dimensional algebra over the field $\Phi = Z/J(Z)$.] Furthermore, we must have some connection between the radical J of R and $J(T)$ so that the structure Theorem IX.1.5, applied to the ring T and the T-module R, will give us some meaningful information about the structure of R. We therefore assume that for each pair $(\bar{R}_i, \bar{R}_k)$ of simple direct summands of $\bar{R} = \bar{R}_1 \oplus \cdots \oplus \bar{R}_q$, the tensor product $\bar{R}_i \otimes_\Phi \bar{R}_k{}^*$ is again simple, where $\bar{R}_k{}^*$ is the opposite ring of $\bar{R}_k$, i.e., $\bar{R}_k{}^*$ is the ring with the same additive structure as $\bar{R}_k$ but with multiplication defined by $a * b = ba$. In particular, if the rings $\bar{R}_i$ have Φ as common center, by Theorem VI.I.5, the hypothesis is fulfilled. Finally we also need $\bigcap J^n = (0)$.

Under these hypotheses on R, R is a noetherian Z-module by Theorem VII.1.30. Since each ideal of R is a Z-module, the lattice of ideals of R and hence the lattice of T-submodules of R, is noetherian.

Theorem 3. Let R be a ring with identity and Z a subring contained in the center of R with $1 \in Z$. Let J' be its radical. Let R be a finitely generated Z-module and J its radical. Then

1. $J' \subseteq Z \cap J$.
2. $J' = Z \cap J$ if Z is the center of R.

Proof. Assume $J'R \not\subseteq J$. Then there exists a maximal left ideal M_0 in R such that $J'R \not\subseteq M_0$. Therefore $R = J'R + M_0$. Let $R = Za_1 \oplus \cdots \oplus Za_N$. It follows that $R = J'a_1 \oplus \cdots \oplus J'a_N + M_0$. Hence there exist elements z_i in J' and $m_0 \in M_0$ such that $(1 - z_1)\, a_1 = z_2a_2 + \cdots + z_Na_N + m_0$. Let z_1' be the quasi-inverse of z_1. Multiplying both sides of the foregoing equality by $(1 - z_1')$ we get that a_1 is contained in $Za_2 + \cdots + Za_N + M_0$ and hence $R = J'a_2 + \cdots + J'a_N + M_0$. Proceeding in the same fashion we finally obtain $R = M_0$ contradicting the maximality of M_0. 2. Let Z be the center of R and let $z \in Z \cap J$. Let $(1 - z')(1 - z) = 1$. We must show that z' is in Z. Let $a \in R$. Then $(a(1 - z') - (1 - z')a)(1 - z) = a(1 - z')(1 - z) - (1 - z')(1 - z)a = a - a = 0$. ∎

To establish the relationship between the radical $J(R)$ of R and the radical $J(T)$ of its multiplication ring T, let

$$\mathfrak{L}_A = \{\mathfrak{L}_a \mid a \in A\} \qquad \text{and} \qquad \mathfrak{R}_A = \{\mathfrak{R}_a \mid a \in A\}$$

where A is an ideal of R. Since multiplication is associative in R

$$\mathfrak{L}_x \mathfrak{R}_y = \mathfrak{R}_y \mathfrak{L}_x \tag{6}$$

and consequently the subgroup

$$W = \mathfrak{L}_J \mathfrak{R}_R + \mathfrak{L}_R \mathfrak{R}_J \tag{7}$$

of Hom(R, R) is an ideal of T. We shall show that W is the radical of T. To this end we remark first that the R-ideal $J'R$ generated by the radical J' of Z is contained in the radical J of R by Theorem 3. Since by hypothesis R is a finitely generated Z-module, the radical ring $J/J'R$ is a finite-dimensional algebra over the field $\Phi = Z/J'$ and hence is nilpotent. But then by (6), $W/J'T$ is also nilpotent. On the other hand by Theorem 3 the ideal $J'T$ is contained in the radical $J(T)$. Therefore

$$W \subseteq J(T) \tag{8}$$

Let $\bar{e}_i$ be the identity of the simple direct summand $\bar{R}_i$ of R/J. By hypothesis $\{\bar{e}_1, ..., \bar{e}_q\}$ may be lifted to an orthogonal family of idempotents $\{e_1, ..., e_q\}$ in R with $e_1 + \cdots + e_q = 1$. In T the elements

$$E_{ik} = \mathfrak{L}_{e_i} \mathfrak{R}_{e_k}, \qquad i, k = 1, ..., q \tag{9}$$

form a family of orthogonal idempotents and the ring T may be decomposed as a direct sum of the additive subgroups $\sum_{ik} \oplus T_{ik}$ and U:

$$T = \sum_{i,k} \oplus T_{ik} + U \tag{10}$$

where $\sum_{i,k} \oplus T_{ik}$ is a ring direct sum of the subrings $T_{ik} = E_{ik}TE_{ik}$ and $U = \sum_{(i,k) \neq (j,l)} E_{ik}TE_{jl}$ is an additive subgroup of the radical $J(T)$.

The factor ring of T_{ik} by its ideal

$$W_{ik} = E_{ik}WE_{ik} = \mathfrak{L}_{J_i} \mathfrak{R}_{R_k} + \mathfrak{L}_{R_i} \mathfrak{R}_{J_k} \tag{11}$$

is artinian since $J'T \subseteq J(T)$. In general however, T_{ik}/W_{ik}, $i, k = 1, ..., q$ is not simple. The hypothesis that for each pair $(\bar{R}_2, \bar{R}_k)$, $\bar{R}_i \otimes_\Phi \bar{R}_k^*$ is simple guarantees the simplicity of T_{ik}/W_{ik} by the following arguments:

For $x \in R_i$ denote the coset $x + W_i$ by $\bar{x}$ and for $y \in R_k$ let $\bar{y}$ denote $y + W_k$. Then by Theorem V.3.2 the mapping

$$\eta: \sum \bar{x} \otimes \bar{\bar{y}} \rightarrow \mathfrak{L}_x \mathfrak{R}_y + W_{ik} \tag{12}$$

is an Φ-algebra epimorphism of the tensor product $R_i/W_i \otimes_\Phi (R_k/W_k)^*$ onto the factor ring T_{ik}/W_{ik} since for all $c \in F_i$

$$\mathfrak{L}_x{}^\alpha \mathfrak{R}_y \equiv \mathfrak{L}_x \mathfrak{R}^\alpha{}_y (\text{mod } W_{ik}) \tag{13}$$

By the simplicity of the tensor product $R_i/W_i \otimes_\Phi (R_k/W_k)^*$ it follows that the epimorphism η is either an isomorphism or the zero homomorphism. Hence if $E_{ik} \neq 0$, T_{ik}/W_{ik} is a simple artinian ring. Therefore the ring T/W, which is isomorphic to $\sum_{i,k} \otimes T_{ik}/W_{ik}$, is either semisimple artinian or 0. This proves that W contains the radical $J(T)$ of T and thus together with (8), we have

$$W = J(T)$$

Finally, the last hypothesis of Theorem 1.5, namely $\bigcap W^n R = (0)$, holds for the ring T and its module R since

$$W^n R = (\mathfrak{L}_J \mathfrak{R}_R + \mathfrak{L}_R \mathfrak{R}_J)^n R = J^n$$

The first two assertions of the following theorem are therefore a consequence of Theorem 1.5:

Theorem 4. Hypotheses: Let R be an arithmetic ring with identity 1 and artinian factor ring by its radical J. Let Z be a local noetherian subring contained in the center of R and let R be a finitely generated Z-module. Let the identity elements of the simple direct summands R_i in $R/J = \bar{R}_i \oplus \cdots \oplus \bar{R}_q$ be liftable to an orthogonal family of idempotents of R with sum 1. For each pair $(\bar{R}_i, \bar{R}_k)$ let $\bar{R}_i \otimes_\Phi \bar{R}_k^*$ be simple, where Φ is the field $Z/J(Z)$.

Notation: Let the mutually orthogonal idempotents $e_1, \ldots, e_q$ be the identity elements modulo J of the rings $\bar{R}_i$, $i = 1, \ldots, q$. Set $R_i = e_i R e_i$ and $J_i = e_i J e_i = J(R_i)$. Then we know that R is the direct sum as a group of the additive subgroups $R_1 \oplus \cdots \oplus R_q$ and U:

$$R = (R_1 \oplus \cdots \oplus R_q) + U \tag{14}$$

where $R_1 \oplus \cdots \oplus R_q$ is a ring-direct sum of the rings R_i and U is an additive subgroup of the radical J of R.

Conclusions: 1. The radical W of the multiplication ring T of R is $W = \mathfrak{L}_J \mathfrak{R}_R + \mathfrak{L}_R \mathfrak{R}_J$ and T/W is artinian.

2. Each ideal A of R is uniquely expressible as a group-direct sum

$$A = \sum_{1 \leqslant i,k \leqslant q} e_i A e_k \tag{15}$$

of the (R_i, R_k)-bimodules $e_i A e_k$. Moreover, $e_i A e_k = J_i{}^s R J_k{}^t$, where s and t are nonnegative integers or ∞ (naturally, we set $J_i{}^\infty = (0)$). For a given fixed pair (i, k), the set $J_i{}^s R J_k{}^t$ for varying s and t forms a chain $\mathcal{C}_{ik}$ under set-theoretic inclusion.

3. The $\cup$-irreducible elements of the lattice $V(R)$ of ideals of R are of the form

$$R J_i{}^s R J_k{}^t R \tag{16}$$

Under multiplication of ideals, this set of ideals of R together with the zero ideal forms a semigroup γ with the property that the partially ordered semigroup $V(\gamma)$ of semigroup ideals of γ, partially ordered by inclusion, is isomorphic to the partially ordered semigroup $V(R)$ of ideals of R. ∎

The first assertion of Part 3, namely that γ is a semigroup, follows from Theorem 1.8 and the fact that the product $RJ_i{}^s RJ_k{}^t R \cdot RJ_j{}^u RJ_l{}^v R$ is obtained by applying T to the (R_i, R_l)-bimodule $J_i{}^s RJ_k{}^t RJ_j{}^u RJ_l{}^v$ contained in the chain $\mathcal{C}_{il}$, or, in other words, by multiplying $J_i{}^s RJ_k{}^t RJ_j{}^u RJ_l{}^v$ on the right and on the left by R. The isomorphism between $V(\gamma)$ and $V(R)$ follows from the fact that in a distributive lattice each element has at most one irredundant representation by $\cup$-irreducible elements (Theorem 1.6).

If we drop the hypothesis that $R_i \otimes R_k$ be simple, Theorem 4 is no longer valid, as the following counterexample shows: Let α be a rational number with $\sqrt{\alpha} \notin \mathbb{Q}$. Set $R_i = \mathbb{Q}e_i + \mathbb{Q}b_i$, $b_i{}^2 = \alpha e_i$, $i = 1, 2$, $R_1 \approx R_2 \approx \mathbb{Q}(\sqrt{\alpha})$. The tensor product $R_1 \otimes_{\mathbb{Q}} R_2 = P$ contains the orthogonal, central idempotents $F_1 = (1/2)(e_1 \otimes e_2) + (1/2\alpha)(b_1 \otimes b_2)$ and $F_2 = (e_1 \otimes e_2) - F_1$. P is therefore the direct sum of the rings F_1P, and F_2P, each of which is isomorphic to $\mathbb{Q}(\sqrt{\alpha})$. Set $R = R_1 \oplus R_2 + J$ where $J^2 = (0)$, $e_2Re_1 = 0$, and $J = e_1Re_2 = F_1J \oplus F_2J$. In the direct sum decomposition of J, $F_1J = \mathbb{Q}c_1 + \mathbb{Q}c_2$ and $F_2J = \mathbb{Q}d_1 + \mathbb{Q}d_2$. Multiplication of nontrivial pairs is defined by

$$\begin{array}{llll} b_1c_1 = c_2 & c_1b_2 = c_2 & b_1d_1 = -d_2 & d_1b_2 = d_2 \\ b_1c_2 = \alpha c_1 & c_2b_2 = \alpha c_1 & b_1d_2 = -\alpha d_1 & d_2b_2 = \alpha d_1 \end{array}$$

The lattice $V(R)$ of ideals of R consists of R, $R_1 + J$, $R_2 + J$, $J = e_1Re_2$, F_1J, F_2J, and (0); and is therefore distributive. On the other hand, since $J_1 = J_2 = 0$, the ideals $RJ_i{}^s RJ_k{}^t R = (0)$ when $s + t > 0$. Hence the product of the now $\cup$-irreducible ideals $R_1 + J$ and $R_2 + J$, in this order, is the $\cup$-irreducible ideal J.

The $\cup$-irreducible ideals $RJ_i^s RJ_k^t R$ of Theorem 4 are expressible in a form which is independent of the idempotents e_i : The ideals

$$E_i = RJ_i^0 R = RR_i R = Re_i Re_i R = Re_i R, \qquad i = 1, ..., q \tag{17}$$

are the only idempotent $\cup$-irreducible ideals of R. From $J_i = e_i J e_i$ it follows that

$$RJ_i R = Re_i RJRe_i R = E_i J E_i \tag{18}$$

Now

$$RJ_i^s R = (RJ_i R)^s \tag{19}$$

since $J_i RJ_i = J_i e_i Re_i J_i = J_i^2$. This implies

$$RJ_i^s R = (E_i J E_i)^s \tag{20}$$

whence

$$RJ_i^s RJ_k^t R = (E_i J E_i)^s (E_k J E_k)^t \tag{21}$$

This proves:

Theorem 5. Under the hypotheses of Theorem 4, R has exactly q idempotent $\cup$-irreducible ideals $E_1, ..., \phi_q$. Moreover each $\cup$-irreducible ideal $RJ_i^s RJ_k^t R$ is expressible in terms of these ideals and J in the form $N_i^s N_k^t$ where $N_i = E_i J E_i$ and $N_i^0 = E_i$. ▌

The structure of $V(R)$ is reflected in the structure of the lattice of semigroup ideals of the semigroup γ since by Theorem 4, $V(R)$ is isomorphic to $V(\gamma)$. We shall therefore investigate the semigroup γ more thoroughly below.

Since the $J_i^s RJ_k^t$ form a chain $\mathcal{C}_{ik}$, the two members $J_i RR_k$ and $R_i RJ_k$ of this chain are comparable. Suppose

$$J_i RR_k \subseteq R_i RJ_k \tag{22}$$

whence $W_{ik} R = J_i RR_k$. It then follows by the induction hypothesis $W_{ik}^n R = J_i^n RR_k$ and (22) that $W_{ik}^{n+1} R = J_i^{n+1} RR_k + J_i^n RJ_k = J_i^{n+1} RR_k$. Therefore $\mathcal{C}_{ik}$ is the "left chain" consisting of the $W_{ik}^s R = J_i^s RR_k$, $s = 0, 1, ...$. Similarly, if $J_i RR_k \subseteq R_i RJ_k$, $\mathcal{C}_{ik}$ is the "right chain" consisting of the $R_i RJ_k^s$, $s = 0, 1, ...$.

The chains $\mathcal{C}_{ik}$ of (R_i, R_k)-bisubmodules of $e_i Re_k$ in the semigroup γ correspond by Theorem 5 to the chains K_{ik} of ideals of R generated by these bimodules. In the case of a left chain these are the ideals $N_i^s E_k$ and in the case of a right chain, the ideals $E_i N_k^s$. Each is either (0) or $\cup$-irreducible.

Moreover the exponent s of an ideal different from (0) in K_{ik} is uniquely determined. In fact from

$$N_i{}^s E_k = N_i{}^t E_k \quad \text{with} \quad s < t \tag{23}$$

it follows that

$$N_i{}^s E_k = N_i^{t-s} N_i{}^s E_k = N_i^{m(t-s)} N_i{}^s E_k$$

for all natural numbers m, whence, because of $N_i \subseteq J$, $N_i{}^s \subseteq J^r$ for all r. Therefore $N_i{}^s E_k = (0)$.

If, in addition, R is prime,

$$N_i E_k = E_i N_k$$

i.e. for such rings we need not differentiate between left and right chains. Proof: If $i = k$, obviously

$$K_{ii} = \{N_i{}^s \mid s = 0, 1, \ldots\}$$

since $N_i E_i = N_i = E_i N_i$. Now let R be a prime ring. The $\cup$-irreducible ideals of R (without the zero ideal) form a semigroup; denote this again by γ.

Since the product $E_i E_j E_i$ lies in K_{ii}, there exists an exponent $f(iji)$ such that

$$E_i E_j E_i = N_i^{f(iji)} \tag{24}$$

Assume K_{ij} is a left chain. Then for some exponent p we have

$$E_i N_j = N_i{}^p E_j \tag{25}$$

We must prove $p = 1$. From (24) and (25) it follows that

$$N_i^{f(iji)} E_j = E_i E_j E_i E_j = E_i N_j^{f(jij)} = N_i^{pf(jij)} E_j$$

whence

$$f(iji) = pf(jij) \tag{26}$$

Analogously, $f(jij) = rf(iji)$ for a suitable exponent r if K_{ji} is also a left chain. Therefore in this case $f(iji) = prf(iji)$ and hence $p \cdot r = 1$ and $p = 1$. If, however, K_{ji} is a right chain, say

$$E_j N_i{}^r = N_j E_i$$

we infer that $p = r$ from

$$N_i^{p+f(iji)} = E_i N_j E_i = N_i^{f(iji)+r}$$

Now setting

$$E_j N_i E_j = N_j{}^t$$

we obtain

$$E_jN_i^{pt} = E_jN_i^{rt} = N_j{}^tE_i = E_jN_iE_jE_i = E_jN_i^{1+f(iji)}$$

whence, by (26)

$$pt = 1 + f(iji) = 1 + pf(jij)$$

This proves $p = 1$. We deal with the case that K_{ij} is a right chain analogously.

We collect these results:

Theorem 6. Let the arithmetic ring R be prime and let it satisfy the hypotheses of Theorem 4. Hence, in particular R/J is artinian. Then, with the notation of Theorem 5, the semigroup γ of its $\cup$-irreducible ideals consists of the disjoint chains

$$K_{ik} = \{N_i{}^sE_k \mid s = 0, 1, \ldots\}$$

$i, k = 1, \ldots, q$. Moreover

$$N_iE_k = E_iN_k \quad \blacksquare \tag{27}$$

The product $E_iE_jE_k$ is a member of the chain K_{ik}. In the case R is prime, set

$$E_iE_jE_k = N_i^{f(ijk)}E_k \tag{28}$$

The exponents $f(ijk)$ are "structure constants" of the semigroup γ in the sense that multiplication in γ is entirely determined by them. We have

$$N_h{}^sE_iN_j{}^tE_k = N_h^{s+t}E_hE_iE_jE_k = N_h^{s+t+f(hij)+f(hjk)}E_k \tag{29}$$

The associativity of multiplication, or, more precisely, the equality $(E_hE_iE_j)\,E_k = E_h(E_iE_jE_k)$ implies

$$f(hij) + f(hjk) = f(hik) + f(ijk) \tag{30}$$

Substituting $h = 1$ in this and solving for $f(ijk)$ we get

$$f(ijk) = f(1ij) + f(1jk) - f(1ik) \tag{31}$$

This suggests that we introduce the so-called "sandwich" matrix

$$\pi\colon (i,j) \to (ij) = f(1ij), \qquad (i,j) \in I \times I \tag{32}$$

where $I = \{1, 2, \ldots, q\}$ and the entries (ij) of π are nonnegative integers. In terms of this notation, (31) becomes

$$f(ijk) = (ij) + (jk) - (ik) \tag{33}$$

By (29) the matrix π describes multiplication in γ by

$$N_h{}^s E_i N_j{}^t E_k = N_h^{s+t+(hi)+(ij)+(jk)-(hk)} E_k \tag{34}$$

The matrix π satisfies the following conditions:

0. $(ii) = 0,\ i = 1, \ldots, q.$
1. $(ij) + (jk) \geqslant (ik),\ i, j, k = 1, \ldots, q.$
2. $(kj) + (ji) > 0,\ i \neq j.$

Condition 0 follows from

$$E_1 E_i = E_1 E_i E_i = N_1^{f(1ii)} E_i = N_1^{(ii)} E_i$$

Condition 1 merely states that, because of (33), $f(ijk)$ i.e. the exponent in

$$E_i E_j E_k = N_i^{f(ijk)} E_k$$

is nonnegative. The last condition is equivalent to the fact that there are no idempotents in γ besides $E_1, \ldots, E_q$.

Within the framework of the theory of semigroups, we may describe the semigroup γ, in the case of a prime ring, as follows:

Theorem 7. Let I be a set and G the infinite cyclic group generated by w. G contains the subsemigroup $C = \{w^s \mid s = 0, 1, \ldots\}$ with $1 = w^0$. Set $M = G \times I \times I$ is a G-operand under

$$w^p(w^s, i, j) = (w^{p+s}; i, j) \tag{35}$$

that is to say

$$c_1 c_2 A = c_1(c_2 A) \qquad \text{for all} \quad c_1, c_2 \in G \quad \text{and} \quad A \in G \times I \times I \tag{36}$$

A semigroup structure on M is said to be compatible with its G-structure if

$$c(AB) = (cA)B = A(cB) \qquad \text{for all} \quad c \in G, \quad A, B \in M \tag{37}$$

Such a semigroup structure may be defined with the helfp of an "exponential sandwich matrix"

$$\pi: (i, j) \to (ij) \in \mathbb{N} \cup 0, \quad (i, j) \in I \times I \tag{38}$$

as follows:

$$(w^s; h, i)(w^t, j, k) = (w^{s+t+(ij)}; h, k)$$

Conditions 0, 1, 2 above for π are equivalent to the $E_i = (1; i, i)$, $i \in I$ being idempotent and the C-suboperand

$$S = \bigcup_{i,j} CE_iE_j$$

of M being a subsemigroup containing no further idempotents. For multiplication of elements of S we have

$$w^sE_hE_iw^tE_jE_k = w^{s+t+(hi)+(ij)+(jk)-(hk)}E_hE_k \tag{39}$$

Proof. By verification.

Definition 2. A semigroup S is *prime* and *quasiuniserial* if it is isomorphic to a semigroup of the type described in Theorem 7.

Formula (34) and the discussion following Theorem 6 show that the mapping which takes the $\cup$-irreducible ideal $N_i{}^sE_k$ to the element $w^sE_iE_k$ of S is an isomorphism. In other words:

Theorem 8. Let R be a prime, arithmetic ring satisfying the hypotheses of Theorem 4. Then the semigroup γ of $\cup$-irreducible ideals of R is prime and quasiuniserial and has only a finite number of idempotents. ▮

Conversely, given a prime, quasiuniserial semigroup S with only finitely many idempotents we may construct a prime arithmetic ring R satisfying the hypotheses of Theorem 4 and with semigroup γ of $\cup$-irreducible ideals isomorphic to S.

To this end, let $S = S_\pi$ be a semigroup, the structure of which is described in terms of a matrix π as in Theorem 7. The set I need not, for the present, be finite. Let Φ be a field. Then, because of (38), the semigroup algebra $\Phi[S]$ is a $\Phi[w]$-algebra, where $\Phi[w]$ denotes the ring of polynomials in the indeterminate w with coefficients from Φ. The ring $\Phi^* = \Phi[[w]]$ of formal power series in w with coefficients from Φ is also a $\Phi[w]$-algebra. We may therefore form the tensor product

$$\Phi^*[S] = \Phi[S] \otimes_{\Phi[w]} \Phi^* \tag{40}$$

The elements of $\Phi^*[S]$ are of the form

$$\sum_{i,j\in I} \alpha_{ij}E_iE_j\,, \qquad \alpha_{ij} \in \Phi^* \tag{41}$$

The semigroup S may be embedded in the Φ^*-algebra $\Phi^*[S]$ by mapping $w^t E_k E_l$ to that element in $\Phi^*[S]$ for which $\alpha_{kl} = w^t$ and $\alpha_{ij} = 0$ for all other pairs (i, j). Later we shall show that $\Phi^*[S]$ has an identity in case the index set I is finite. However, we shall not assume this for the present. Until we do, we must distinguish between algebra and ring ideals in $\Phi^*[S]$. In the following, ideal will mean algebra ideal unless otherwise stated.

Let A be an ideal of the semigroup S. The image of A under the embedding of S in $\Phi^*[S]$ generates the Φ^*-module $\Phi^*[A]$ which is an ideal in the algebra $\Phi^*[S]$. The mapping

$$\psi\colon A \to \Phi^*[A], \qquad A \text{ an ideal of } S \tag{42}$$

is a $\cap$-homomorphism of the lattice $V(S)$ of ideals of S to the lattice $V(R)$ of ideals of $R = \Phi^*[S]$. Moreover, ψ also preserves multiplication:

$$\psi(AB) = \psi(A)\,\psi(B)$$

In the other direction, we define the mapping

$$\phi\colon A^* \to \begin{cases} A^* \cap S & \text{if } A^* \neq (0) \\ \varnothing & \text{if } A^* = (0) \end{cases} \tag{43}$$

where A^* is an ideal of R. Clearly ϕ is a $\cap$-homomorphism of $V(R)$ to $V(S)$. We shall show that ψ is an isomorphism of $V(S)$ onto $V(R)$ with ϕ as inverse. This is an immediate consequence of the following lemma:

Lemma 1. If the element $w^t E_h E_k$ of the semigroup $S = S_\pi$ occurs with nonzero coefficient in the element

$$x = \sum_{i,j} \sum_{s=0}^{\infty} c_{ijs} w^s E_i E_j$$

of $R = \Phi^*[S]$, i.e. if $c_{hkt} \neq 0$, then the principal ideal $(x)'$ generated by x in R contains $w^t E_h E_k$. ∎

The lemma will be better understood if we first introduce a partial ordering $\leqslant$ in the semigroup S as follows:

Lemma 2. Let (a) denote the set SaS in the semigroup $S = S_\pi$. Then $\leqslant$, defined by

$$a \leqslant b \Leftrightarrow a \in (b) \tag{44}$$

is a partial ordering on S which is compatible with multiplication.

Proof. $a = w^s E_i E_j = E_i w^s E_i E_j \cdot E_j$ implies

$$\mathrm{F}_1. \quad a \in (a) \qquad \text{for all} \quad a \in S$$

whence $\leqslant$ is reflexive. It is also transitive since $a \leqslant b \leqslant c$ implies $a \in (b)$, $b \in (c)$, which in turn gives $b = xcy$. Thus $a \in (b) = (xcy) \subseteq (c)$. To prove the antisymmetry of $\leqslant$, we note that if $a \leqslant b \leqslant c$, there exist elements $x, y, u, v \in S$ such that

$$a = xby \qquad \text{and} \qquad b = uav \tag{45}$$

Then from $a = E_h a E_k$ and $b = E_i b E_j$ it follows that

$$E_h a E_k = a = E_h x u E_h a E_k v y E_k \tag{46}$$

By Property 2 of the matrix π, i.e. since $(hk) + (kh) > 0$ for $h \neq k$, (45) implies $E_h x u E_h = E_h$ and $E_k v y E_k = E_k$. Now since $b = E_i b E_j$ and $a = E_h a E_k$ we may assume that $x = E_h x E_i$ and $u = E_i u E_h$ in (45). Then, again because of Property 2 of π, it follows from $E_h = xu = E_h x E_i u E_h$ that $h = i$ and $x = u = E_h$. Similarly we may show $j = k$ and $v = y = E_k$. Hence by (45) $a = E_h b E_k = E_i b E_j = b$. The antisymmetry of $\leqslant$ which we have proved may also be expressed:

$$\mathrm{F}_3. \quad (a) = (b) \Leftrightarrow a = b \qquad \text{for all} \quad a, b \in S.$$

Finally the compatibility of $\leqslant$ with multiplication in S is a consequence of a weak form of commutativity of S, namely of

$$\mathrm{F}_2. \quad (ab) = (a)(b) \qquad \text{for} \quad a, b \in S.$$

In fact from F_2 and $a \leqslant c$ it follows that

$$ab \in (c)\, b \subseteq (c)(b) = (cb) \qquad \text{for all} \quad b \in S$$

whence $ab \leqslant cb$. Similarly $ba \leqslant bc$.

Proof of (F_2). $w^s E_h E_i \cdot w^r E_m E_n \cdot w^t E_j E_k = w^z \cdot w^s E_h E_i \cdot w^t E_j E_k$ is equivalent to

$$\begin{aligned} s + r + t + (hi) + (im) + (mn) + (nj) + (jk) - (hk) \\ = z + s + t + (hi) + (ij) + (jk) - (hk) \end{aligned}$$

This equation for z has a nonnegative integer solution since, by Property 1 of π, the inequality

$$(im) + (mn) + (nj) \geqslant (ij)$$

holds.

We now proceed with the proof of Lemma 1. Among all the elements of S which occur with nonzero coefficient in x, choose one, say $w^p E_u E_v$, which is maximal relative to $\leqslant$. Then the ring ideal $(x)'$ contains the element

$$y = E_u x E_v = \sum_s \sum_{i,j} c_{ijs} w^s E_i E_j E_v$$

$$= \sum_s \sum_{i,j} c_{ijs} w^{s+(ui)+(ij)+(jv)-uv)} E_u E_v \tag{47}$$

$w^p E_u E_v$ is still maximal in the set of all elements of S which occur with nonzero coefficient in y; for otherwise there exist $(i, j) \neq (u, v)$ and an exponent s such that

$$p \geqslant s + (ui) + (ij) + (jv) - (uv) \qquad \text{and} \qquad c_{ijs} \neq 0$$

This however, implies

$$w^p E_u E_v = w^{p-s} E_u \cdot w^s E_i E_j \cdot E_v < w^s E_i E_j$$

contradicting the maximality of $w^p E_u E_v$ in x. This proves that the coefficient of $E_u E_v$ in y is an element of the form $\gamma = w^p(c_{uv}\nu + \cdots)$ in Φ^*. Hence there is an element δ in Φ^* of the form $c_{uvp}^{-1} w^0 + \cdots$ such that $\delta\gamma = w^p$. But then the element

$$\delta E_u y = w^p E_u E_v$$

is in the ideal $(x)'$ of R since $y \in (x)'$. Finally since $w^p E_u E_v \in (x)'$, all elements of the form $w^{p+t} E_u E_v$, $t \geqslant 0$ are contained in $(x)'$.

Lemma 1 shows that each ideal A^* of the Φ^*-algebra $R = \Phi^*[S]$ is generated as a Φ^*-module by those elements of S which occur with nonzero coefficients in elements of A^*. Hence each ideal of R is the image under the mapping ψ in (42) of some ideal A of S. More precisely

$$A^* = \Phi^*[A^* \cap S] = \psi(A^* \cap S) = \psi\phi A^*$$

The $\cup$-homomorphism ψ is therefore a lattice isomorphism of the lattice $V(S)$ of ideals of S onto the lattice $V(R)$ of ideals of R. Since ψ is also multiplicative, ψ, together with its inverse ϕ is an isomorphism of the lattice ordered semigroup of ideals; for again by Lemma 1,

$$\psi(A \cdot (B \cup C)) = \psi(A) \cdot (\psi B) + \psi(A)\, \psi(C) \tag{48}$$

The lattice operations of $V(S)$ are set theoretic intersection and set theoretic union. Therefore $V(S)$ is a D^*-distributive lattice (cf. Section IX.1), i.e.

$$A \cup \bigcap_{\lambda \in \Lambda} B_\lambda = \bigcap_{\lambda \in \Lambda} (A \cup B_\lambda) \qquad \text{for every subset } \Lambda \text{ of } I \tag{D^*}$$

On the other hand, the lattice isomorphism ψ is complete, i.e. ψ maps the intersection of an arbitrary family of ideals of S to the intersection of their images in R, since the same is trivially true for the inverse ϕ of ψ by (43). Hence the lattice $V(R)$ is also D^*-distributive.

This proves:

Theorem 9. Let S be a prime, quasiuniserial semigroup and Φ a field. Then, with the notation of Theorem 7 for S, the semigroup algebra $\Phi[S]$ is an algebra over the ring $\Phi[w]$ of polynomials in w. Let $\Phi^* = \Phi[[w]]$ be the ring of formal power series in w. Then the tensor product

$$R = \Phi^*[S] = \Phi[S] \otimes_{\Phi[w]} \Phi^* \tag{49}$$

is a Φ^*-algebra with D^*-distributive lattice of ideals $V(R)$. Moreover the mapping

$$\psi: A \to \Phi^*[A] \qquad \text{for ideals } A \text{ of } S \tag{50}$$

together with its inverse

$$\phi: A^* \to A^* \cap S \qquad \text{for ideals } A^* \text{ of } R \tag{51}$$

is a complete isomorphism of the lattice ordered semigroup $V(S)$ of ideals of the semigroup S onto $V(R)$, the lattice ordered semigroup of ideals of R. ∎

If the number of idempotents in S is finite, $R = \Phi^*[S]$ satisfies all the hypotheses of Theorem 4. To show this, we first determine the radical of R.

The idempotents $E_1, ..., E_q$ of S are maximal relative to the partial ordering $\leqslant$ introduced in Lemma 2. The remaining elements of S form the ideal $N = S - \{E_1, ..., E_q\}$ in S. This ideal generates in $R = \Phi^*[S]$ the ideal $\psi N = \Phi^*[N]$. We show

$$\bigcap N^n = \varnothing \tag{52}$$

Proof: In a product $a_1 a_2 \cdots a_{q+1}$ of $q + 1$ factors from N either some factor is contained in wS or each factor is of the form $E_i E_j$ with $i \neq j$. In the latter case some idempotent, say E_h, occurs in the above product in two non-adjoining positions: $\cdots E_h \cdots E_k \cdots E_h \cdots$, $h \neq k$. It follows from this and Condition 2 for π, i.e. from $(hk) + (kh) > 0$, that the product $a_1 a_2 \cdots a_{q+1}$ lies in wS. Thus $N^{\alpha+1} \subseteq wS$, whence there exists a mapping $T: m \to T(m)$ with

$$N^n \subseteq w^m S \qquad \text{for} \quad n \geqslant T(m) \tag{53}$$

and therefore

$$\bigcap N^n \subseteq \bigcap w^m S = \varnothing$$

Since the isomorphism ψ is multiplicative it follows from (53) and (52) that

$$(\psi N)^n \subseteq w^m R \qquad \text{for} \quad n \geqslant T(m) \tag{54}$$

and

$$(\psi N)^n = (0). \tag{55}$$

By definition, $R = \Phi^*[S]$ consists of elements of the form $z = \sum_{s=0}^{\infty} c_s w^s$ with $c_s = \sum_{i,j=1}^{q} \gamma_{ijs} E_i E_j$, $\gamma_{ijs} \in \Phi^*$. Now if z is contained in ψN we have by (54)

$$z^n \in w^m R \qquad \text{for all} \quad n \geqslant T(m)$$

Therefore the element $z_1 = -z - z^2 - z^3 - \cdots \in R$ is an element for which $z + z_1 - z_1 z = 0$, proving that z, and hence the ideal ψN, is quasiregular. Therefore ψN is contained in the radical $J = J(R)$. Conversely, from the isomorphism

$$R/\psi N \approx \sum_{i=1}^{q} \oplus \Phi_i\, , \qquad \Phi_i \approx \Phi$$

it follows that $R/\psi N$ is semisimple. This proves

$$J = J(R) = \psi N \tag{56}$$

Idempotents of R/J may be lifted to idempotents of R. Proof: (cf. Jacobson). Let $u \in R$ and $u^2 \equiv u (\text{mod } J)$ and let us suppose that the desired idempotent e in R is expressible in the form $e = u + x - 2ux$ where x is an element which commutes with u. Then $e^2 = e$ is equivalent to

$$u^2 - u + r(u^2 - u)(x^2 - x) + x^2 - x = 0 \tag{57}$$

Since we do not know beforehand whether R has an identity, we adjoin one to the ϕ^*-algebra $R = \Phi^*[S]$ and then obtain in R^1 from (57) the condition

$$u^2 - u + (1 + 4(u^2 - u))(x^2 - x) = 0 \tag{58}$$

Solving formally for $x^2 - x$ we get

$$x^2 - x = -\sum_{n=0}^{\infty} (-4)^n (u^2 - u)^{n+1} \tag{59}$$

The series on the right represents an element z of R since, by (54)

$$(u^2 - u)^n \in w^m R \qquad \text{for all} \quad n \geqslant T(m)$$

Obviously z commutes with u. The equation

$$x^2 - x = z$$

has a solution in R: For if the characteristic char $\Phi \neq 2$, we may solve formally for x and obtain $x = \frac{1}{2}(1 - (1 + 4z)^{1\backslash 2})$. This expression may be expanded formally in powers of z to obtain

$$\tfrac{1}{2}(1 - (1 + 4z)^{1/2}) = \sum_{n=1}^{\infty} \frac{1}{2n - 1} \binom{2n - 1}{n} (-z)^n \tag{60}$$

and the right-hand side of (60) is a solution in R. If, however, char $\Phi = 2$, (59) reduces to

$$x^2 - x = u^2 - u = z$$

and in this case too the infinite series in (60) with integer coefficients is a solution.

Now we prove the existence of an identity in R using the following result due to R. Baer:

Theorem 10. Let $J = J(R)$ be the radical of the ring R and suppose J contains no idempotent ideals different from 0. For each ideal A of R, let $RA = A = AR$. Let idempotents be liftable modulo J. Then if R/J has an identity, so does R.

Proof. Let $e + J$ be the identity of R/J where e is an idempotent of R. Let

$$l(e) = \{x - xe \mid x \in R\} \qquad \text{and} \qquad r(e) = \{x - ex \mid x \in R\}$$

be the left (resp. right) annihilator of e. We must show that $l(e) = r(e) = (0)$. Now $N = l(e) \cdot r(e)$ is an ideal of R. Define the ideals Q_l and Q_r by

$$Q_l = \{x \in R \mid Rx \subseteq N\} \qquad \text{and} \qquad Q_r = \{x \in R \mid xR \subseteq N\}$$

We show these ideals are equal: By hypothesis and definition $Q_l = RQ_l \subseteq N$ and $Q_r = Q_rR \subseteq N$. On the other hand, $RN \subseteq N$ and $NR \subseteq N$ since N is an ideal. Therefore

$$N \subseteq Q_l \qquad \text{and} \qquad N \subseteq Q_r$$

This proves

$$Q_l = N = Q_r$$

Since e is an idempotent, $R = Re + l(e)$, whence $R \cdot r(e) = l(e) \cdot r(e) = N$. It follows that $r(e) \subseteq Q_l = N$. Similarly we show $l(e) \subseteq Q_r = N$. This implies $N^2 = N \cdot N \supseteq l(e)\, r(e) = N \supseteq N^2$ and hence $N^2 = N$. Now by hypothesis, $e + J$ is the identity of R/J. Therefore $l(e)$ and $r(e)$—and hence also the ideal N—are contained in the radical J of R. Hence $N = (0)$ and therefore $l(e) = r(e) = (0)$. ∎

This theorem may be applied to the Φ^*-algebra $R = \Phi^*[S]$ if S contains only a finite number of idempotents since in this case:

α. R/J is a direct sum of fields
β. idempotents are liftable modulo J
γ. $RA^* = (\psi S)(\psi A) = \psi(SA) = \psi A = A^* = \psi(AS) = (\psi A)(\psi S) = A^*R$
for all ideals A^* of $R = \Phi^*[S]$.

We have thus shown that the Φ^*-algebra $R = \Phi^*[S]$ is a semiperfect ring and that its ring ideals are algebra ideals. The subring $\Phi^* = \Phi_1{}^*$ is contained in the center of R and is a local noetherian ring. Moreover R is finitely generated over Φ^*. We show finally that the last hypothesis of Theorem 4 is also satisfied: Under the embedding of S in $R = \Phi^*[S]$, the idempotents $E_1, \ldots, E_q$ of the semigroup S map to the identity elements $\bar{E}_1, \ldots, \bar{E}_q$ of the simple direct summands $\bar{R}_i = \bar{E}_i\bar{R}\bar{E}_i$ of $\bar{R} = R/J$. Each $\bar{R}_i$ is isomorphic to Φ. Hence $\bar{R}_i \otimes_\Phi \bar{R}_k{}^* \approx \Phi$ and is therefore simple.

Combining these results with Theorem 8 we obtain the following characterization:

Theorem 11. 1. Let R be a prime arithmetic ring satisfying the hypotheses of Theorem 4. The $\cup$-irreducible ideals of R form a prime quasiuniserial semigroup γ with only finitely many idempotents.

2. Let S be a prime quasiuniserial semigroup with only finitely many idempotents and let Φ be a field. Then the Φ^*-algebra (cf. Theorem 9) $R = \Phi^*[S] = \Phi[S] \otimes_{\Phi[w]} \Phi^*$ over the ring Φ^* of formal power series in w is a prime, arithmetic, semiperfect ring whose factor ring by its radical is a direct sum of fields each isomorphic to Φ.

3. The lattice ordered semigroups of ideals of the semigroups γ and S of Parts 1 and 2 and the lattice ordered semigroup of ideals of R are isomorphic to one another:

$$V(R) \approx V(\gamma) \approx V(S) \quad \blacksquare$$

In rings R satisfying the hypotheses of Theorem 11, every (two-sided) ideal is principal by Theorem IX.1.5. A prime, quasiuniserial semigroup S is commutative if and only if $q = 1$ since

$$E_1E_2E_1 = w^{(12)+(21)}E_1 \neq E_1{}^2E_2 = E_1E_2$$

Nevertheless we may prove, using property (F_2) of S, i.e. $(a)(b) = SaS \cdot SbS = SabS = (ab)$ for all $a, b \in S$ and Lemma 1, that, just as in the case of commutative rings,

$$(x)'(y)' = (xy)' \tag{61}$$

where $(x)'$ and $(y)'$ are the principal ideals generated by x and y in $R = \Phi^*[S]$, S a prime, quasiuniserial, semigroup with only finitely many idempotents.

Even if S contains an infinite number of idempotents, the converse of Theorem 9 holds (cf. Behrens 1970).

The semigroups $S = S_\pi$ introduced earlier on have the following purely semigroup theoretic properties, besides properties F_1, F_2, and F_3 proved in Lemma 2:

F_4. Given $a \neq a^2$ and $b \in S$, there exists a natural number n such that $a^n \leqslant b$ ($\leqslant$ is the relation defined in Lemma 2).

F_5. All idempotents in S are primitive, i.e. $E^2 = E$, $F^2 = F$ and $E = EF = FE$ imply $E = F$.

F_6. Let E_i, $i \in I$ be the idempotents of S. Then the subsemigroup E_iSE_k of S has exactly one maximum (with respect to $\leqslant$) element among those elements of E_iSE_k different from the maximum E_iE_k.

In the theory of semigroups it may be shown that, Properties F_1 to F_6 characterize the semigroups S_π (cf. Behrens [6]).

The theory of the lattice ordered semigroup of ideals in a ring R and in a semigroup S, is called the arithmetic of R and S, respectively. By Theorem II, the arithmetic of R is the same as the arithmetic of the prime quasiuniserial semigroups. The latter is given in Behrens [10].

Noetherian Ideal Theory in Nonassociative Rings

CHAPTER X

A nonassociative ring R differs from an associative ring in that multiplication is no longer assumed associative. Under multiplication, R forms a groupoid in the sense of the following:

Definition 1. A *groupoid* is a nonempty set S together with an operation, i.e. a mapping of $S \times S$ to S. The image of (a, b) under this mapping is often denoted by $a \cdot b$ or just simply by ab.

Definition 2. A *nonassociative ring* is a set R with two operations addition $(+)$ and multiplication $(\cdot)$ satisfying

1. R is an abelian group under $+$
2. R is a groupoid under $\cdot$
3. For any three elements $a, b, c \in R$,

$$a \cdot (b + c) = a \cdot b + a \cdot c \qquad \text{and} \qquad (b + c) \cdot a = b \cdot a + c \cdot a$$

Comparing this definition with that given in Definition I.1.4 we note that R is an associative ring if R is, in addition, a semigroup under $\cdot$. Left, right, and two-sided ideals in R are defined in exactly the same way as before.

In this chapter R will denote a nonassociative ring. From the definition, we note that this does not exclude the possibility that R be associative.

Definition 3. R is *left noetherian* if R satisfies the maximal condition on left ideals, i.e. if each nonempty family $\mathcal{F}$ of left ideals, partially ordered by set theoretic inclusion, contains a maximal left ideal in $\mathcal{F}$. We define *right noetherian* rings analogously. In this chapter a ring R will be called *noetherian* if R satisfies the maximal condition on (two-sided) ideals.

Analogous to the situation in the artinian case (cf. Definition II.3.4), R is left noetherian if and only if every strictly ascending chain

$$B_1 \subset B_2 \subset B_3 \cdots$$

of left ideals B_i stops after finitely many steps.

In classical ideal theory of rings which are both commutative and associative, the starting point is the representation of an arbitrary ideal as an intersection of $\cap$-irreducible ideals. In an analogous fashion we have:

Theorem 1. In a noetherian ring R every ideal is the intersection of finitely many $\cap$-irreducible ideals, i.e. ideals I with the property that if $I = A \cap B$, then either $I = A$ or $I = B$.

Proof. Suppose the theorem is false and let $\mathcal{F}$ be the family of ideals which are not expressible as the intersection of finitely many $\cap$-irreducible ideals. Then $\mathcal{F} \neq \varnothing$. Let M be a maximal element of $\mathcal{F}$. Then $M = A \cap B$, $A \subset M$ and $B \subset M$. Hence $A \notin \mathcal{F}$ and $B \notin \mathcal{F}$. Therefore both A and B (and hence M) are expressible as the intersection of finitely many $\cap$-irreducible ideals. It follows that $M \notin \mathcal{F}$, a contradiction. ∎

In the classical case the $\cap$-irreducible ideals Q are primary, i.e. for all $a, b \in R$,

$$ab \in Q \quad \text{and} \quad a \notin Q \Rightarrow b \in r(Q)$$

Here r is the mapping of the set of ideals in R into itself defined by

$$r\colon A \to r(A) = \{b \in R \mid (b)^n \subseteq A \text{ for some } n \in \mathbb{N}\}$$

(cf. Theorem 15 below); $r(A)$ is called the *(primary) radical* of A.

In general, the representation of an ideal as the intersection of $\cap$-irreducible ideals is not unique. However, in the classical case we can prove that the intersection of primary ideals with the same prime radical is again a primary ideal with the same radical. Therefore in the representation of an ideal A as the intersection of $\cap$-irreducible ideals, we collect together those ideals having the same radical. Finally we omit any redundant ideals from this representation, i.e. we omit any ideal containing the intersection of the others, to obtain an irredundant representation of A as the intersection of finitely many primary ideals belonging to different radicals. Moreover, if

$A = \bigcap_{i=1}^{n} Q_i = \bigcap_{j=1}^{m} Q_j'$ are two such representations of A we can show that $m = n$ and $r(Q_i) = r(Q_i')$ after a permutation of the indices.

The ring $\mathbb{Z}$ of rational integers is noetherian. The ideals (p^n) generated by powers of primes are the primary ideals of $\mathbb{Z}$ and $r(p^n) = (p)$. In this case the intersection of two primary ideals with different radicals is equal to their product. An analogous statement holds also for Asano orders (cf. Theorems VII.2.11 and VII.2.12).

The classical theory described above was generalised by Lesieur and Croisot [1963] to associative noncommutative rings using the concept of the tertiary radical of an ideal. In the commutative case the tertiary and primary radicals coincide. Finally Kurata [1965] further generalized the theory to cover the nonassociative case.

In the definition of the tertiary radical $t(A)$ of an ideal A and the associated concept of a tertiary ideal we are guided by the classical case. In fact "tertiary" plays the same role in this situation that "primary" plays in the classical case. Therefore if we are to imitate the method outlined above we must ensure that a $\cap$-irreducible ideal is tertiary or contrapositively, we define the concept of a tertiary ideal in such a way as to ensure that a non-tertiary ideal A is easily seen to be $\cap$-reducible. Thus we would like to be able to conclude from

$$(a)(b) \subseteq A, \qquad a \notin A, \quad b \notin t(A) \tag{1}$$

that A has a representation as the intersection of two ideals properly containing A. In (1), (a) is of course the smallest ideal containing a. These considerations lead us to the following:

Definition 4. 1. The *tertiary radical* $t(A)$ of an ideal A of a nonassociative ring R is the set

$$t(A) = \{b \in R \mid (A : (b)) \cap (c) \subseteq A \Rightarrow c \in A \text{ for all } c \in R\} \tag{2}$$

where $(A : (b)) = \{x \in R \mid (x)(b) \subseteq A\}$.

2. An ideal Q is *tertiary* if

$$(a)(b) \subseteq Q \qquad \text{and} \qquad a \notin Q \quad \text{imply} \quad b \in t(A)$$

It is not clear at this stage that $t(A)$ is an ideal but we shall show this later on.

Theorem 2. 1. Every $\cap$-irreducible ideal is tertiary.

2. If A is an ideal of R, $A \neq R$ and $b \in t(A)$, then $(A : (b)) \supset A$.

3. $A \subseteq t(A)$.

Proof. 1. Let A be a nontertiary ideal. Then there exist elements a and $b \in R$ satisfying (1); $b \notin t(A)$ implies the existence of $c \notin A$ such that

$$(A\colon (b)) \cap (c) \subseteq A \tag{3}$$

Since the lattice of ideals of R is modular and $A \subseteq (A\colon (b))$, it follows from (3) that

$$A = (A\colon (b)) \cap ((c) + A) \tag{4}$$

Because of (1), $(A\colon (b)) \supset A$ and since $c \notin A$, $(c) + A \supset A$.

2. If $(A\colon (b)) = A$ and $b \in t(A)$ then for all $c \in R$ $(A\colon (b)) \cap (c) = A \cap (c) \subseteq A$ and hence $c \in A$ by the definition of $t(A)$. Therefore $R = A$, a contradiction.

3. Let $a \in A$. Then $(A\colon (a)) = R$. Therefore for all $c \in R$ $(A\colon (a)) \cap (c) \subseteq A$ implies $(c) \subseteq A$ and hence $c \in A$ proving that $a \in t(A)$. ▮

The tertiary radical $t(A)$ of an ideal A is a subset of the ring R which is closed under multiplication by elements of R from the right and from the left as we may easily verify from the definition. To prove that $t(A)$ is a subgroup of the additive group of R we first recall the definition of an essential extension of a module (cf. Definition VII.3.3):

Definition 5. Let T be an associative ring, M a left T-module and N a submodule of M. M is an *essential extension* of N if every nonzero submodule U of M intersects N nontrivially, i.e. $N \cap U \neq (0)$.

As in Definition VI.1.2 we define the multiplication ring of a nonassociative ring R:

Definition 6. Let R be a nonassociative ring. The subring of the endomorphism ring of the abelian group $\{R, +\}$, generated by the left multiplications $\mathfrak{L}_a\colon x \to ax$, the right multiplications $\mathfrak{R}_a\colon x \to xa$ and the identity endomorphism, is called the *multiplication ring* of R and is denoted by $T(R)$.

Just as in the associative case, the ideals of R are the T-submodules of the T-module R. We remark also that T is an associative ring even if R is not. As in the associative case, given an ideal A of R we may form the set of cosets $x + A$ and define an addition and multiplication in terms of the representatives of the cosets. The factor ring R/A is the homomorphic image of R under

$$x \to x + A$$

(cf. Section II.4). Moreover, R/A is a T-module.

Theorem 3. Let T be the multiplication ring of R. Let $\bar{x}$ denote the coset $x + A$. The tertiary radical of A is the set

$$t(A) = \{b \in R \mid \bar{R} \text{ is an essential extension of } (\bar{0} : (\bar{b})) \text{ as a } T\text{-module}\} \quad (5)$$

Proof. $\bar{R} = R/A$ is an essential extension of $(\bar{0} : (\bar{b}))$ if and only if for each nonzero ideal $\bar{C}$ of $\bar{R}$, $(\bar{0} : (\bar{b})) \cap \bar{C} \neq \bar{0}$. This condition is equivalent to the same condition for principal ideals $\bar{C} = (\bar{c}) \neq \bar{0}$. But

$$t(A) = \{b \mid c \notin A \Rightarrow (A : (b)) \cap (c) \nsubseteq A\} = \{b \mid \bar{c} \neq \bar{0} \Rightarrow (\bar{0} : (\bar{b})) \cap (\bar{c}) \neq (\bar{0})\}$$

and the assertion is proved. ∎

The following result is valid in general:

Theorem 4. Let the T-module M be an essential extension of its submodules U_1 and U_2. Let $U' \supseteq U_1$. Then M is an essential extension of $U_1 \cap U_2$ and U'.

Proof. Let $V \neq (0)$ be a submodule of M. Since $U_2 \cap V \neq (0)$, it follows that $(U_1 \cap U_2) \cap V = U_1 \cap (U_2 \cap V) \neq (0)$. Clearly $V \cap U_1 \neq (0)$ implies $U' \cap V \neq (0)$. ∎

Theorem 5. $t(A)$ is an ideal of R.

Proof. By Theorem 3 it is sufficient to prove that if $\bar{R}$ is an essential extension of $(\bar{0} : (\bar{b}_1))$ and $(\bar{0} : (\bar{b}_2))$ then it is also an essential extension of $(\bar{0} : (\overline{b_1 - b_2}))$. Clearly $(\bar{0} : (\bar{b}_1 - \bar{b}_2)) \supseteq (\bar{0} : (\bar{b}_1) \cap (\bar{0} : (\bar{b}_2))$. Therefore by Theorem 4, $\bar{R}$ is an essential extension of $(\bar{0} : (\bar{b}_1 - \bar{b}_2))$. ∎

In the classical case it can be shown that the primary radical of the intersection of two ideals is equal to the intersection of the primary radicals of these ideals (cf. Theorem 16 below). In the general case this is no longer true. However, with the weaker result of Theorem 6 we shall prove in Theorem 8 that the intersection of two tertiary ideals with the same radical is again a tertiary ideal belonging to this radical.

Theorem 6. $t(A_1) \cap t(A_2) \subseteq t(A_1 \cap A_2)$

Proof. Let $b \in t(A_1) \cap t(A_2)$ and let c be an element of R such that

$$(c) \cap (A_1 \cap A_2 : (b)) \subseteq A_1 \cap A_2 \quad (6)$$

We must show that $c \in A_1 \cap A_2$. The left-hand side of (6) is equal to

$$(c) \cap (A_1 : (b)) \cap (A_2 : (b))$$

Since this is contained in A_2 and since $b \in t(A_2)$ it follows, from the definition of the tertiary radical of A_2 applied to $(c) \cap (A_1 : (b))$, that $(c) \cap (A_1 : (b))$ is contained in A_2 and therefore also in $(A_2 : (b))$. The left-hand side of (6) is therefore equal to $(c) \cap (A_1 : (b))$. Since it is also contained in A_1 and as $b \in t(A_1)$ by hypothesis, c is in A_1. Similarly we prove that c is in A_2. ∎

Theorem 7. Let P and $Q \neq R$ be ideals of R and suppose

1. $P \subseteq t(Q)$ and
2. $AB \subseteq Q$, $A \not\subseteq Q \Rightarrow B \subseteq P$ for all ideals A, B of R. Then Q is tertiary and $t(Q) = P$.

Proof. By Definition 4 of a tertiary ideal and the fact that $P \subseteq t(Q)$, Q is tertiary. We must now show that each element $b \in t(Q)$ lies in P. By Theorem 2(2) there exists $x \in (Q: (b)) \backslash Q$. Then $(x)(b) \subseteq Q$ and $x \notin Q$. Hence by Part 2, $b \in P$. ∎

Theorem 8. If Q_1 and Q_2 are tertiary with $t(Q_1) = t(Q_2)$, then $Q_1 \cap Q_2$ is also tertiary and $t(Q_1 \cap Q_2) = t(Q_1)$.

Proof. Let A and B be ideals of R such that $AB \subseteq Q_1 \cap Q_2$ and $A \not\subseteq Q_1 \cap Q_2$. Assume without loss of generality that $A \not\subseteq Q_1$. Then

$$B \subseteq t(Q_1) = t(Q_2) = t(Q_1) \cap t(Q_2)$$

By Theorem 6, $t(Q_1) \cap t(Q_2) \subseteq t(Q_1 \cap Q_2)$ proving that $Q_1 \cap Q_2$ is tertiary. Moreover the hypotheses of Theorem 7 are satisfied with $Q = Q_1 \cap Q_2$ and $P = t(Q_1) = t(Q_2)$. ∎

Let A be an ideal of R. We have shown that $A = Q_1 \cap \cdots \cap Q_n$, Q_i $\cap$-irreducible and hence tertiary by Theorem 2.1. Theorem 8 enables us to lump together those Q_i with the same radical to obtain a representation of A as the intersection of tertiary ideals belonging to different radicals. It is then natural to normalize this representation in the sense of the following:

Definition 7. A representation $A = Q_1 \cap \cdots \cap Q_n$ of an ideal A as the intersection of tertiary ideals Q_i is said to be *irredundant* if no Q_i contains the intersection of the Q_j, $j \neq i$. The representation is said to be *normal* if it is irredundant and, in addition, $t(Q_i) \neq t(Q_j)$, $i \neq j$.

To prove that such representations are, in a certain sense, unique, we must first prove the following two theorems:

Theorem 9. Let $Q \neq R$ be a tertiary ideal and let b be an element of R. Then

$$(Q: (b)) \supset Q \Leftrightarrow b \in t(Q)$$

Proof. The sufficiency of the condition has already been proved in Theorem 2(2) for arbitrary ideals. The necessity follows from the definition of a tertiary ideal: $b \notin t(Q)$ and $(x)(b) \subseteq Q$ imply $x \in Q$. ▮

Theorem 10. Let $A = Q \cap X = Q' \cap X'$ and suppose Q and Q' are tertiary with $t(Q) \neq t(Q')$. Then $A = X \cap X'$.

Proof. Suppose, without loss of generality, that $t(Q) \not\subseteq t(Q')$ and let $b \in t(Q) \backslash t(Q')$. It is sufficient to show

$$(X \cap X') \cap (Q: (b)) \subseteq Q \tag{7}$$

since then, by definition of $t(Q)$, it would follow that

$$X \cap X' \subseteq Q \qquad \text{and hence} \quad A \subseteq X \cap X' \subseteq Q \cap X = A$$

Since $X \subseteq (X: (b))$,

$$\begin{aligned}(X \cap X') \cap (Q: (b)) &\subseteq (X: (b)) \cap X' \cap (Q: (b)) \\ &= ((Q \cap X): (b)) \cap X' = ((Q' \cap X'): (b)) \cap X' \\ &= (Q': (b)) \cap (X': (b)) \cap X' \\ &= (Q': (b)) \cap X' = Q' \cap X'\end{aligned}$$

The last equality above follows from Theorem 9 and the fact that $b \notin t(Q')$. Since $Q' \cap X' = A \subseteq Q$, the inclusion in (7) is proved. ▮

The following two theorems claim the existence and the essential uniqueness of the normal representations of the ideals in R. As an immediate consequence of Theorems 1, 2, and 8 we have:

Theorem 11. In a noetherian nonassociative ring every ideal has a normal representation as an intersection of finitely many teriary ideals. ▮

Examples may be found which show that we cannot claim uniqueness of normal tertiary representation. The following theorem however asserts that the radicals of the tertiary ideals occurring in the representation are uniquely determined.

Theorem 12. Let

$$A = Q_1 \cap \cdots \cap Q_n = Q_1' \cap \cdots \cap Q_m' \tag{8}$$

be two normal representations of an ideal A of a noetherian nonassociative ring R as the intersection of tertiary ideals Q_i and Q_j'. Then $n = m$ and after renumbering, $t(Q_i) = t(Q_i')$, $i = 1, ..., n$.

Proof. Assume by way of contradiction that

$$t(Q_1) \neq t(Q_j'), \qquad j = 1, ..., m \tag{9}$$

Applying Theorem 10 to $A = Q_1 \cap (Q_2 \cap \cdots \cap Q_n) = Q_1' \cap (Q_2' \cap \cdots \cap Q_n')$ we get

$$A = Q_2 \cap \cdots \cap Q_n \cap Q_2' \cap \cdots \cap Q_m' \tag{10}$$

Since $t(Q_1) \neq t(Q_2')$ we may again apply Theorem 10 to

$$A = Q_1 \cap (Q_2 \cap \cdots \cap Q_n) = Q_2' \cap (Q_2 \cap \cdots \cap Q_n \cap Q_3' \cap \cdots \cap Q_m')$$

to obtain

$$A = Q_2 \cap \cdots \cap Q_n \cap Q_3' \cap \cdots \cap Q_m' \tag{11}$$

The inequality $t(Q_1) \neq t(Q_3')$ leads in a similar fashion to

$$A = Q_2 \cap \cdots \cap Q_n \cap Qj' \cap \cdots \cap Q_m'$$

Proceeding in this manner we finally get

$$A = Q_2 \cap \cdots \cap Q_n$$

contradicting the irredundancy of the representation in (8). Therefore there exists an index j lying between 1 and m such that $t(Q_1) = t(Q_j')$. In a similar fashion we may show the existence of an index k such that $t(Q_2) = t(Q_k')$. We observe that $k \neq j$ since the radicals of the tertiary ideals Q_j' in the representation $A = Q_1' \cap \cdots \cap Q_m'$ are distinct from one another by hypothesis. In this manner we finally obtain the inequality $n \leqslant m$, and after renumbering, the equalities $t(Q_i) = t(Q_i')$, $i = 1, ..., n$. Since the hypotheses of the theorem are symmetric with respect to the Q_i and the Q_j', it follows that $m = n$. ▌

The radical $t(Q)$ of a tertiary ideal Q need not be prime (cf. Definition III.2.4). Nevertheless we have the following:

Theorem 13. If R is an associative commutative ring and Q a tertiary ideal of R, then $t(Q)$ is a prime ideal.

Proof. Let x and y be elements not in $t(Q)$. Then since multiplication is associative in R

$$\begin{aligned}(Q\colon (x)(y)) &= \{a \in R \mid (a) \cdot ((x)(y)) \subseteq Q\} \\ &= \{a \in R \mid (a)(x) \subseteq (Q : (y))\}\end{aligned}$$

Since $y \notin t(Q)$, $(Q\colon (y)) = Q$ by Theorem 9.
Hence

$$\begin{aligned}(Q\colon (x)(y)) &= \{a \in R \mid (a) \subseteq (Q : (x))\} \\ &= \{a \in R \mid a \in Q\}\end{aligned}$$

where the final equality is again a consequence of Theorem 9. Since the condition of Theorem 9 is also sufficient, it follows that $(x)(y) = (xy) \not\subseteq t(Q)$.∎

In certain types of associative noetherian rings there is a connection between the primary radical

$$r(A) = \{b \in R \mid (b)^n \subseteq A \text{ for some } n \in \mathbb{N}\} \tag{12}$$

and the tertiary radical $t(A)$:

Theorem 14. Let R be an associative noetherian ring with the property that the tertiary radical of any ideal is prime. Then for every ideal A of R, $r(A) \subseteq t(A)$.

Proof. Let $A = Q_1 \cap \cdots \cap Q_m$ be a normal representation of A as the intersection of tertiary ideals Q_i. It follows from $(b)^n \subseteq A \subseteq Q_i \subseteq t(Q_i)$, $i = 1, \ldots, m$ that $b \in t(Q_i)$ since by hypothesis the tertiary radical $t(Q_i)$ is prime. Hence by Theorem 6

$$b \in t(Q_1) \cap \cdots \cap t(Q_m) \subseteq t(Q_1 \cap \cdots \cap Q_m) = t(A) \quad ∎$$

If in addition R is commutative, i.e. in the classical case, the two radicals coincide:

Theorem 15. Let R be an associative, commutative, noetherian ring. Then the primary and tertiary radicals of an ideal A coincide.

Proof. Because of Theorems 13 and 14 we have only to show that $t(A) \subseteq r(A)$. Suppose there is an element $b \in t(A) \backslash r(A)$. Since R satisfies the ascending chain condition on ideals, there exists an exponent $k \in \mathbb{N}$ such that

$$(A : (b)^k) = (A : (b)^{k+1}) \tag{13}$$

This implies that

$$\begin{aligned}(A : (b)) \cap (b^k) &= \{xb^k \mid x \in R,\, xb^{k+1} \in A\} \\ &= \{xb^k \mid x \in R,\, xb^k \in A\} = A\end{aligned}$$

Since $b \in t(A)$, by the very definition of $t(A)$, $b^k \in A$ proving that $b \in r(A)$, a contradiction. ∎

A stronger version of Theorem 6 may also be proved in the classical case:

Theorem 16. If R is associative, commutative, and noetherian,

$$r(A_1) \cap r(A_2) = r(A_1 \cap A_2)$$

for all ideals A_1, A_2 of R.

Proof. $a \in r(A_1 \cap A_2) \Rightarrow a^n \in A_1 \cap A_2$ for some natural number n. Therefore $a \in r(A_1) \cap r(A_2)$. ∎

Theorem 17. If R is an associative, commutative, and noetherian ring, then

$$r(A) = \bigcap P \tag{14}$$

where the intersection is taken over all prime ideals P of R containing A.

Proof. Here $x^n \in A \subseteq P \Rightarrow x \in P$. Hence $r(A) \subseteq \bigcap P$. Conversely, the radicals $r(Q_i)$ of the primary ideals Q_i in the normal representation $A = Q_1 \cap \cdots \cap Q_m$ are prime ideals P_i, say. By Theorem 16, $r(A) = P_1 \cap \cdots \cap P_m$. Hence $P \subseteq r(A)$. ∎

Finally, we prove one more result which has already been used in the proof of Theorem IX.2.2.

Theorem 18 (Krull). Let R be an associative, commutative, and noetherian ring with identity and let A be an ideal contained in the Jacobson radical $J(R)$. Then

$$T = \bigcap_n A^n = (0) \tag{15}$$

Proof. Set $T = \bigcap A^n$ and let

$$TA = Q_1 \cap \cdots \cap Q_m \tag{16}$$

be a normal representation of the ideal TA as an intersection of primary ideals Q_i, $i = 1, ..., m$. It follows that $TA \subseteq Q_i$. Let $P_i = r(Q_i)$. If $A \not\subseteq P_i$, then $T \subseteq Q_i$ since Q_i is primary. On the other hand if $A \subseteq P_i$, then a suitable power of A is contained in Q_i since A, as an ideal of a noetherian ring, is finitely generated and from $(a)^n = (b)^m = (0)$ it follows by the binomial theorem that $(a + b)^{n+m} = (0)$. Hence by (15), $T \subseteq Q_i$. It follows that $T \subseteq \cap Q_i$ and consequently

$$TA = T \tag{17}$$

Suppose $T \neq (0)$. Let $\{d_1, ..., d_s\}$ be a system of generators of T of minimal length s. By (17) there exist elements $a_1, ..., a_s$ in A satisfying

$$d_1a_1 + \cdots + d_sa_s = d_s \tag{18}$$

This implies

$$d_s(1 - a_s) = d_1a_1 + \cdots + d_{s-1}a_{s-1} \in (d_1, ..., d_{s-1}) \tag{19}$$

Since $a_s \in J(R)$, a_s has a quasiinverse c. Therefore $(1 - a_s)(1 - c) = 1$. It follows from (19) by multiplying both sides by $(1 - c)$ that $d_s \in (d_1, ..., d_{s-1})$. Hence $T = (d_1, ..., d_{s-1})$ contradicting the minimality of s. ∎

Orders in Semisimple Artinian Rings

CHAPTER XI

The ring of rational integers is noetherian with the field of rationals as quotient field. From a ring theoretic standpoint, commutative fields are precisely the commutative simple artinian rings. It is therefore natural to generalize the idea of a quotient ring of an integral domain to noncommutative rings and then to attempt to characterize those rings with semisimple artinian quotient rings. The reader will find it instructive to compare the subsequent investigation with Section VII.2 on Asano orders.

Definition 1. An element a of a ring R is a *left zero divisor* if there exists $x(\neq 0) \in R$ such that $ax = 0$. Similarly we define a *right zero divisor*. An element $a \in R$ is said to be *regular* if a is neither a left nor a right zero divisor.

Definition 2. The ring Q is a (*left*) *quotient ring* of its subring R if

α. Q contains an identity.
β. Every regular element $a \in R$ is a *unit* in Q, i.e. there exist x and $y \in Q$ such that $ax = 1$ and $ya = 1$.
γ. Every element of Q is of the form $a^{-1}b$, where a and b are elements of R and a is regular.

In this case we also say that R is a (*left*) *order* in Q.

Given a regular element $a \in R$ and an arbitrary element $b \in R$, the product ba^{-1} is in Q. Therefore there exist elements a_1 and b_1 in R, a_1 regular, such that $ba^{-1} = a_1^{-1}b_1$. In other words if R has a quotient ring, then given arbitrary elements $a, b \in R$, a regular, a and b have a common left multiple

$a_1 b = b_1 a$ in R, where a_1 is also regular. The following theorem asserts that this condition is also sufficient that R possess a quotient ring, provided R contains regular elements.

Theorem 1 (Ore). Let R be a ring containing regular elements and satisfying *Ore's condition*, i.e. given a pair of elements (a, b) in R, a regular, there exists a pair (a_1, b_1) in R, a_1 regular, such that $a_1 b = b_1 a$. Then there exists a ring Q and a monomorphism ψ of R into Q such that Q is a (left) quotient ring of $\psi(R)$. ▮

The necessity of Ore's condition has already been proved above. The sufficiency will be proved at the end of this chapter.

To calculate with quotients $a^{-1}b$ in Q, the existence of "common denominators" is important. The following Lemma 1 is a repetition of Theorem VII.2.9 which we include here for completeness sake.

Lemma 1. Let Q be a quotient ring of R and $a_1, \ldots, a_n$ regular elements of R. Then there exist regular elements $a, b_1, \ldots, b_n$ in R such that $a_i^{-1} = a^{-1} b_i$, $i = 1, \ldots, n$.

Proof (*by induction on n*). For $n = 1$ we may take $a = a_1^2$, $b_1 = a_1$. By induction there exist regular elements $a_0, b_1', \ldots, b_{n-1}'$ such that $a_i = a_0^{-1} b_i'$, $i = 1, \ldots, n - 1$. By Ore's condition there exist b and b_n in R, b_n regular, such that $b_n a_n = b a_0$. In this, $b = b_n a_n a_0^{-1}$, and as a product of three regular elements it is also regular. Therefore so is $a = b_n a_n = b a_0$ and a satisfies $a_0^{-1} = a^{-1} b$, $a_n^{-1} = a^{-1} b_n$. Finally, setting $b_i = b b_i'$ for $i = 1, \ldots, n - 1$ we have

$$a_i^{-1} = a_0^{-1} b_i' = a^{-1} b b_i' = a^{-1} b_i \quad ▮$$

The left ideals H of Q are mapped to left ideals of R by

$$\phi\colon H \to H \cap R$$

In the following Theorem 2 we shall give a characterization within R of those left ideals of R which occur as images of left ideals of Q under ϕ. We shall also show that ϕ is a $\cap$-semilattice isomorphism. In the particular case that Q is a semisimple artinian ring we obtain a complemented lattice of left ideals which is, however, only a $\cap$-subsemilattice of the lattice of left ideals of R.

1. The mapping ϕ is one-one.

Proof: Let H_1 and H_2 be left ideals of Q with $\phi H_1 = \phi H_2$ and let x_1 be an element of H_1. Then $x_1 = b_1^{-1} a_1$ where $a_1, b_1 \in R$ and b_1 is regular. Hence

$$a_1 = b_1 x_1 \in H_1 \cap R = H_2 \cap R \subseteq H_2 \tag{1}$$

Therefore H_2 contains $x_1 = b_1^{-1}a_1$. Similarly we show $H_2 \subseteq H_1$.

2. The mapping ϕ is a complete $\cap$-semilattice homomorphism, i.e.

$$\phi \bigcap_{i\in I} H_i = \bigcap_{i\in I} \phi H_i \tag{2}$$

This is clear from the definition $\phi H_i = H_i \cap R$. Since $R = \phi Q$, the set $\{\phi H \mid H \text{ is a left ideal of } Q\}$ *forms a lattice* $\mathfrak{L}'$ provided we define the join of two ideals ϕH_1 and ϕH_2 by

$$\phi H_1 \cup \phi H_2 = \bigcap \{\phi H \mid \phi H_1 + \phi H_2 \subseteq \phi H\}$$

3. The mapping ϕ is a lattice isomorphism of the lattice of all left ideals of Q onto $\mathfrak{L}'$.

Proof. In Q the join of two left ideals L_1 and L_2 is also the meet of all left ideals containing both L_1 and L_2. Since ϕ is one-one by (1), it follows that ϕ preserves joins.

4. The left ideals of R belonging to $\mathfrak{L}'$ may be characterized by the following property: $X \in \mathfrak{L}'$ if and only if, given elements $a, b \in R$, a regular,

$$ab \in X \Rightarrow b \in X \tag{3}$$

Proof. Here $ab \in X = \phi H = H \cap R \subseteq H$ implies $b = a^{-1}ab \in H$. Conversely, given a left ideal X of R, the set

$$QX = \{b^{-1}x \mid b \in R,\ b \text{ regular},\ x \in X\} \tag{4}$$

is the left ideal in Q generated by X; for the left ideal in Q generated by X consists of elements expressible as finite sums of the form

$$\sum a_i^{-1}c_ix_i, \qquad a_i, c_i \in R,\ a_i \text{ regular},\ x_i \in X$$

and by Lemma 1 the a_i^{-1} occurring in this sum are expressible as $a^{-1}b_i$. Thus the above sum may be written in the form

$$a^{-1} \sum b_ic_ix_i = a^{-1}x$$

where $x \in X$. We must now show that if $X \in \mathfrak{L}'$,

$$\phi(QX) = X \tag{5}$$

Clearly $X \subseteq \phi(QX) = QX \cap R$. Conversely, if $b^{-1}x \in QX \cap R$, $b \cdot b^{-1}x = x \in X$. Therefore by the condition in (3), $b^{-1}x \in X$. We collect these results:

Theorem 2. Let Q be a quotient ring of the ring R. Then the mapping

$\phi: H \to H \cap R$, H a left ideal of Q, is a lattice isomorphism of the lattice of all left ideals of Q onto the lattice $\mathfrak{L}'$ of those ideals X of R for which the following implication holds:

$$\text{if } a, b \in R, a \text{ regular, then } ab \in X \Rightarrow b \in X \tag{6}$$

$\mathfrak{L}'$ is a complete $\cap$-subsemilattice of the lattice of all left ideals of R. The join $X_1 \cup X_2$ of two elements in $\mathfrak{L}'$ is defined to be the intersection of all left ideals of $\mathfrak{L}'$ containing both X_1 and X_2. ∎

Theorem 3. Let Q be a quotient ring of the ring R. Let the left ideal X of R be a left annihilator, i.e. let X be such that there exists a subset M of R for which $X = l(M) = \{c \in R \mid cM = 0\}$. Then $\phi(QX) = X$. Therefore $X \in \mathfrak{L}'$.

Proof. Clearly $QXM = (0)$. On the other hand it follows from $a^{-1}bM = (0)$ that $bM = (0)$ and hence $b \in X$ showing that $a^{-1}b \in QX$. Therefore QX is the left annihilator of M in Q. This implies that $\phi(QX) = QX \cap R = X$. ∎

Now in the particular case that Q is semisimple artinian, the left ideals of Q form a complemented lattice satisfying the minimal condition (Theorems IV.1.2 and IV.1.4). Since every semisimple artinian ring is left noetherian, this lattice also satisfies the maximal condition. By Theorem 3, this implies that the left annihilators in an order R in Q satisfy the maximal condition.

If Q is semisimple artinian a further chain condition holds in R, namely, each family of independent left ideals is finite. We call a finite set $\{X_1, ..., X_n\}$ of left ideals *independent* if $X_j \cap \sum_{i \neq j} X_i = (0)$ for each $j = 1, 2, ...$. An arbitrary set of left ideals is said to be independent if each finite subset is. Now let $\{X_1, ..., X_n\}$ be an independent set of left ideals of R. Then the set $\{QX_1, ..., QX_n\}$ of left ideals of Q is also independent. Proof: Assume by way of contradiction that

$$0 \neq b \in QX_j \cap \sum_{i \neq j} QX_i \tag{7}$$

By Lemma 1, b is expressible in the form

$$b = a^{-1}c_j = \sum_{i \neq j} a^{-1}c_i, \qquad \text{where} \quad c_k \in X_k \tag{8}$$

From this it follows that $0 \neq c_j = \sum_{i \neq j} c_i$, contradicting the independence of $\{X_1, ..., X_n\}$. Therefore every set of independent left ideals of R is finite. Finally we prove that there are no nilpotent ideals in an order R in a semisimple artinian ring Q; for assume that the ideal N of R is nilpotent, say

$N^m = (0)$, $N^{m-1} \neq (0)$, $m \geqslant 2$. Then QNQ is an ideal in Q different from (0) and by Theorem III.3.4 it is a direct sum of some of the simple direct summands of Q. The sum of the identity elements of these direct summands is an idempotent e belonging to the center of Q and generating QNQ:

$$QNQ = eQ = Qe$$

As an element of QNQ it is expressible in the form

$$e = \sum q_i n_i r_i\,, \qquad q_i\,, r_i \in Q \quad \text{and} \quad n_i \in N$$

By Lemma 1 we may express q_i in the form $a^{-1}b_i$ and hence

$$e = a^{-1} \sum b_i n_i r_i = a^{-1} \sum x_i r_i \qquad \text{where} \quad x_i = b_i n_i \in N$$

Therefore $ea = ae$ is an element of NQ. This implies

$$N^{m-1}ea \subseteq N^m Q = 0 \cdot Q = (0)$$

Since a is regular it follows that

$$eN^{m-1} = N^{m-1}e = (0)$$

and hence

$$(N^{m-1}Q)^2 \subseteq NQN^{m-1}Q \subseteq QeN^{m-1} = (0)$$

Thus $N^{m-1}Q$ is a nilpotent right ideal of Q different from zero since $N^{m-1} \neq (0)$, contradicting the fact that Q is semisimple artinian.

We have thus proved that R is semiprime in the following sense:

Definition 3. A ring R is *semiprime* if R contains no nilpotent ideals.

If in the above discussion we strengthen the hypotheses and assume that Q is a simple artinian ring, then the order R in Q is *prime*, i.e. (0) is a prime ideal, or, otherwise stated, the product of two nonzero ideals is nonzero.

Proof. Let $B(\neq (0))$ be an ideal of R. Then $QBQ = Q$ contains the identity 1 of Q. Therefore there exist elements $b \in B$, $q \in Q$ and a regular in R such that $1 = a^{-1}bq$. Now let z be an element of the left annihilator $l(B)$ of B. Since $a = bq$, $za = zbq \in zBq = (0)$. Therefore $z = 0$ since a is regular, proving that $l(B) = (0)$. Hence if A and B are ideals in R and $AB = (0)$, then either $A = (0)$ or $B = (0)$.

We state these results in the following:

Theorem 4. Let R be an order in the semisimple artinian ring Q. Then R is a semiprime ring satisfying *Goldie's two chain conditions*:

1. the set of left ideals of R which are annihilators of subsets of R satisfies the maximal condition.

2. every independent set of left ideals of R is finite.

The ring R is prime if and only if Q is simple. ▮

The converse of this theorem also holds, namely, *a semiprime ring satisfying Goldie's two chain conditions possesses a semisimple artinian quotient ring.*

The lattice of left ideals of a semisimple artinian ring is modular, complemented and satisfies the minimal condition. By Theorem 2 this lattice is mapped onto a lattice $\mathfrak{L}'$ of left ideals in the order R in Q by means of $\phi: H \to H \cap Q$.

$\mathfrak{L}'$ is a $\cap$-subsemilattice of the lattice of all left ideals of R. Therefore, to prove the converse of Theorem 4 by attempting to recover the lattice $\mathfrak{L}'$, it is natural to look for those ideals of R which possess a maximal semicomplement in $\mathfrak{L}$, the lattice of left ideals of R, in the sense of the following definition to ${}_R R$:

Definition 4. Let M be an R-module. A submodule K of M is a *semicomplement of the submodule* U of M if $K \cap U = 0$. K is a *maximal semicomplement of* U if K is maximal in the set of all semicomplements of U. The submodule K of M is a *maximal semicomplement* if K is a maximal semicomplement of some submodule of M.

There is a connection between maximal semicomplements and essential extensions:

Definition 5. (cf. Definition VII.3.3). The submodule V of M is an *essential extension in M of the submodule U of M* if every nonzero submodule V' of V intersects U nontrivially, i.e. if $V' \cap U \neq (0)$. U is said to be *closed* in M if U has no essential extension in M other than itself.

Theorem 5. Let U be a submodule of the R-module M. Then U has a maximal semicomplement K. This in turn has a maximal semicomplement E containing U. E is then a maximal essential extension of U in M and is therefore closed. On the other hand, every closed submodule E of M is a maximal semicomplement in M. Finally if U_1 and U_2 are maximal semicomplements with U_1 a maximal semicomplement of U_2, then U_2 is also a maximal semicomplement of U_1.

Proof. 1. The existence of K follows from Zorn's lemma since $U \cap (0) = (0)$. Applying Zorn's lemma again and because $U \cap K = (0)$ we obtain E containing U, maximal subject to $E \cap K = (0)$. We must now show that very submodule $V \neq (0)$ of E intersects U nontrivially. Assume $V \cap U = (0)$. Let $K' = K + V$. Since $V \subsetneqq E$ and $E \cap K = (0)$, $K' \neq K$.

Therefore $U \cap K' \neq (0)$. On the other hand let $x = v + k \in (V + K) \cap U$, $v \in V$, $k \in K$. Then $k = x - v \in E \cap K = (0)$ since $V \subseteq E$ and $x \in U \subseteq E$. Therefore $k = 0$ and $x = v \in V \cap U = (0)$. This implies that $U \cap K' = (0)$, a contradiction. Hence E is an essential extension of U. Assume now that E_1 is an essential extension of U properly containing E. Then $E_1 \cap K \neq (0)$ since E is a maximal semicomplement of K. On the other hand

$$(E_1 \cap K) \cap U = E_1 \cap (K \cap U) = E_1 \cap (0) = (0),$$

a contradiction. Therefore E is a maximal essential extension of U in M and hence closed.

2. Now let E be a closed submodule of M and K a maximal semicomplement of E in M. Then there exists a maximal semicomplement E' of K in M containing E. As was shown in Part 1, E' is an essential extension of E and hence, by the maximality of E, $E' = E$.

3. By Parts 1 and 2, the closed submodules of M are precisely the maximal semicomplements in M. Therefore if U_1 and U_2 are two maximal semicomplements with U_1 a maximal semicomplement of U_2, then U_2, as a maximal essential extension of itself, is a maximal semicomplement of U_1. ▮

Theorem 5 will now be applied to the left ideals L of a semiprime ring R satisfying Goldie's first chain condition. From the definition of an essential extension it is clear that every essential extension of L in ${}_RR$ is contained in the set

$$\hat{L} = \{a \in R \mid (a' \mid \cap L \neq (0) \text{ for all } 0 \neq a' \in (a \mid\} \tag{9}$$

where $(a \mid$ denotes the left ideal of R generated by a. In fact if L' is an essential extension of L and $a \in L'$ then for all $0 \neq a' \in (a \mid$, $(0) \neq (a' \mid \subseteq L'$ and hence $(a' \mid \cap L \neq (0)$. If the set $\hat{L}$ is a left ideal of R then it is the unique maximal essential extension of L in R. Now clearly $R\hat{L} \subseteq \hat{L}$ but in general $\hat{L}$ is not a subgroup of the additive group of the ring R. However if R is semiprime and satisfies Goldie's first chain condition, $\hat{L}$ is indeed a left ideal. To prove this we first need the following:

Lemma 2. Let x be an element of the semiprime ring R satisfying Goldie's first chain condition. Then the left annihilator $l(x)$ possesses a semicomplement different from (0).

Proof. If $x = 0$, the lemma is trivially true. Assume therefore that $x \neq 0$. Let $l(z)$ be maximal in the family $\{l(p) \mid 0 \neq p \in R \text{ and } l(p) \subseteq l(x)\}$. We assert that $RzR \neq (0)$. For suppose not. Then $RzR = (0)$ implies $Rz \subseteq l(R)$. But $(l(R))^2 \subseteq l(R)\, R = (0)$. Therefore since R is semiprime $l(R) = (0)$ and hence

$Rz = (0)$. But as above the right annihilator of R is (0). Hence $z = 0$, contradicting $z \neq 0$ and proving our assertion. Now if $zRx = (0)$ then $zR \subseteq l(x) \subseteq l(z)$ and hence $(RzR)^2 = (0)$ contradicting $RzR \neq (0)$. Thus $zRx \neq (0)$.

Let r be an element in R such that $zrx \neq 0$. By the maximality of $l(z)$, $l(z) = l(zrx)$ or, in other words

$$yzrx = 0 \Rightarrow yz = 0 \tag{10}$$

Now $Rzr \neq (0)$ since, as remarked above, the right annihilator of R is (0). Therefore there exists $v \in R$ such that $vzr \neq (0)$. This element generates a semicomplement $(vzr \mid$ of $l(x)$. Proof: Let $c \in (vzr \mid \cap\, l(x)$. Then $c = v'zr$ for some element $v' \in (v)$ and $v'zrx = cx = 0$. Hence $v'z = 0$ by (10) proving that $c = 0$. ∎

Theorem 6. Let L be a left ideal in the semiprime ring R satisfying Goldie's first chain condition. Then the set

$$\hat{L} = \{a \in R \mid (a' \mid \cap\, L \neq (0) \text{ for all } 0 \neq a' \in (a \mid\} \tag{11}$$

is a left ideal of R. Moreover, $\hat{L}$ is the (unique) maximal essential extension of L in ${}_RR$.

Remark. If we assume the theory of injective modules developed in Section VII.3, then by VII.3.10, $\hat{L}$ is the injective hull of L in ${}_RR$.

Proof of Theorem 6. We must show that $\hat{L}$ is a subgroup of the additive group of R. Let x and y be elements of $\hat{L}$ and set $d = x - y$. We must show that for each element $0 \neq d' \in (d \mid$, the principal left ideal $(d' \mid$ intersects L nontrivially. Now $d' = x' - y'$ for some $x' \in (x \mid$ and $y' \in (y \mid$. By Lemma 2 the left annihilator $l(d')$ of d' possesses a semicomplement $C \neq (0)$. Let $0 \neq c \in C$. Since $C \cap l(d') = (0)$, $cd' = c(x' - y') \neq 0$. Without loss of generality, let $cx' \neq 0$. Since by hypothesis $x \in \hat{L}$, $(cx' \mid$ intersects L nontrivially. Let $0 \neq c'x' \in (cx' \mid \cap\, L$, where $c' \in (c \mid$. If $c'y' = 0$, $0 \neq c'x' = c'(x' - y') = c'd' \in (d' \mid \cap\, L$. On the other hand if $c'y' \neq 0$, $(c'y' \mid \cap\, L \neq (0)$ since $y \in L$. Let $0 \neq c''y' \in (c'y' \mid \cap\, L$ where $c'' \in (c' \mid$. Then $c''x' \in L$ since $c''x' \in (c'x' \mid \subseteq L$. Therefore, since $c'' \in C$ and $C \cap (d' \mid = (0)$, $0 \neq c''(x' - y') \in L \cap (d' \mid$. ∎

The left ideal L of R is therefore closed as a submodule of the module ${}_RR$ if and only if $\hat{L} = L$. This may also be stated as follows:

Theorem 7. Let L be a left ideal of R and let $\hat{L}$ be as in (11). Then $\hat{L} = L$ if and only if for all $b \in R$

$$b \notin L \Rightarrow \text{there exists } 0 \neq b' \in (b \mid \text{such that } (b' \mid \cap L = (0) \qquad (12)$$

The proof is almost trivial: If $\hat{L} = L$, $b \notin L \Rightarrow b \notin L$. Hence there exists $b' \in (b \mid$ such that $(b' \mid \cap L = (0)$ by the very definition of $\hat{L}$. Conversely, if $L \subset \hat{L}$ let $a \in \hat{L} \setminus L$. Then the implication in (12) is not valid for this element. ∎

The following Theorem 8 implies that the closed left ideals of R form a complete $\cap$-semilattice:

Theorem 8. Let $\{L_i \mid i \in I\}$ be a family of left ideals of R, each L_i closed in the sense of definition (12) of Theorem 7. Then the set theoretic intersection $D = \bigcap L_i$ is also closed.

Proof. If $b \notin D$ there is an index i_0 for which $b \notin L_{i_0}$. Thus by (12) there exists an element $b' \in (b \mid$ such that $(b' \mid \cap L_{i_0} = (0)$. Since $D \cap L_i \subseteq L_{i_0}$, $(b' \mid \cap D = (0)$. Therefore $\hat{D} = D$. ∎

From this and the fact that

$$L_1 \subseteq L \Rightarrow \hat{L}_1 \subseteq \hat{L}, \qquad L_1 \text{ and } L \text{ left ideals of } R \qquad (13)$$

we may infer that the mapping

$$L \to \hat{L} \qquad (14)$$

is a closure operator in the ring R. Indeed, by (13) and Theorem 8 we have:

Theorem 9. Let R be a semiprime ring satisfying Goldie's first chain condition. Then the closure $\hat{L}$ of a left ideal L of R may also be defined by $\hat{L} = \bigcap_{i \in I} L_i$ where $\{L_i \mid i \in I\}$ is the set of closed ideals containing L.

Proof. $L \subseteq \hat{L}_i = L \Rightarrow \hat{L} \subseteq \hat{L}_i$ by (13), and $\widehat{\bigcap L_i} = \bigcap L_i$ by Theorem 8. ∎

Since the ring R itself is clearly a closed left ideal, Theorem 9 enables us to define the join of two elements of the complete $\cap$-semilattice of closed left ideals of R in the following way: Let A and B be closed left ideals of R and set

$$A \cup B = \bigcap \{L_i \mid L_i \supseteq A + B, \hat{L}_i = L_i\} \qquad (15)$$

More generally, if $\{A_\lambda \mid \lambda \in \Lambda\}$ is any collection of closed left ideals set

$$\bigcup A_\lambda = \bigcap \{L_i \mid A_\lambda \subseteq L_i \text{ for all } \lambda \in \Lambda\} \tag{16}$$

We see immediately that *the set of closed ideals under set-theoretic intersection and Definitions* (15) *and* (16) *for the join form a lattice* $\mathfrak{L}$ (cf. Definitoin II.3.2). $\mathfrak{L}$ is moreover a complete lattice. The join of two left ideals in $\mathfrak{L}$ is not in general their sum but rather the closure of their sum. We therefore have:

Theorem 10. Let R be a semiprime ring satisfying Goldie's first chain condition. The closed left ideals of R form a complete lattice $\mathfrak{L}$ under the two following operations: The meet is the ordinary set theoretic intersection while the join of two left ideals A and B is defined by

$$A \cup B = \widehat{A + B} = \bigcap \{L_i \mid A + B \subseteq L_i\,,\ \hat{L}_i = L_i\} \tag{17}$$

Proof. The second equality in (17) is merely a corollary to Theorem 9. ▮

The lattice $\mathfrak{L}$ has all the properties which the lattice $\mathfrak{L}'$ of Theorem 2 has. Recall that $\mathfrak{L}'$ is the lattice of those left ideals of the order R in a semisimple artinian ring Q which have the property stated in (6). Later we shall show that actually $\mathfrak{L} = \mathfrak{L}'$ (compare the discussion following Theorem 5). We now set out to prove these properties.

Theorem 11. In a semiprime ring R satisfying Goldie's first chain condition the lattice $\mathfrak{L}$ of closed left ideals is modular and complemented.

Proof of the modularity of $\mathfrak{L}$. Let A, B, $C \in \mathfrak{L}$ and suppose $A \subseteq C$. Since

$$A \subseteq C \Rightarrow A \cup (B \cap C) \subseteq (A \cup B) \cap C$$

is true in any lattice we have only to show that $(A \cup B) \cap C \subseteq A \cup (B \cap C)$. Let $x \in (A \cup B) \cap C = (\widehat{A + B}) \cap C$. By definition of the closure, $x \in \widehat{A + B}$ implies that for each $0 \neq x' \in (x\,|$, $(x'\,| \cap (A + B) \neq (0)$. Let

$$0 \neq x'' \in (A + B) \cap (x'\,|.$$

Then $x'' = a + b$, $a \in A$, $b \in B$ and $(x''\,| \subseteq (x'\,| \subseteq (x\,| \subseteq C$. It follows that $b = x'' - a \in C$ since $A \subseteq C$. Therefore $b \in B \cap C$ and $0 \neq x'' = a + b$ lies in $A + (B \cap C)$. Hence we have shown that $(x'\,|$ intersects $A + (B \cap C)$ nontrivially for each $0 \neq x' \in (x\,|$. Therefore $x \in A \cup (B \cap C)$.

Proof that $\mathfrak{L}$ *is complemented.* Let $L \in \mathfrak{L}$. By Theorem 5 applied to the

R-module ${}_R R$, L is a maximal semicomplement of some ideal $L' \in \mathfrak{L}$ which is itself a maximal semicomplement of L. We must show that

$$L \cup L' = \widehat{L + L'} = R$$

Assume $L \cup L' \subset R$. Then by Theorem 5 there exists $(0) \neq T \in \mathfrak{L}$ such that $(L \cup L') \cap T = (0)$. By the modularity of $\mathfrak{L}$

$$\begin{aligned} L \cap (L' \cup T) \subseteq L \cap (L' \cup L) \cap (T \cup L') &= L \cap (((L' \cup L) \cap T) \cup L') \\ &= L \cap L' = (0) \end{aligned} \tag{18}$$

since $(L' \cup L) \cap T = (0)$. But this implies that $L' \cup T (\supset L')$ is a semicomplement of L contradicting the maximality of L'. Therefore $L \cup L' = R$. ▮

To prove that $\mathfrak{L}$ satisfies the minimal condition we need the hypothesis that every independent family of left ideals of R be finite.

Theorem 12. Let R be a semiprime ring satisfying Goldie's two chain conditions. Then the lattice $\mathfrak{L}$ of closed left ideals of R is modular, complemented, and satisfies the minimal condition.

Proof. Let

$$L_1 \supset L_2 \supset \cdots \tag{19}$$

be a descending chain of closed left ideals. Since $\mathfrak{L}$ is complemented, for each L_{i+1} there exists a closed left ideal C_{i+1} such that

$$C_{i+1} \cap L_{i+1} = (0) \qquad \text{and} \qquad C_{i+1} \cup L_{i+1} = R$$

Then $D_{i+1} = L_i \cap C_{i+1}$ is a relative complement of L_{i+1} in L_i since $D_{i+1} \cap L_{i+1} \subseteq C_{i+1} \cap L_{i+1} = (0)$ and

$$D_{i+1} \cup L_{i+1} = (L_i \cap C_{i+1}) \cup L_{i+1} = L_i \cap (C_{i+1} \cup L_{i+1}) = L_i \cap R = L_i$$

by the modularity of $\mathfrak{L}$. From this it follows that $D_i \neq (0)$ provided $L_i \supset L_{i+1}$. For each natural number n, $D_{n-1} \cap D_n \subseteq D_{n-1} \cap L_{n-1} = (0)$. Therefore $D_{n-1} + D_n = D_{n-1} \oplus D_n$, i.e. $D_{n-1} + D_n$ is a direct sum of D_{n-1} and D_n in the lattice of left ideals of R. Now assume inductively that the sum of any $n(\geqslant 2)$ consecutive D_i is direct and consider $D_j + \cdots + D_{j+n}$. By induction we have $D_{j+1} \oplus \cdots \oplus D_{j+n}$. Now $D_{j+1} = L_{j+i-1} \cap C_{j+1}$. Therefore $D_{j+i} \subseteq L_j$, $i = 1, 2, \ldots$. Hence $D_{j+1} \oplus \cdots \oplus D_{j+n} \subseteq L_j$. But $D_j \cap L_j = L_{j-1} \cap C_j \cap L_j = (0)$ and therefore $D_j \oplus D_{j+1} \oplus \cdots \oplus D_{j+n}$ is a direct sum. Since by hypothesis each independent set of left ideals of R is finite it follows that the chain in (19) must be finite. ▮

We have thus shown that under the hypotheses of Theorem 12 the lattice $\mathfrak{L}$ of closed left ideals of R has the same properties as the lattice of left ideals of a semisimple artinian ring Q. In the remainder of this chapter we shall show that in such a ring R the closed left ideals are precisely those satisfying the implication in (6) of Theorem 2 and moreover that Ore's condition (Theorem 1) holds in R. R then possesses a quotient ring Q whose left ideals H are mapped isomorphically by $\phi: H \to H \cap R$ onto the closed left ideals of R, i.e. onto the elements of the lattice $\mathfrak{L} = \mathfrak{L}'$. By Theorem 12 it will then follow that Q is semisimple artinian.

To carry out the program outlined above we first remark that Goldie's first chain condition implies the *minimal condition on right annihilators.* For let

$$r(M_1) \supset r(M_2) \supset \cdots \tag{20}$$

be a descending chain of right annihilators. Then since

$$r(l(r(M))) = r(M) \tag{21}$$

we obtain an ascending chain

$$l(r(M_1)) \subset l(r(M_2)) \subset l(r(M_3)) \subset \cdots$$

of left annihilators. The equality in (21) follows from

$$r(M) \subseteq r(l(r(M)))$$

and the fact that

$$M \subseteq l(r(M)) \Rightarrow r(M) \supseteq r(l(r(M)))$$

In a ring satisfying the hypotheses of Theorem 12, the *minimal condition also holds for left annihilators* since, by the following theorem, left annihilators are closed.

Theorem 13. Let R be a ring satisfying the hypotheses of Theorem 12. Then for any subset M of R, the left annihilator $l(M)$ is closed.

Proof. By Theorem 8 it is sufficient to prove the theorem in the case $M = \{a\}$. By the characterization of closed left ideals given in Theorem 7 we must show that if $b \notin l(a)$ there exists $c \in R$ such that $cb \neq 0$ but $(cb \mid \cap\, l(a) = (0)$. Since $b \in l(a)$, it follows that $ba \neq 0$. By Lemma 2, $l(ba)$ possesses a semicomplement $C \neq (0)$. Let $0 \neq c \in C$. Then $cba \neq 0$ since $C \cap l(ba) = (0)$. Let $x \in (cb \mid \cap\, l(a)$. Then $x = c'b$ for some $c' \in (c \mid \subseteq C$ and $c'ba = 0$. This implies $c' \in l(ba) \cap C = (0)$ proving that $c' = 0$ and hence $x = 0$. ∎

As an immediate consequence of this result we have:

Theorem 14. Let R be a ring satisfying the hypotheses of Theorem 12. Then $r(\hat{L}) = r(L)$ for each left ideal L of R.

Proof. $L \subseteq \hat{L} \Rightarrow r(L) \supseteq r(\hat{L})$. Let $b \in r(L)$. Then from $L \subseteq l(b)$ and Theorem 13 it follows that $\hat{L} \subseteq \widehat{l(b)} = l(b)$. Therefore $\hat{L}b = (0)$ and hence $b \in r(\hat{L})$. ▮

Theorem 15. Under the hypotheses of Theorem 12, an element $a \in R$ is regular if and only if $\widehat{Ra} = R$. If fact, $l(a) = (0)$ already implies the regularity of a.

Proof. We first show that

$$\widehat{Rx} = \widehat{Ry} = R \Rightarrow \widehat{Rxy} = R \tag{22}$$

For $L \neq (0)$ a left ideal of R, set $L' = \{r \in R \mid ry \in L\}$. Clearly $l(y) \subseteq L'$. On the other hand $(0) \neq Ry \cap L = L'y$. Therefore $l(y) \neq L'$. Choose $b \in L' \backslash l(y)$. Since $\widehat{l(y)} = l(y)$ there exists by Theorem 7, $0 \neq b' \in (b)$ such that $(b' | \cap l(y) = (0)$. Now set $L'' = \{r \in R \mid rx \in (b' |\}$.
Then $L''x = Rx \cap (b' | \neq (0)$ since $\widehat{Rx} = R$. Now $b' \in (b | \subseteq L'$ and hence

$$(0) \neq L''xy \subseteq Rxy \cap (b' | y \subseteq Rxy \cap L$$

Moreover,

$$\widehat{Ra} = R \Rightarrow a \quad \text{regular} \tag{23}$$

since by Theorem 14,

$$r(a) \subseteq r(Ra) = r(\widehat{Ra}) = r(R) = (0)$$

To show that $(a) = (0)$ consider the ascending chain

$$l(a) \subseteq l(a^2) \subseteq l(a^3) \subseteq \cdots$$

of left annihilators. By Goldie's first chain condition there exists n for which $l(a^n) = l(a^{n+1})$. Let $x \in Ra^n \cap l(a)$. Then $x = ya^n$ and $0 = xa = ya^{n+1}$. From this it follows that $y \in l(a^{n+1}) = l(a^n)$ and hence $x = 0$. By (22) and since $Ra = R$, $Ra^n = R$. This however contradicts $Ra^n \cap l(a) = (0)$ if $l(a) \neq (0)$. To prove the last assertion of Theorem 15 it is sufficient to prove the implication

$$l(c) = (0) \Rightarrow \widehat{Rc} = R \tag{24}$$

Suppose by way of contradiction that $Rc \cap L = (0)$ for some left ideal $L \neq (0)$. Form the left ideals Lc^n, $n = 0, 1, 2, \ldots$. Each one of these ideals is different from zero, since $l(c) = (0)$ and $Lc^m = (0)$ imply $L = (0)$. Moreover they form a linearly independent set of ideals; for suppose

$$a_{n+1}c^{n+1} = a_0 + a_1c + \cdots + a_nc^n, \qquad a_i \in L$$

Then

$$a_0 = a_{n+1}c^{n+1} - a_1c - \cdots - a_nc^n \in Rc \cap L = (0)$$

Thus

$$a_{n+1}c^{n+1} = a_1c + \cdots + a_nc^n$$

But since $l(c) = 0$,

$$a_{n+1}c^n = a_1 + \cdots + a_nc^{n-1}$$

As above, $a_1 = 0$. Proceeding in this fashionwe eventually arrive at $a_{n+1} = 0$. Thus Lc^i, $i = 0, 1, \ldots$ is an infinite set of linearly independent left ideals, contradicting Goldie's second chain condition. ▮

This characterization of regular elements allows us to prove that in a ring R satisfying the hypotheses of Theorem 12, the lattice $\mathfrak{L}$ of closed left ideals of R is contained in the lattice $\mathfrak{L}'$ mentioned in Theorem 2. In other words, for each $X \in \mathfrak{L}$,

$$a \text{ regular and } ab \in X \Rightarrow b \in X \tag{25}$$

Proof. Let $\hat{X} = X$, let a be regular and let $ab \in X$. It follows that $Ra \subseteq (X : b)$ and by Theorem 15, $R = \widehat{Ra} \subseteq \widehat{(X : b)}$. Hence $\widehat{(X : b)} = R$. Assume $b \notin X = \hat{X}$. By Theorem 7 there exists $0 \neq b' \in (b \mid$ such that $(b' \mid \cap X = (0)$. In the case $b' = yb$ for some element $y \neq 0$ in R, then

$$(y \mid b \cap X = (yb \mid \cap X = (b' \mid \cap X = (0)$$

and hence

$$(y \mid \cap (X : b) = (0) \qquad \text{contradicting} \quad \widehat{(X : b)} = R$$

If, however, $0 \neq b' = zb$ for some integer z, then $R \cdot zb \cap X \subseteq (b' \mid \cap X = (0)$. Therefore $Rz \cap (X : b) = (0)$ again contradicting $\widehat{(X : b)} = R$. Hence

$$\mathfrak{L} \subseteq \mathfrak{L}' \tag{26}$$

The proof that equality holds in (26) for prime rings satisfying Goldie's two chain conditions may be based on the following:

Theorem 16. Let R be a prime ring satisfying Goldie's two chain conditions and let L be a left ideal with $\hat{L} = R$. Then L contains a regular element.

Proof. Since left annihilators are closed and the lattice $\mathfrak{L}$ of closed left ideals in R satisfies the minimal condition, there exists an element $a \in L$ such that $l(a)$ is minimal in the family of left annihilators of elements of L. Assume a is not regular and hence by Theorem 14, $Ra \neq R$. Then there exists a left ideal $K \neq (0)$ of R such that $Ra \cap K = (0)$. Furthermore, since, by hypothesis, $\hat{L} = R$, $K \cap L \neq (0)$ and therefore K may be chosen so that $K \subseteq L$. Now let $k \in K$ and $x \in l(a + k)$. Then $x(a + k) = 0$ and hence $xa = -xk \in Ra \cap K = (0)$. Thus $x \in l(a) \cap l(k)$ proving that $l(a + k) \subseteq l(a) \cap l(k) \subseteq l(a)$. But $a + k \in L$ and $l(a)$ is minimal subject to the conditions set out above. Therefore $l(a) = l(a + k)$ and $l(a) = l(a) \cap l(k)$. This implies that $l(a) \subseteq l(k)$ or, in other words, $l(a) \cdot k = 0$ for all $k \in K$. Since by hypothesis R is prime it follows from $l(a) \cdot K = (0)$ and $K \neq (0)$ that $l(a) = (0)$. Therefore by (24), $Ra = R$, contradicting our earlier assumption that $Ra \neq R$. ▮

We may now prove that $\mathfrak{L} = \mathfrak{L}'$ for prime rings R satisfying Goldie's chain conditions and furthermore that such rings possess a quotient ring Q. This follows from the following more general theorem, applied to Theorem 16:

Theorem 17. Let R be semiprime satisfying Goldie's two chain conditions and let each left ideal $L \subseteq R$ for which $\hat{L} = R$ contain a regular element. Then the closed left ideals of R are precisely those for which the implication

$$a, b \in R, \qquad a \text{ regular } ab \in X \Rightarrow b \in X \tag{27}$$

is valid. Moreover R possesses a quotient ring Q.

Proof. 1. In (26) we showed $\mathfrak{L} \subseteq \mathfrak{L}'$. Suppose now that the left ideal X is not closed. We must show that $X \notin \mathfrak{L}'$. Let $b \in \hat{X} \backslash X$. We first prove that $\widehat{(X : b)} = R$. Assume by way of contradiction that for some left ideal $L \neq (0,)$ $L \cap (X : b) = (0)$. Then since $b \neq 0$ there exists $c \in L$ such that $cb \neq 0$. Since $b \in \hat{X}$ it follows that $(cb \mid \cap X \neq (0)$ and hence $(0) \neq (c \mid \cap (X : b)$ contradicting our assumption that $L \cap (X : b) = (0)$. Since $\widehat{(X : b)} = R$, by hypothesis there is a regular element $a \in (X : b)$. Now for the pair a, b we have a regular, $ab \in X$ but $b \in X$. Hence $X \notin \mathfrak{L}'$.

2. Let a and b be elements of R with a regular. By Theorem 1 we need only show that this pair satisfies Ore's condition, i.e. we must show that there exist elements a_1 and b_1 with a_1 regular such that $a_1 b = b_1 a$. By the regularity of a it follows that $\widehat{Ra} = R$. Consider now the left ideal $L = \{c \in R \mid cb \in Ra\}$. We show $\hat{L} = R$. If X is a left ideal for which $Xb = (0)$ then $X \subseteq L$. If on the other hand $Xb \neq (0)$, then since $\widehat{Ra} = R$, $Xb \cap Ra \neq (0)$.

In this case let $0 \neq xb \in Ra$, where $x \in X$. Then $0 \neq x \in L$ and $X \cap L \neq (0)$ proving that $\hat{L} = R$. By hypothesis L therefore contains a regular element a_1. By definition of L, Ra contains $a_1 b$. Therefore there exists $b_1 \in R$ with $a_1 b = b_1 a$. ∎

All that we claimed has now been proved for prime rings satisfying Goldie's two chain conditions. In fact by Theorem 16 such rings fulfill the hypotheses of Theorem 17 and it therefore follows that the mapping

$$\phi: H \to H \cap R$$

of Theorem 2 is a monomorphism of the lattice of left ideals H of the quotient ring Q onto the lattice $\mathfrak{L}' = \mathfrak{L}$ of closed left ideals of R. By Theorem 12 this lattice is complemented and satisfies the minimal condition. Therefore Q is semisimple artinian. (The stronger condition that R be prime actually enables us to conclude that Q is simple by Theorem 4.)

To prove the same results for semiprime rings satisfying Goldie's two chain conditions it is sufficient to prove that each left ideal L of R for which $\hat{L} = R$ contains a regular element. This we now proceed to do.

Since by Theorem 13 the left annihilators of subsets of such rings are closed, there exists an ideal P of R which is minimal in the family of left annihilators of left ideals of R. Considered as a subring of R, P satisfies the conclusion of the following:

Lemma 3. P is a prime ring satisfying Goldie's two chain conditions.

Proof. 1. Let $T \neq (0)$ be a left ideal of the ring P. If $RT = (0)$ then $\{x \mid Rx = (0)\}$ is a nonzero nilpotent ideal of R, contradicting semiprimeness of R. Therefore $RT \neq (0)$. Assume $PT = (0)$. Then since P is an ideal of R, $P \cdot RT = PR \cdot T \subseteq PT = (0)$. It follows from $RT \subseteq RP \subseteq P$ that $(RT)^2 \subseteq P \cdot RT = (0)$ and hence $RT = (0)$ contradicting what we have just proved. Thus PT is a nonzero left ideal of R contained in T.

2. It follows immediately from this that P satisfies Goldie's second chain condition since the independence of the left ideals $\{T_1, ..., T_q\}$ of P implies the independence of the left ideals $\{PT_1, ..., PT_q\}$ of R.

3. Let T_1 and T_2 be left ideals of P which are the left annihilators in P of the subsets M_1 and M_2, respectively. Since $T_i = l(M_i) \cap P$, then $T_1 \subset T_2$ implies $l(M_1) \subset l(M_2)$. Therefore a strictly increasing sequence of left annihilators in P gives rise to a strictly increasing sequence of left annihilators in R. Hence P satisfies Goldie's first chain condition.

4. Let A and B be ideals with $AB = (0)$ and $A \neq (0)$. Since $PB \subseteq B$, $APB = (0)$ and hence $A \subseteq l(PB) \cap P$. But $l(PB) \cap P$ is the intersection of

two annihilator ideals and is therefore itself an annihilator ideal. By the minimality of P, $l(PB) \cap P = P$ and hence $P \subseteq l(PB)$. From this inclusion it follows that $(PB)^2 \subseteq P \cdot PB = (0)$. But PB is a left ideal of R. Therefore $PB = (0)$. By Part 1 above this implies $B = (0)$. Hence P is a prime ring. ▮

With this result we base the proof of the following theorem on Theorem 16:

Theorem 18. Let R be a semiprime ring satisfying Goldie's two chain conditions and let L be a left ideal of R such that $\hat{L} = R$. Then L contains a regular element.

Proof. 1. Let $S = P_1 \oplus \cdots \oplus P_n$ be a maximal direct sum of minimal annihilator ideals (minimal in the sense of Lemma 3). Then the closure of S considered as a left ideal of R is R:

$$\hat{S} = R \tag{28}$$

Proof of (28): Let $K \neq (0)$ be a left ideal of R. Assume by way of contradiction that $S \cap K = (0)$. Then since $SK \subseteq S \cap K$, $SK = (0)$ and hence $K \subseteq r(S)$. On the other hand $S \cap r(S) = (0)$ since R is semiprime. Now $r(S) \cdot S \subseteq r(S) \cap S = (0)$ and hence $r(S) \subseteq l(S)$. Similarly $r(S) \subseteq l(S)$. Since $(0) \neq K$ is contained in $r(S) = l(S)$, the left annihilator $l(S)$ is nonzero and therefore contains a minimal annihilator ideal, say P_{n+1}. Since $S \cap l(S) = (0)$, $P_1 \oplus \cdots \oplus P_n \oplus P_{n+1}$ is a direct sum of annihilator ideals contradicting the maximality of $P_1 \oplus \cdots \oplus P_n$. Hence there is no left ideal $K \neq (0)$ of R for which $S \cap K = (0)$. In other words, $\hat{S} = R$.

2. Now let L be a left ideal of R such that $\hat{L} = R$. We show that for each $i = 1, \ldots, n$ the closure of $L \cap P_i$ considered as a left ideal of the ring P_i is P_i. Proof: Let $T \neq (0)$ be a left ideal of P_i. Then, as in the proof of Lemma 3, $P_i T$ is a nonzero left ideal of R contained in T. Since the closure of L is R by hypothesis,

$$(0) \neq L \cap P_i T \subseteq (L \cap P_i) \cap P_i T \subseteq (L \cap P_i) \cap T$$

Thus the closure of $L \cap P_i$ in P_i is P_i. Now by Lemma 3, P_i is a prime ring satisfying Goldie's two chain conditions. We may therefore apply Theorem 16 to the left ideal $L \cap P_i$ of the ring P_i to obtain a regular element p_i in $L \cap P_i$. With these p_i, $i = 1, \ldots, n$ we form an element contained in L which is regular in the whole ring R. Let $p = p_1 + \cdots + p_n$. Assume $l(p) \neq (0)$. Then since $\hat{S} = R$, $l(p) \cap S \neq (0)$. Let $0 \neq q = q_1 + \cdots + q_n \in l(p) \cap S$ where $q_i \in P_i$. Since the sum of the ideals P_i is direct,

$$0 = qp = q_1 p_1 + \cdots + q_n p_n .$$

But $q_i p_i \in P_i$ and $l(p_i) \cap P_i = (0)$. Therefore $q_i = 0$ and $q = 0$, a contradiction. Hence $l(p) = (0)$ and by Theorem 15 this implies p is regular. ▌

As a consequence of this result, Theorem 17 is applicable to semiprime rings satisfying Goldie's two chain conditions. Therefore we have finally:

Theorem 19 (Goldie). A ring R is an order in a semisimple artinian ring Q if and only if R is semiprime and satisfies Goldie's two chain conditions (cf. Theorem 4). Moreover, Q is simple if and only if R is prime (cf. Theorem 4). ▌

A left noetherian ring R, i.e. a ring R satisfying the maximal condition on left ideals, clearly satisfies both of Goldie's chain conditions. Let W be a maximal nilpotent left ideal of R. Since the sum of two nilpotent left ideals is again nilpotent $[L_1{}^m = L_2{}^n = (0) \Rightarrow (L_1 + L_2)^{m+n-1} = (0)]$, W is the maximum of all nilpotent left ideals of R. W is moreover an ideal since for each $c \in R$, Wc is nilpotent. W is a generalization of the *Wedderburn–Artin radical* and a special case of the *Baer lower radical*. The factor ring R/W has no nonzero nilpotent left ideals:

Proof. Let $\bar{L}^n = (\bar{0})$ where $\bar{L}$ is a left ideal of $\bar{R} = R/W$. Then $L^n \subseteq W$. But $W^m = (0)$ for some m. Hence $L^{nm} \subseteq W^m = (0)$ and $L \subseteq W$. Therefore $\bar{L} = (\bar{0})$. ▌

In other words:

Theorem 20. Let W be the maximum of all nilpotent left ideals of the left noetherian ring R. Then R/W is an order in a semisimple artinian ring Q. ▌

To complete our investigations we still have to show that Ore's condition (cf. Theorem 1) on a ring R is sufficient to ensure that R possess a (left) quotient ring Q. In essence the method of constructing Q from R is very similar to that used to construct the rationals from the integers. We form a certain set of pairs of elements of R and define an equivalence relation on this set. In this case, however, if Q is a (left) quotient ring of R and if $b^{-1}a = d^{-1}c$ it does not necessarily follow that either $da = cb$ or $bc = da$. Because of this difficulty we define our equivalence relation as follows:

Let R be a ring satisfying Ore's condition and let a, b, c, d be elements of R with b and d regular. We say the pair (a, b) is in the relation $\sim$ to the pair (c, d) if whenever (x, y) is a pair of regular elements for which $xb = yd$ then also $xa = yb$. We describe this situation symbolically by

$$(x, y) \to (a, b) \sim (c, d). \tag{29}$$

By Ore's condition, given (b, d) as above, there exists (x, y) with y regular such that $xb = yd$. We first show that x is also regular. Since y and d are both regular so is yd. Then $l(yd) = (0)$ implies $l(x) = 0$. For rings satisfying the hypotheses of Theorem 12 this already implies that x is regular by Theorem 15. However, we want to present Ore's theorem in full generality and so we make no assumption about R other than that it satisfies Ore's condition. By Ore's condition there exists a pair (p, q) with q regular such that $pb = gd$. Since $xb = yd$ is regular there exists a pair (f, g), g regular, such that $fxb = gpb$. Since b is not a right zero divisor it follows that $fx = gp$. Thus fx is regular since gp is and hence x is not a left zero divisor.

To determine whether (a, b) is in the same equivalence class (c, d) it is sufficient to check that for some pair (x, y) of regular elements, $xb = yd$ implies $xa = yc$. For in this case, if (x_1, y_1) is another pair of regular elements with $x_1 y = y_1 d$, there exists a regular pair (f, g) with $fy = gy_1$. Then $gx_1 b = gy_1 d = fyd = fxb$ and hence $gx_1 = fx$. But then $gx_1 a = fxa = fyc = gy_1 c$ from which we infer that $x_1 a = y_1 c$.

The reflexivity and symmetry of the relation $\sim$ is easily proved. We now prove transitivity: Let

$$(x, y) \to (a, b) \sim (c, d) \qquad \text{and} \qquad (u, v) \to (c, d) \sim (e, f)$$

By Ore's condition we can find (p, q) such that $pu = qy$. It is then easy to check that

$$(r, s) \to (a, b) \sim (e, f)$$

where $r = qx$ and $s = pv$.

The relation $\sim$ is therefore an equivalence relation and gives rise to a decomposition of the set $\{(a, b) \mid b \text{ regular}\}$ into equivalence classes. Denote the class to which (a, b) belongs by a/b. It is natural to try to define the sum of the two classes a/b and c/d by

$$\frac{a}{b} + \frac{c}{d} = \frac{xa + yc}{yd} \tag{30}$$

where (x, y) is a pair of regular elements for which $xb = yd$. We must of course show that the right-hand side of (30) is independent of (x, y) and the representatives (a, b) and (c, d) of the classes a/b and c/d, respectively. To arrive at the definition of the product of two of these equivalence classes assume first that R has a quotient ring Q. Then $a/b = a^{-1}b$ and the product $b^{-1}ad^{-1}c$ is obtained by expressing ad^{-1} in the form $w^{-1}z$ and substituting to get

$$b^{-1}ad^{-1}c = b^{-1}w^{-1}zc = (wb)^{-1}zc = \frac{zc}{wb}$$

These heuristic considerations lead us to the following definition of the product:

$$\frac{a}{b}\cdot\frac{c}{d}=\frac{zc}{wb} \qquad \text{where} \quad wa = zd \quad \text{and} \quad w \text{ is regular} \tag{31}$$

Again we must show that the right-hand side of (31) is independent of (w, z) and also of the representatives of the respective equivalence classes.

If for example (w', z') is such that $w'a = z'd$ we must show that $mw'b = nwb$ implies $mz'c = nzc$. Now it follows from the regularity of b that $mw' = nw$ and hence $mz'd = mw'a = nwa = nzd$ showing that $mz' = nz$. This proves

$$(m, n) \to (zc, wb) \sim (z'c, w'b)$$

To show that the right-hand side of (31) is independent of the respective representatives of the factors, let $a/b = a'/b'$ and $c/d = c'/d'$. Then

$$(x, y) \to (a, b) \sim (a', b') \qquad \text{and} \qquad (u, v) \to (c, d) \sim (c', d')$$

Let

$$wa = zd \qquad \text{and} \qquad w'a' = z'd'$$

Thus

$$\frac{a}{b}\cdot\frac{c}{d}=\frac{zc}{wb} \qquad \text{and} \qquad \frac{a'}{b'}\cdot\frac{c'}{d'}=\frac{z'c'}{w'b'}$$

Since both wb and $w'b'$ are regular, there exists a regular pair (p, p') such that

$$pwb = p'w'b'$$

We show that

$$(p, p') \to (zc, wb) \sim (z'c', w'b')$$

Since (pw, x) is regular, there exists a regular pair (m, n) such that

$$mpw = nx$$

Then $nyb' = nxb = mpwb = mp'w'b'$ and hence $ny = mp'w'$. Therefore $mp'w'a' = nya' = nxa = mpwa$ and so $p'w'a' = pwa$ proving that $p'z'd' = pzd$. Since u is regular there is a pair (r, s) with r regular such that $rpz = su$. Then $rp'z'd = rpzd = sud = svd'$ and hence $rp'z' = sv$. This implies $rp'z'c' = svc' = suc = rpzc$. Since r is regular it follows that $p'z'c' = pzc$.

In a similar fashion we may prove that the right-hand side of (30) is independent of the choice of the "common denominator" and of the choice of representatives of the respective summands.

Furthermore we can show that the equivalence classes under addition as defined in (30) and multiplication as defined in (31) form a ring Q.

The associativity of multiplication, for example, may be proved as follows: For the product

$$\left(\frac{a}{b} \cdot \frac{c}{d}\right) \cdot \frac{e}{f}$$

let $wa = zd$. Then

$$\left(\frac{a}{b} \cdot \frac{c}{d}\right) \cdot \frac{e}{f} = \frac{zc}{wb} \cdot \frac{e}{f}$$

For the product $(zc/wb) \cdot (e/f)$ let $u \cdot zc = v \cdot f$. Then

$$\left(\frac{a}{b} \cdot \frac{c}{d}\right) \cdot \frac{e}{f} = \frac{ve}{uwb}$$

On the other hand it follows from $uz \cdot c = v$ that

$$\frac{a}{b} \cdot \left(\frac{c}{d} \cdot \frac{e}{f}\right) = \frac{a}{b} \cdot \frac{ve}{uzd}$$

Since $wa = zd$, $u^2wa = u \cdot uzd$ and hence

$$\frac{a}{b} \cdot \left(\frac{c}{d} \cdot \frac{e}{f}\right) = \frac{uve}{uuw} = \frac{ve}{uwb}$$

The ring Q possesses an identity, namely, the class c/c, c regular. Proof:

$$\frac{a}{b} \cdot \frac{c}{c} = \frac{zc}{wb} \qquad \text{where} \quad wa = zc$$

By hypothesis there exists a regular pair (x, y) such that $xwb = yb$. Thus $xw = y$ and so $xzc = xwa = ya$, i.e.

$$(x, y) \to (zc, wb) \sim (a, b)$$

Similarly we may show that c/c is a left identity in Q.

The ring R may be embedded in Q by means of the mapping

$$\eta\colon a \to \frac{ca}{c}, \qquad a \in R$$

Here η is independent of the choice of the regular element c in R since the equality

$$\frac{ca}{c} = \frac{da}{d}$$

follows from the implication $xc = yd \Rightarrow xca = yda$. η is multiplicative since, from $wca = zc$, it follows that

$$\frac{ca}{c} \cdot \frac{cb}{c} = \frac{zcb}{wc} = \frac{wcab}{wc} = \frac{cab}{c}$$

The inverse b^{-1} of the image $\eta b = cb/c$ of the element $b \in R$ is c/cb since from $c \cdot cb = c \cdot cb$ follows

$$(\eta b)b^{-1} = \frac{cb}{c} \cdot \frac{c}{cb} = \frac{cc}{cc} = \frac{c}{c}$$

Finally $b^{-1}a = a/b$ for all pairs (a, b) with b regular. Proof: From $c \cdot c = c \cdot c$ it follows that

$$\frac{c}{cb} \cdot \frac{ca}{c} = \frac{c \cdot ca}{c \cdot cb} = \frac{ca}{cb} = \frac{a}{b}$$

Rings of Continuous Functions

CHAPTER XII

1. Biregular right algebras

In the first nine chapters we have studied rings by representing them as rings of linear mappings of some right K-module M into itself. In this representation, the product ab of two elements a and b of R corresponded to the K-linear mapping of M into itself obtained by the successive application of b and a, in that order:

$$(ab)\, m = a(bm) \qquad \text{for all} \quad m \in M$$

Rings may also be studied from a different point of view. Starting with a set $\mathscr{P}$ of elements (called points) we consider the set R of functions from $\mathscr{P}$ to some set W. If W has an algebraic structure—for example a commutative group structure written additively—we may make R into an abelian group by defining

$$a + b\colon P \to a(P) + b(P) \qquad \text{for} \quad P \in \mathscr{P}, \quad a, b \in R$$

In case W is a ring, this additive group of functions forms a ring with the product ab defined by

$$ab\colon P \to a(P)\, b(P) \qquad \text{for} \quad P \in \mathscr{P}$$

If the ring W contains an identity 1, the function

$$e\colon P \to 1 \qquad \text{for all} \quad P \in \mathscr{P}$$

is an identity for the ring R of functions.

In Chapter III, to describe the structure of semisimple rings more precisely, we introduced the concept of a subdirect sum (Section III.2). There we had a set of functions on an index set I, the index set of the primitive ideals P_i of R. Of course as an indexing set we may take the set $\mathscr{P}$ of primitive ideals of the ring R. We then get the representation

$$R \approx \sum_{P \in \mathscr{P}}{}_s\, R/P$$

of the semisimple ring R as a subdirect sum of the primitive rings R/P, where an element $a \in R$ corresponds to the function a^* defined by

$$a^*(P) = a + P \in R/P \qquad \text{for all} \quad P \in \mathscr{P}$$

The sum $a + b$ and product ab of a and b in R correspond to the sum and product of the functions a^* and b^* defined by

$$(a^* + b^*)(P) = a^*(P) + b^*(P), \qquad (a^* \cdot b^*)(P) = a^*(p) \cdot b^*(P)$$

There are similarities between this and the situation described at the beginning of this section. Nevertheless, there are certain differences. At the beginning of the section we had a set of functions defined on a set $\mathscr{P}$ with values in a single ring W. Here the value of a^* at $P \in \mathscr{P}$ is in the ring R/P which depends on P, so that we may no longer speak of "the" ring W containing all the values of the functions a^*, b^*, ... Even in the case that all the factor rings R/P, $P \in \mathscr{P}$ are isomorphic to one another (and hence to a ring K) we do not in general obtain any interesting results. Indeed, in such a case we may choose for each $P \in \mathscr{P}$ an isomorphism

$$\psi_P\colon a + P \to \psi_P(a + P), \qquad a + P \in R/P \tag{1}$$

of R/P onto K and then replace a^* by the function a' defined on $\mathscr{P}$ with values in K given by

$$a'(P) = \psi_P(a + P) \qquad \text{for} \quad P \in \mathscr{P} \tag{2}$$

Again we obtain a ring of functions isomorphic to the original ring R. However, the freedom of choice we have for the isomorphisms ψ_P makes it in general impossible for the functions a' to possess "desirable" properties. The situation would be somewhat different if it were possible to choose the isomorphisms in a "natural way."

The following investigation of rings A of functions a', b', ... defined on a set $\mathscr{P}$ of points P with values in a ring K containing an identity 1 will give a rough idea of the structure of a semisimple ring R with "natural isomorphisms" of R/P to K. First we make A into a right K-module by defining

$$(a'k)(P) = [a'(P)]\,k \qquad \text{for all} \quad a' \in A,\ k \in K, \text{ and } P \in \mathscr{P} \tag{3}$$

Clearly $a'(k_1 + k_2) = a'k_1 + a'k_2$, $(a_1' + a_2')\,k = a_1'k + a_2'k$ and $[a'(k_1k_2)](P) = [a'(P)](k_1k_2) = [a'(P)\,k_1]\,k_2 = ([a'k_1)\,k_2)(P)$ proving that A is indeed a right K-module. Moreover,

$$(a'b')\,k = a'(b'k) \qquad \text{for all} \quad a', b' \in A \text{ and } k \in K \tag{4}$$

since

$$((a'b')\,k)(P) = [a'(P)\,b'(P)]\,k = a'(P) \cdot ([b'k](P))$$

for all $P \in \mathscr{P}$. In otherwords, A is a right algebra over K in the sense of the following:

Definition 1. Let A and K be rings and let K contain an identity. Then A is a *right algebra over K* if A is a right K-module satisfying the further condition

$$(ab)\,k = a(bk) \qquad \text{for} \quad a, b \in A \text{ and } k \in K \tag{5}$$

In Chapter I an algebra was only defined over *commutative* rings Φ. Clearly an algebra in that sense is a right algebra with the further condition that $(ab)\,k = (ak)\,b$ for all $k \in \Phi$. In the definition of a right algebra given above, we do not exclude the possibility that K be noncommutative.

The ring A of functions on a set $\mathscr{P}$ with values in a ring K with identity contains many *central idempotents* e, i.e. idempotents in the center Z of A.

Indeed, to each subset S of $\mathscr{P}$ the so-called *characteristic function* g_S *of the set* S defined by

$$g_S(P) = \begin{cases} 1 & \text{if} \quad P \in S \\ 0 & \text{if} \quad P \notin S \end{cases} \tag{6}$$

is a central idempotent in A as may be easily seen by checking

$$g_S(P)\,g_S(P) = g_S(P)$$
$$a'(P) \cdot g_S(P) = g_S(P)\,a'(P) \qquad \text{for all} \quad a' \in A$$

In particular A contains the characteristic function $e_{a'}$ of the *support*

$$T(a') = \{P \mid P \in \mathscr{P},\ a'(P) \neq 0\} \tag{7}$$

of the function a'. In addition to being an idempotent, $e_{a'}$ satisfies

$$a' = a' \cdot e_{a'} \tag{8}$$

whence

$$a' \in (e_{a'}) \qquad \text{which implies} \quad (a') \subseteq (e_{a'}) \tag{9}$$

Here (x) denotes the principal ideal in A generated by x, i.e. the smallest ideal containing x.

The inclusion in (9) raises the important question of whether the principal ideal (a') actually coincides with the principal ideal $(e_{a'})$. For an element $k \in K$ and $a' \in A$ define

$$T(a', k) = \{P \mid P \in \mathscr{P},\ a'(P) = k\} \tag{10}$$

This is the set of points at which a' takes on the value k. $T(a', k)$ can, of course, be empty. In each case A contains the characteristic function $g_{T(a',k)}$ of this set. Therefore the principal ideal (a') contains the function $a' \cdot g_{T(a',k)}$ with values

$$[a' g_{T(a',k)}](P) = \begin{cases} k & \text{for} \quad P \in T(a', k) \\ 0 & \text{otherwise} \end{cases} \tag{11}$$

If $T(a', k)$ contains the point P and $k \neq 0$, then $k = a'(P)$ is an element of K different from 0. *Now if we take K to be simple*, the principal ideal generated by k is the whole of K. In particular, the identity 1 of K is expressible in the form

$$1 = \sum_{1 \leqslant j \leqslant n} k_j k l_j \tag{12}$$

where k_j and l_j are suitable elements of K. The functions $g_{T(a',k)} k_j$ and $g_{T(a',k)} l_j$ are contained in A. The principal ideal (a') in A generated by a' contains therefore the function

$$\sum_{1 \leqslant j \leqslant n} [g_{T(a',k)} k_j][a' g_{T(a',k)}][g_{T(a',k)} l_j] \tag{13}$$

which takes on the value 1 on $T(a', k)$ and 0 elsewhere. Hence $g_{T(a',k)} \in (a')$. All this is valid provided $T(a', k) \neq \varnothing$ and $k \neq 0$.

The characteristic function $e_{a'}$ of the support $T(a')$ of the function a' is the sum of the functions $g_{T(a',k)}$ for all $k \neq 0$. Even in the case that a' takes on infinitely many values we may still define this sum without difficulty. However, in order to ensure that this sum is contained in (a') *we must, because of the very definition of an ideal, make the hypothesis that a' take on only finitely many different values in K.* Then $e_{a'}$ is contained in (a') and, together with (9) we have

$$(a') = (e_{a'}) \tag{14}$$

This discussion suggests that instead of considering the whole ring A of functions from $\mathscr{P}$ to K, we should rather look at the subset A_1 consisting of those functions which take on only finitely many different values in K. We see immediately that A_1 is a subring of A which is even a right algebra over K. In A_1 each principal ideal (a') is generated by a *central idempotent, i.e. an idempotent contained in the center of the ring*. In particular, the characteristic functions introduced above belong to A_1 since they take on at most the two values 0 and 1.

Definition 2. A ring different from 0 is *biregular* if each principal ideal of the ring is generated by a central idempotent. Our discussion above proves:

Theorem 1. Let $\mathscr{P}$ be a set and K a simple ring, with identity. Let A be the ring of functions, from $\mathscr{P}$ to K and let A_1 be the subring of A consisting of those functions which take on only finitely many values in K. Then A_1 is biregular. ▮

For biregular right algebras R over a simple ring K with identity, the problem of choosing the isomorphisms ψ_P of (1) in a "natural way" can be handled more easily since, as we shall see, we can define in a "natural way" a monomorphism of K into R/P for each primitive ideal P of R. We first need the following theorem on biregular rings:

Theorem 2. Hypothesis: Let R be a biregular ring. Conclusions:

1. R is semisimple.
2. R is primitive if and only if R is simple with identity.
3. The primitive ideals P of R are precisely the maximal ideals of R.
4. Each factor ring and each direct summand of R is also biregular.

Proof. 1. Let $a \in J$. The principal ideal (9) is generated by an idempotent e. Since $e \in J$ there exists $z \in R$ such that $e + z - ze = 0$. Multiplying on the right by e and using the fact that $e^2 = e$, we get $e = e + ze - ze = 0e = 0$. Therefore $a = 0$.

2. Let R be primitive and M a faithful irreducible R-module. Since R is biregular, there exists a central idempotent $e \neq 0$ in R. The subgroup eM of the additive group of M is an R-submodule of M since $ReM = eRM \subseteq eM$. Hence since $e \neq 0$ and M is faithful, $eM = M$. It follows from $[x - xe]\,e = 0$ for all $x \in R$ that

$$(0) = [x - xe]\,eM = [x - xe]\,M$$

whence $x = xe = ex$. Therefore e is the identity of R proving that R contains only one nonzero central idempotent, namely, the identity. Since each principal ideal is generated by a central idempotent it follows that R is simple. The converse has already been proved in Theorem II.7.8.

3. The primitive ideals P and R are precisely those for which R/P is primitive. Hence Part 3 follows directly from Part 2.

4. If $\bar{a} \in R/I$ and $(a) = (e_a)$, then $(\bar{a}) = (\bar{e}_a)$, where $\bar{e}_a$ is still a central idempotent of R/I. Now suppose $R = R_1 \oplus R_2$ and let $a_1 \in R_1$. Since $R_2 a_1 R_2 = 0$, it follows that there exists $e_1 = e_1{}^2$ in the center of R_1 for which $(a_1) = (e_1)$. ∎

Each right ideal W of a biregular right algebra R over the ring K is a right K-module since by (5) and (8)

$$ak = (a \cdot e_a)\, k = a \cdot (e_a k) \qquad \text{for} \quad a \in R,\, k \in K$$

Hence, for every ideal P of R, R/P can be turned into a right K-module by defining

$$[a + P]\, k = ak + P \qquad \text{for} \quad a \in R,\, k \in K \tag{15}$$

Moreover, with this definition, R/P is even a right algebra over K since it is easily checked that

$$[(a + P) \cdot (b + P)]\, k = (a + P)([b + P]\, k)$$

By Theorem 2, R/P is biregular and if P is primitive, R/P is a simple ring with identity $e_P + P$. These facts enable us to prove the following:

Theorem 3. Let R be a biregular right algebra over the simple ring K with identity. Let the ideal P of R be primitive and let $e_P + P$ be the identity of R/P. Then the mapping

$$\chi_P \colon k \to (e_P + P)\, k \qquad \text{for} \quad k \in K \tag{16}$$

is a monomorphism of K into R/P. We call χ_P the *natural monomorphism of K into R/P.*

Proof. The additivity of χ_P is clear since R/P is a right K-module. To prove that χ_P is multiplicative we need Property (5) of the right algebra R/P:

$$\begin{aligned}(e_P + P)(k_1 k_2) &= [(e_P + P)\, k_1]\, k_2 = [((e_P + P)\, k_1)(e_P + P)]\, k_2 \\ &= ([e_P + P]\, k_1) \cdot ([e_P + P]\, k_2)\end{aligned}$$

The kernel of the homomorphism χ_P is an ideal of the simple ring K. Since

$$\chi_P(1) = e_P + P \neq P$$

$\ker \chi_P = (0)$, i.e. χ_P is a monomorphism. ∎

The natural monomorphism of K into R/P need not be an epimorphism for each primitive ideal P of R, i.e. given $a \in R$ there does not necessarily exist $k \in K$ for which

$$a + P = (e_P + P)\,k \tag{17}$$

However, we shall need the condition $\chi_P(K) = R/P$ for all primitive ideals P of the biregular (and hence semisimple) right algebra R over K, if we want to interpret the elements a' of the subdirect sum

$$R \approx \sum_{P \in \mathscr{P}}{}_{s} R/P \tag{18}$$

as functions from $\mathscr{P}$ to K. Therefore we shall make this hypothesis for all $P \in \mathscr{P}$ in the following

Theorem 4. Hypotheses: Let K be a simple ring with identity and let R be a biregular right algebra over K. For each primitive ideal P of R let the natural monomorphism χ_P of K into R/P defined in Theorem 3 be an epimorphism. Conclusions: Let $\mathscr{P}$ be the set of primitive ideals P of R and $e_P + P$ the identity of R/P. For each $a \in R$ define the mapping a' from $\mathscr{P}$ to K by

$$a + P = (e_P + P)\,a'(P) \qquad \text{for all} \quad P \in \mathscr{P} \tag{19}$$

Then the mapping

$$\phi\colon a \to a' \qquad \text{for} \quad a \in R$$

is an isomorphism of R onto a right K-subalgebra R' of the right K-algebra of functions from $\mathscr{P}$ to K.

Proof. All that is left to prove is that ϕ is a monomorphism: Since $P + P = P$ we have

$$\begin{aligned} a + b + P = (a + P) + (b + P) &= (e_P + P)\,a'(P) + (e_P + P)\,b'(P) \\ &= (e_P + P)(a'(P) + b'(P)) \end{aligned}$$

Furthermore, from $RP \subseteq P$ and $PR \subseteq P$ it follows that

$$\begin{aligned} a \cdot b + P = (a + P)(b + P) &= [(e_P + P)\,a'(P)][(e_P + P)\,b'(P)] \\ &= (e_P + P)(a'(P)\,b'(P)) \end{aligned}$$

Finally, for each $k \in K$, $ak + P = (a + P)\,k = (e_P + P)\,a'(P)\,k$ by (19). The mapping ϕ is therefore a right K-algebra homomorphism. The element

$a \in R$ is mapped under ϕ to the zero function if and only if $a \in P$ for all $P \in \mathscr{P}$, i.e. if and only if

$$a \in \bigcap_{P \in \mathscr{P}} P$$

But by Theorem 2, R is semisimple and hence $\bigcap_{P \in \mathscr{P}} P = (0)$ since the radical of a ring is the intersection of its primitive ideals. ∎

As a first step in the characterization of R' in the ring of functions we prove:

Theorem 5. The ring $R' = \phi R$ of functions on $\mathscr{P}$, as defined in Theorem 4, contains, together with a function a', the characteristic function e_a' of its support.

Proof. By (7) the support of the function a' is the set

$$T(a') = \{P \mid P \in \mathscr{P}, a'(P) \neq 0\}$$

Let $a' = \phi a$ and $(a) = (e_a)$, where e_a is a central idempotent. Then the following equivalences hold:

$$P \notin T(a') \Leftrightarrow a'(P) = 0 \Leftrightarrow a \in P \Leftrightarrow e_a \in P \Leftrightarrow e_a'(P) = 0$$

The central idempotent $e_a + P \neq \bar{0}$ of the simple ring R/P is its identity $e_P + P$. It follows that the value of the function $e_a' \in R'$ at P, as defined in (19), is the identity 1 of K. Therefore e_a' is the characteristic function of $T(a')$. ∎

In the next two sections we shall impose a topological structure on $\mathscr{P}$ and K which will enable us to characterize the functions in $R' = \phi R$ among all functions from $\mathscr{P}$ to K in terms of elementary topological properties. In preparation for this, the following investigation will shed light on the part played by the central idempotents.

Each central idempotent e of a ring R gives rise to a representation of R as a direct sum:

$$R = A \oplus B_e \qquad \text{where} \quad A = eR \quad \text{and} \quad B_e = \{x - ex \mid x \in R\} \qquad (20)$$

Proof. Since e is central, $A = eR = Re$ and B_e are ideals. Each element x of R is expressible in the form $x = ex + (x - ex)$. Hence $R = A + B_e$. If $x \in A \cap B_e$, $x = ye = z - ze$ for some $x, y \in R$. Therefore $x = yee = (z - ze)\, e = ze - ze = 0$ and (20) is proved.

Now let $R = A \oplus B$, A and B ideals of R, and let $\mathscr{P}$ be the set of primitive ideals P of R. Then $\mathscr{P}$ is the disjoint union of the two subsets

$$\mathscr{A} = \{P \mid P \in \mathscr{P}, P \supseteq A\} \qquad \text{and} \qquad \mathscr{B} = \{P \mid P \in \mathscr{P}, P \supseteq B\} \tag{21}$$

i.e.

$$\mathscr{P} = \mathscr{A} \cup \mathscr{B}, \qquad \mathscr{A} \cap \mathscr{B} = \varnothing \tag{22}$$

Proof. No P contains both A and B since $P \neq R$. From $A \cap B = (0)$ it follows that $AB = (0) \in P$, and, since each P is prime by Theorem III.2.6, either $A \subseteq P$ or $B \subseteq P$. For all rings a decomposition $R = A \oplus B$ gives rise to a decomposition $\mathscr{A} \cup \mathscr{B}$ of $\mathscr{P}$. If R is semisimple we even have

$$A = \bigcap_{P \in \mathscr{A}} P \qquad \text{and} \qquad B = \bigcap_{P \in \mathscr{B}} P \tag{23}$$

Proof. Each primitive ideal of R/A is the image under the natural homomorphism of R to R/A of a primitive ideal of R containing A. Since $R = A \oplus B$, the ring R/A, which is isomorphic to B, has no quasiregular ideal other than (0). Hence R/A is semisimple and its zero ideal is the intersection of its primitive ideals $\bar{P} = P/A$, where, as noted above, P is a primitive ideal of R containing A. Hence $A = \bigcap_{P \supseteq A} P$. The second equality in (23) is proved similarly.

This result which we restate for convenience in Part 1 of the following theorem may be applied to the particular decomposition of a semisimple ring R arising from a central idempotent to yield the second assertion of this theorem.

Theorem 6. 1. Let $R = A \oplus B$ be a representation of the semisimple ring R as a direct sum of ideals A and B. Then A is the intersection of all primitive ideals of R containing A. Similarly for B.

2. In particular, if e is a central idempotent of the semisimple ring R and if

$$R = eR \oplus \{x - ex \mid x \in R\}$$

is the resulting decomposition of R relative to e, then eR is the intersection of all primitive ideals of R containing e and $\{x - ex \mid x \in R\}$ is the intersection of all primitive ideals not containing e. ∎

In Section 3 we shall see that Theorem 6 and the fact that in a biregular ring each principal ideal (a) is generated by a central idempotent e_a provide the key to the topological characterization of those functions which constitute the ring R' of Theorem 4. Indeed, by Theorem 5, if a' is an element of

R' so is the characteristic function e_a' of its support $T(a')$. The function e_a' takes on only the values 0 and 1. If we endow the set $\{0, 1\}$ with the discrete topology, then the topology on the set $\mathscr{P}$ of primitive ideals of R is so defined as to make e_a' continuous (defined in Definition XII.3.1). This means that both the set of points $P \in \mathscr{P}$ at which e_a' is 0 and the set of points $P \in \mathscr{P}$ at which e_a' is 1 are closed. By definition of the function e_a', the set of points at which e_a' is zero consists of those $P \in \mathscr{P}$ containing e_a and the set of points at which e_a' is 1 of those $P \in \mathscr{P}$ not containing e_a. By the second assertion of Theorem 6 these sets are, respectively, the set of $P \in \mathscr{P}$ containing e_aR and the set of $P \in \mathscr{P}$ containing $\{x - e_ax \mid x \in R\}$. Thus the continuity of the function e_a' means that both the set of $P \in \mathscr{P}$ containing e_aR and the set of $P \in \mathscr{P}$ containing $\{x - e_ax \mid x \in R\}$ are closed relative to the topology to be defined on $\mathscr{P}$. In the next section we shall show that the set $\mathscr{P}$ of primitive ideals P of a ring R may be endowed with a topological structure in such a way that the closed sets $\mathscr{A}$ of $\mathscr{P}$ are precisely those consisting of primitive ideals P containing some ideal A of R (cf. Theorem IX.2.7).

2. The structure space of a ring

The structure space $\mathscr{P}$ of a biregular ring R (to be defined later) has a number of topological properties which will play an important role in the next section in the characterization of such rings as rings of continuous functions. The general topological concepts required will be developed in the first part of this section. The reader with some background in topology may omit this part and begin with Theorem 7.

Definition 1. A *closure operator H on a set $\mathscr{P}$* is a mapping of the lattice $V(\mathscr{P})$ of subsets of $\mathscr{P}$ into itself satisfying the following conditions:

H_1. $H\varnothing = \varnothing$
H_2. $\mathscr{A} \subseteq H\mathscr{A}$
H_3. $HH\mathscr{A} = H\mathscr{A}$
H_4. $H(\mathscr{A} \cup \mathscr{B}) = H\mathscr{A} \cup H\mathscr{B}$

for $\varnothing, \mathscr{A}, \mathscr{B} \in V(\mathscr{P})$.

Theorem 1. Hypothesis: Let H be a closure operator on the set $\mathscr{P}$. Call a subset $\mathscr{A}$ of $\mathscr{P}$ *closed* if $H\mathscr{A} = \mathscr{A}$ and call it *open* if it is the complement $\complement\mathscr{B} = \mathscr{P}\backslash\mathscr{B}$ of a closed set $\mathscr{B}$.

Conclusions: 1. $H\mathscr{A}$ is the intersection of all closed subsets of $\mathscr{P}$ containing $\mathscr{A}$.

2. The union of a finite number and the intersection of an arbitrary number of closed sets are closed.

3. The family $\mathscr{F}(P)$ of open subsets of $\mathscr{P}$ containing $\{P\}$ forms a family of basic neighborhoods of P in the sense of Section II.2. The system $\{\mathscr{F}(P) \mid P \in \mathscr{P}\}$ therefore endows $\mathscr{P}$ with a topology.

Proof. 1. If $\mathscr{A} \subseteq \mathscr{B}$, then $H\mathscr{A} \subseteq H\mathscr{B}$, since

$$\mathscr{A} \subseteq \mathscr{B} \Rightarrow \mathscr{A} \cup \mathscr{B} = \mathscr{A} \Rightarrow H\mathscr{A} = H(\mathscr{A} \cup \mathscr{B}) = H\mathscr{A} \cup H\mathscr{B}$$
$$\Rightarrow H\mathscr{A} \subseteq H\mathscr{B}$$

Therefore, if $\mathscr{A} \subseteq \mathscr{B}$ and $\mathscr{B}$ is closed, $H\mathscr{A} \subseteq H\mathscr{B} = \mathscr{B}$ proving that $H\mathscr{A}$ is contained in the intersection of all closed sets containing $\mathscr{A}$. But $H\mathscr{A}$ is itself a closed set containing $\mathscr{A}$ by (H_3) and (H_2). Thus Part 1 is proved.

2. The first assertion shows that the intersection of an arbitrary number of closed sets is closed. By induction it follows from H_2 and H_3 that the union of a finite number of closed sets is also closed.

3. We must prove properties U_1, U_2, and U_3 of Section II.2 for the family $\mathscr{F}(P)$. Since $\mathscr{P} = \mathscr{P} \backslash \varnothing = C\varnothing$ is in $\mathscr{F}(P)$, $\mathscr{F}(P) \neq \varnothing$. All $\mathscr{U}(P) \in \mathscr{F}(P)$ contain P by definition. Hence

U_1. $\mathscr{F}(P) \neq \varnothing$ and $P \in \mathscr{U}(P)$ for all $\mathscr{U}(P) \in \mathscr{F}(P)$. ▮

If $C\mathscr{A}$ and $C\mathscr{B}$ belong to $\mathscr{F}(P)$, $C\mathscr{A} \cap C\mathscr{B} = C(\mathscr{A} \cup \mathscr{B})$ is the complement of a closed set $\mathscr{A} \cup \mathscr{B}$ not containing P. It therefore belongs to $\mathscr{F}(P)$. This is actually more than is required in U_2:

U_2. Given $\mathscr{U}(P)$ and $\mathscr{V}(P)$ there exists $\mathscr{W}(P) \in \mathscr{F}(P)$ such that $\mathscr{W}(P) \subseteq \mathscr{U}(P) \cap \mathscr{V}(P)$. If the point Q is contained in the basic neighborhood $C\mathscr{A} \in \mathscr{F}(P)$ then $C\mathscr{A} \in \mathscr{F}(Q)$ and of course $C\mathscr{A} \subseteq C\mathscr{A}$. This proves:

U_3. Given $Q \in \mathscr{U}(P)$, there exists $\mathscr{V}(Q) \in \mathscr{F}(Q)$ such that $\mathscr{V}(Q) \subseteq \mathscr{U}(P)$.

Recall that a neighborhood of a point P is a subset of $\mathscr{P}$ containing a basic neighborhood of P.

In Section II.2 we defined the closure of a subset $\mathscr{A}$ of $\mathscr{P}$ to be the set of points $P \in \mathscr{P}$ for which every neighborhood $\mathscr{U}(P)$ of P intersects $\mathscr{A}$ nontrivially. We can show (cf. Franz [1960]) that the closure, in this sense, of a subset $\mathscr{A}$ of $\mathscr{P}$ is precisely $H\mathscr{A}$ and that the mapping which takes a subset to its closure satisfies H_1 to H_4. In other words: The topology on $\mathscr{P}$ may be defined in terms of a closure operator instead of specifying the families $\mathscr{F}(P)$ of basic neighborhoods.

Not very topological space is "Hausdorff." The following is a list of the so-called "separation axioms":

Definition 2. Let $\mathscr{P}$ be a topological space.

1. $\mathscr{P}$ is a *T_0-space* if for every pair of distinct points of $\mathscr{P}$ there is a neighborhood of at least one of them not containing the other.

2. $\mathscr{P}$ is a *T_1-space* if for every pair of distinct points of $\mathscr{P}$ there is a neighborhood of each not containing the other.

3. $\mathscr{P}$ is a *T_2-space* (also called a Hausdorff space) if for each pair of distinct points of $\mathscr{P}$ there is a pair of nonintersecting neighborhoods containing the given points.

4. $\mathscr{P}$ is *totally disconnected* if for each pair of distinct points P and Q of $\mathscr{P}$, there exist two open subsets $\mathscr{A}$ and $\mathscr{B}$ of $\mathscr{P}$ with the following properties:

$$P \in \mathscr{A}, \qquad Q \in \mathscr{B}, \qquad \mathscr{A} \cup \mathscr{B} = \mathscr{P}, \qquad \mathscr{A} \cap \mathscr{B} = \varnothing$$

The expression "totally disconnected" derives from the concept of *connectedness.* We call a topological space *connected* if there do not exist two nonempty open sets $\mathscr{A}$ and $\mathscr{B}$ such that

$$\mathscr{A} \cup \mathscr{B} = \mathscr{P} \qquad \text{and} \qquad \mathscr{A} \cap \mathscr{B} = \varnothing$$

In Theorem 6 we shall come across an even stronger separation axiom, the 0-dimensionality. To explain it we need the concept of local compactness which depends for its definition on the idea of a cover.

Definition 3. A family $\mathscr{F}$ of subsets of a topological space $\mathscr{P}$ is said to *cover* $\mathscr{P}$ if each point of $\mathscr{P}$ is contained in at least one member of $\mathscr{F}$. $\mathscr{F}$ is called a *cover* of $\mathscr{P}$. If each member of $\mathscr{F}$ is open, we call $\mathscr{F}$ an *open cover.*

Theorem 2. Let $\mathscr{P}$ be a topological space. Then the following two statements are equivalent:

1. Each open cover $\mathscr{F}$ of $\mathscr{P}$ contains a finite subfamily $\mathscr{F}'$ which also covers $\mathscr{P}$.

2. Each family $\mathscr{G}$ of closed subsets of $\mathscr{P}$ which has empty intersection contains a finite subfamily with empty intersection.

Proof. Suppose Part 1 holds and let $\mathscr{G} = \{\mathscr{A}_i \mid i \in I\}$ be a family of closed sets $\mathscr{A}_i$ with

$$\bigcap_{i \in I} \mathscr{A}_i = \varnothing$$

Taking complements of both sides we get

By Part 1 there exists a finite set $\{i_1, ..., i_n\} \subseteq I$ such that

$$\bigcup_{1 \leqslant k \leqslant n} C\mathscr{A}_{i_k} = \mathscr{P}$$

since the $C\mathscr{A}_i$ are open. It follows that

$$\bigcap_{1 \leqslant k \leqslant n} \mathscr{A}_{i_k} = C \bigcup_k C\mathscr{A}_{i_k} = C\mathscr{P} = \varnothing$$

Dually, Part 1 follows from Part 2. ▮

Definition 4. A topological space $\mathscr{P}$ in which either Part 1 or Part 2 of Theorem 2 holds is said to be *compact.*

A topological space is said to be locally compact (cf. Definition 5 below) if every point of the space is contained in a closed neighborhood which is compact relative to the "induced" topology.

Theorem 3. Let $\mathscr{S}$ be a subset of the topological space $\mathscr{P}$ and suppose the topology on $\mathscr{P}$ is defined by the closure operator H. Then the mapping

$$H_{\mathscr{S}}: \mathscr{A} \to H_{\mathscr{S}}\mathscr{A} = \mathscr{S} \cap H\mathscr{A} \qquad \text{for} \quad \mathscr{A} \subseteq \mathscr{S} \tag{1}$$

also satisfies conditions (H_1) to (H_4) and therefore endows $\mathscr{S}$ with a topological structure, the so-called *induced topological structure.*

Proof. H_1 is trivial. Since $\mathscr{A} \subseteq H\mathscr{A}$ and $\mathscr{A} \subseteq \mathscr{S}$, then $\mathscr{A} \subseteq \mathscr{S} \cap H\mathscr{A} = H_{\mathscr{S}}\mathscr{A}$ and H_2 is proved. This implies that

$$\begin{aligned} H_{\mathscr{S}}H_{\mathscr{S}}\mathscr{A} &= H_{\mathscr{S}}(\mathscr{S} \cap H\mathscr{A}) = \mathscr{S} \cap H(\mathscr{S} \cap H\mathscr{A}) \\ &= \mathscr{S} \cap HH\mathscr{A} = \mathscr{S} \cap H\mathscr{A} = H_{\mathscr{S}}\mathscr{A} \end{aligned}$$

proving the idempotence H_3 of $H_{\mathscr{S}}$. The following line of equalities proves H_4 :

$$\begin{aligned} H_{\mathscr{S}}(\mathscr{A} \cup \mathscr{B}) &= \mathscr{S} \cap H(\mathscr{A} \cup \mathscr{B}) = \mathscr{S} \cap (H\mathscr{A} \cup H\mathscr{B}) \\ &= (\mathscr{S} \cap H\mathscr{A}) \cup (\mathscr{S} \cap H\mathscr{B}) = H_{\mathscr{S}}\mathscr{A} \cup H_{\mathscr{S}}\mathscr{B} \end{aligned}$$ ▮

Definition 5. 1. A subset $\mathscr{S}$ of the topological space $\mathscr{P}$ is said to be compact if it is compact relative to the induced topology on $\mathscr{S}$.

2. A topological space $\mathscr{P}$ is *locally compact* if each point $P \in \mathscr{P}$ is contained in a closed compact neighborhood of P.

Since the whole space $\mathscr{P}$ is closed, a compact space is locally compact.

By Definition 5, a subset $\mathscr{S}$ of $\mathscr{P}$ is compact if every cover of $\mathscr{S}$ *by open sets of* $\mathcal{S}$ has a finite subcover. No mention is made of how $\mathcal{S}$ behaves with respect to covers of $\mathcal{S}$ *by open sets of* $\mathcal{P}$. The following theorem provides us with this information:

Theorem 4. Let $\mathcal{S}$ be a closed compact subset of $\mathcal{P}$. Let $\mathcal{F} = \{\mathcal{A}_i \mid i \in I\}$ be a family of open sets *of* $\mathcal{P}$ which covers $\mathcal{S}$, i.e.

$$\mathcal{S} \subseteq \bigcup_{i \in I} \mathcal{A}_i$$

Then there exists a finite subset $\{i_1, ..., i_n\} \subseteq I$ for which

$$\mathcal{S} \subseteq \bigcup_{l \leqslant k \leqslant n} \mathcal{A}_{i_k} \tag{2}$$

Proof. Let $\mathcal{A}_i' = \mathcal{S} \cap \mathcal{A}_i$. Now $H_s(\mathcal{S} \cap C\mathcal{A}_i') = \mathcal{S} \cap H(\mathcal{S} \cap C\mathcal{A}_i') = \mathcal{S} \cap (\mathcal{S} \cap C\mathcal{A}_i') = \mathcal{S} \cap C\mathcal{A}_i'$ thus showing that the complement of $\mathcal{A}_i'$ in $\mathcal{S}$ is closed. Therefore $\mathcal{A}_i'$ is open *in* $\mathcal{S}$. Hence the family $\mathcal{F}' = \{\mathcal{A}_i' \mid i \in I\}$ is an open cover of $\mathcal{S}$. Since $\mathcal{S}$ is compact there is a finite subcover $\{\mathcal{A}_{i_k}' \mid k = 1, ..., n\}$. But $\mathcal{A}_{i_k}' \subseteq \mathcal{A}_{i_k}$. Therefore $\mathcal{S} \subseteq \bigcup_{1 \leqslant k \leqslant n} \mathcal{A}_{i_k}$. ∎

The following theorem helps to bring the concept of compactness into sharper focus:

Theorem 5. 1. Each closed subset of a compact space is compact.
2. In a T_2-space (Hausdorff space) a compact subset is closed.

Proof. 1. Let $\mathcal{S}$ be a closed subset of the topological space $\mathcal{P}$ with closure operator H. Let $\mathcal{A}$ be a closed subset of the topological space $\mathcal{S}$, i.e. $\mathcal{A} = H_{\mathcal{S}}\mathcal{A} = \mathcal{S} \cap H\mathcal{A}$. Since $\mathcal{A} \subseteq \mathcal{S}$, $H\mathcal{A} \subseteq H\mathcal{S} = \mathcal{S}$. Therefore $\mathcal{A} = \mathcal{S} \cap H\mathcal{A} = H\mathcal{A}$ and $\mathcal{A}$ is a closed subset of the topological space $\mathcal{P}$. To prove that $\mathcal{A}$ is compact we use the second characterization of compactness given in Theorem 2. Thus let $\mathcal{G} = \{\mathcal{A}_i \mid i \in I\}$ be a family of closed sets of $\mathcal{S}$ with empty intersection. Since by hypothesis $\mathcal{P}$ is compact and the $\mathcal{A}_i$ are closed in $\mathcal{P}$, we can find a subfamily $\{\mathcal{A}_{i_k} \mid k = 1, ..., n\}$ with empty intersection.

2. Let $\mathcal{S}$ be a compact set in the T_2-space $\mathcal{P}$. We show that the complement $C\mathcal{S}$ of $\mathcal{S}$ is open, i.e. we show that given any point $P \in C\mathcal{S}$ there exists a neighborhood $\mathcal{U}_0(P)$ contained entirely in $C\mathcal{S}$. Since $\mathcal{P}$ is Hausdorff, given a point $Q \in \mathcal{S}$ and a fixed point $P \in C\mathcal{S}$, there exist open neighborhoods $\mathcal{V}(Q)$ and $\mathcal{U}_Q(P)$ such that $\mathcal{V}(Q) \cap \mathcal{U}_Q(P) = \varnothing$. As Q varies over $\mathcal{S}$, the neighborhoods $\mathcal{V}(Q)$ cover $\mathcal{S}$. Since $\mathcal{S}$ is compact, a finite number of these cover $\mathcal{S}$, say $\mathcal{V}(Q_1), ..., \mathcal{V}(Q_n)$. The intersection of the

corresponding $\mathscr{U}_{Q_1}(P), \ldots, \mathscr{U}_{Q_n}(P)$ is an open set and is therefore a neighborhood $\mathscr{U}_0(P)$ of P. Since $\mathscr{V}(Q) \cap \mathscr{U}_Q(P) = \varnothing$ for all $Q \in \mathscr{S}$, it follows that

$$\mathscr{S} \cap \mathscr{U}_0(P) \subseteq (\mathscr{V}(Q_1) \cup \cdots \cup \mathscr{V}(Q_n)) \cap \mathscr{U}_0(P) = \varnothing$$

and hence $\mathscr{U}_0(P) \subseteq C\mathscr{S}$. ∎

For a locally compact totally disconnected space we finally prove the following separation property:

Theorem 6. Let $\mathscr{P}$ be a locally compact totally disconnected topological space. Let $\mathscr{V}$ be one of the closed compact subsets of $\mathscr{P}$ containing the point P_0. Let $\mathscr{U}$ be an open subset of $\mathscr{P}$ for which

$$P_0 \in \mathscr{U} \subseteq \mathscr{V}$$

Then there exists a subset $\mathscr{D}'$ of $\mathscr{P}$ which is both open and closed and which satisfies the condition

$$P_0 \in \mathscr{D}' \subseteq \mathscr{U}$$

A topological space with this property at the point P_0 is said to be of *dimension* 0 *at* P_0. If the space is 0-dimensional at each point, it is said to be 0-*dimensional.*

Proof. The set of points contained in the closure of $\mathscr{U}$ but not in $\mathscr{U}$ is called the *boundary* of $\mathscr{U}$:

$$bd\mathscr{U} = H\mathscr{U} \cap C\mathscr{U}$$

Since $HH\mathscr{U} = H\mathscr{U}$ and since $\mathscr{U}$ is open, $bd\mathscr{U}$, as the intersection of two closed sets, is closed. Since $\mathscr{U}$ is contained in $\mathscr{V} = H\mathscr{V}$, $bd\mathscr{U}$ is also contained in $\mathscr{V}$. P_0 is not contained in $bd\mathscr{U}$. For each point $Q \in bd\mathscr{U}$ there exists an open and closed set $\mathscr{W}_Q$ with the property that $Q \in \mathscr{W}_Q$ and $P_0 \in C\mathscr{W}_Q$ since $\mathscr{P}$ is totally disconnected. As a closed subset of the compact set $\mathscr{V}$, $bd\mathscr{U}$ is also compact by Theorem 4. Thus only finitely many $\mathscr{W}_Q$ are needed to cover $bd\mathscr{U}$, say

$$bd\mathscr{U} \subseteq \mathscr{W}_{Q_1} \cup \cdots \cup \mathscr{W}_{Q_s} = \mathscr{W}$$

Moreover,

$$P_0 \in C\mathscr{W}_{Q_1} \cap \cdots \cap C\mathscr{W}_{Q_s} = \mathscr{D}$$

$\mathscr{D}$ is both open and closed since it is the intersection of a finite number of open-and-closed sets $C\mathscr{W}_{Q_1}$. Hence $\mathscr{D} \cap H\mathscr{U}$ is closed. Furthermore since $\mathscr{D} \cap bd\mathscr{U} \subseteq \mathscr{D} \cap \mathscr{W} = \varnothing$, it follows that

$$\mathscr{D} \cap H\mathscr{U} = \mathscr{D} \cap (\mathscr{U} \cup bd\mathscr{U}) = (\mathscr{D} \cap \mathscr{U}) \cup (\mathscr{D} \cap bd\mathscr{U}) = \mathscr{D} \cap \mathscr{U}$$

Therefore $\mathscr{D} \cap H\mathscr{U}$ is also open, being the intersection of the two open sets $\mathscr{D}$ and $\mathscr{U}$. The set $\mathscr{D}' = \mathscr{D} \cap \mathscr{U}$ contains P_0 and satisfies the required properties. ∎

These general topological concepts may now be applied to the so-called structure space $\mathscr{P}$ of a ring with at least one primitive ideal, and hence, in particular, to the structure space of a biregular ring.

Theorem 7. Let $\mathscr{P}$ be the nonempty set of primitive ideals P of the ring R. Let $\mathscr{A}$ be a subset of $\mathscr{P}$ and let the ideal

$$D_{\mathscr{A}} = \bigcap_{P \in \mathscr{A}} P$$

be the intersection of all ideals contained in $\mathscr{A}$. In particular, set $D_{\mathscr{A}} = R$ if $\mathscr{A} = \varnothing$. Let H be the mapping which assigns to each subset $\mathscr{A} \subseteq \mathscr{P}$ the set

$$H\mathscr{A} = \{P \mid P \in \mathscr{P},\ P \supseteq D_{\mathscr{A}}\} \tag{3}$$

Then H satisfies Conditions H_1 to H_4 and therefore endows $\mathscr{P}$ with a topological structure. $\mathscr{P}$ with this topology is called the *structure space of the ring R*.

Proof. 1. $H\varnothing = \varnothing$ since no primitive ideal P contains R.

2. $P \supseteq D_{\mathscr{A}}$ for all $P \in \mathscr{A}$ and hence $\mathscr{A} \subseteq H\mathscr{A}$.

3. By Part 2, $H\mathscr{A} \subseteq HH\mathscr{A}$. Conversely, since the intersection of all $P \in H\mathscr{A}$ contains the ideal $D_{\mathscr{A}}$, we have $HH\mathscr{A} \subseteq H\mathscr{A}$.

4. From $D_{\mathscr{A} \cup \mathscr{B}} = D_{\mathscr{A}} \cap D_{\mathscr{B}}$ it follows that $H(\mathscr{A} \cup \mathscr{B}) \supseteq H\mathscr{A} \cup H\mathscr{B}$. Now let P be a primitive ideal contained in $H(\mathscr{A} \cup \mathscr{B})$. Then

$$P \supseteq D_{\mathscr{A} \cup \mathscr{B}} = D_{\mathscr{A}} \cap D_{\mathscr{B}} \supseteq D_{\mathscr{A}} \cdot D_{\mathscr{B}} \tag{4}$$

since $D_{\mathscr{A}}$ and $D_{\mathscr{B}}$ are ideals. By Theorem III.2.6, P is prime. It therefore follows from (4) that either $D_{\mathscr{A}}$ or $D_{\mathscr{B}}$ is contained in P. Hence either $P \in H\mathscr{A}$ or $P \in H\mathscr{B}$. ∎

Some of the algebraic properties of R are reflected in the topological properties of its structure space $\mathscr{P}$. In general the structure space of a ring is not Hausdorff as the ring $\mathbb{Z}$ of integers shows. Nevertheless we have

Theorem 8. The structure space $\mathscr{P}$ of a ring R is a T_0-space. If all primitive ideals of R are maximal, $\mathscr{P}$ is a T_1-space.

Proof. Suppose $\mathscr{P}$ is not T_0. Then there exists a pair of distinct points

(P, Q) (primitive ideals) with the property that each neighborhood of either one contains the other. It follows that $H\{P\} = H\{Q\}$ for the two singleton sets $\{P\}$ and $\{Q\}$; for suppose $Q \in H\{P\}$, say. Then the open set $CH\{P\}$ contains the point Q but not the point P since by (H_2), $P \in H\{P\}$. But this contradicts the choice of the pair (P, Q). Therefore $H\{P\} = H\{Q\}$ which implies $P = Q$, a contradiction.

2. The closure $H\{P\}$ of the singleton set $\{P\}$ consists of all primitive ideals containing P. Thus if P is maximal, $H\{P\} = \{P\}$. Now let $Q \neq P$. Then Q is contained in the open set $CH\{P\} = C\{P\}$. Similarly $P \in C\{Q\}$. ∎

The following theorem implies, among other things, that *the structure space of a biregular ring R is Hausdorff.*

Theorem 9. The structure space of a biregular ring R is totally disconnected.

Proof. Let the element a of the ring R belong to the primitive ideal P_2 but not to P_1. The central idempotent e for which $(a) = (e)$ (and which exists by the biregularity of R) yields, by the second assertion of Theorem IX.1.6, the direct sum decomposition of R

$$R = eR \oplus \{x - ex \mid x \in R\}$$

where $eR = (e) = (a)$ is the intersection of all primitive ideals containing (a) and $\{x - ex \mid x \in R\}$ is the intersection of all primitive ideals not containing (a). These two sets of primitive ideals $\mathscr{A}$ and $\mathscr{B}$ are therefore both closed and hence also both open since each is the complement of the other. Thus P_2 is in $\mathscr{A}$ and P_1 is in $C\mathscr{A} = \mathscr{B}$. ∎

The structure space of a biregular ring need not be compact. However, it is locally compact (cf. Theorem 11). The compactness of the structure space $\mathscr{P}$ of a ring R follows from the existence of an identity in R. Since this sufficient condition for the compactness of $\mathscr{P}$ plays an important role in the investigation of rings with locally compact structure spaces we prove the following:

Theorem 10. The structure space of a ring with identity 1 is compact.

Proof. Let $\mathscr{G} = \{\mathscr{A}_i \mid i \in I\}$ be a family of closed subsets $\mathscr{A}_i$ of the structure space $\mathscr{P}$ of R. Each $\mathscr{A}_i$ consists of primitive ideals P of R. Let $D_{\mathscr{A}_i}$ denote the intersection of all the ideals P contained in $\mathscr{A}_i$. The intersection $\bigcap \mathscr{A}_i$ consists of those primitive ideals which are contained in each of the $\mathscr{A}_i$.

Suppose $\bigcap_i \mathscr{A}_i = \varnothing$. We must show that there exists a finite subset $\{i_1, ..., i_n\} \subseteq I$ such that $\Sigma_{1 \leqslant k \leqslant n} \mathscr{A}_{i_k} = \varnothing$. Let

$$S = \sum_{i \in I} D_{\mathscr{A}_i}$$

S consists of finite sums of elements from the $D_{\mathscr{A}_i}$:

$$S = \{d_1 + \cdots + d_m \mid d_j \in D_{\mathscr{A}_j}, \{i_1, ..., i_m\} \subseteq I, m \in \mathbb{N}\}$$

since $\bigcap_{i \in I} \mathscr{A}_i = \varnothing$, no primitive ideal P of R belongs to each $\mathscr{A}_i$. The primitive ideals of the factor ring R/S correspond under the natural homomorphism of R onto R/S to the primitive ideals of R containing S as we may easily conclude from Definition II.7.2. Since $\mathscr{A}_i$ is closed, $P \in \mathscr{A}_i$ if and only if $P \supseteq D_{\mathscr{A}_i}$. It therefore follows from $\bigcap \mathscr{A}_i = \varnothing$ that R/S contains no primitive ideals. The radical $\bar{J}$ of R/S is therefore the whole of R/S. Since $1 + S$ is the identity of R/S, and since $\bar{J} = R/S$, it follows that $S = R$ because the identity of the ring R/S is the zero of the ring only in the case that $R/S = (\bar{0})$. In particular, $1 \in S$ and is therefore expressible in the form

$$1 = d_1 + d_2 + \cdots + d_n, \qquad d_j \in D_{\mathscr{A}_{i_j}}, \qquad j = 1, ..., n$$

It follows that

$$1 \in D_{\mathscr{A}_{i_1}} + \cdots + D_{\mathscr{A}_{i_n}}$$

whence

$$R/(D_{\mathscr{A}_{i_1}} + \cdots + D_{\mathscr{A}_{i_n}}) = (\bar{0})$$

Therefore

$$\mathscr{A}_{i_1} \cap \cdots \cap \mathscr{A}_{i_n} = \varnothing \quad \blacksquare$$

A biregular ring may not have an identity. However, it contains so many central idempotents e that its structure space is locally compact.

Theorem 11. The structure space of a biregular ring is locally compact.

Proof. Let P_0 be a point of the structure space $\mathscr{P}$ of the ring R (P_0 is therefore a primitive ideal of R). Let a be an element of R not in P_0. Since R is biregular, the principal ideal (a) is generated by an idempotent e. This idempotent e gives rise to a decomposition

$$R = eR \oplus \{x - ex \mid x \in R\} \tag{5}$$

of R, and this in turn leads to a decomposition of $\mathscr{P}$ into the union of disjoint open-and-closed subsets

$$\mathscr{A} = \{P \mid e \in P\} \qquad \text{and} \qquad \mathscr{B} = \{P \mid x - ex \in P \text{ for all } x \in R\}$$

(Compare with the proof of Theorem 9.) Since a (and hence e) is not contained in P_0, P_0 is a member of $\mathscr{B}$. Therefore $\mathscr{B}$ is a neighborhood of P_0. The set $\mathscr{B}$ consists of all the primitive ideals of $R = eR + B_e$ containing the ideal

$$B_e = \{x - ex \mid x \in R\}$$

These primitive ideals mod B_e form the structure space of the factor ring R/B_e. By (5), R/B_e is isomorphic to eR. The images of the primitive ideals of the set $\mathscr{B}$ under a ring-isomorphism of R/B_e onto eR form the structure space of the ring eR. The element $e \in eR$ is an identity for eR and hence by Theorem 10, its structure space is compact. This implies that $\mathscr{B}$ is also a compact space relative to the induced topology of $\mathscr{P}$ on the subset $\mathscr{B}$ since in all cases the topology is defined in terms of the closure operator (3). ∎

By Theorems 9 and 11 the structure space $\mathscr{P}$ of a biregular ring is totally disconnected and locally compact. This, together with Theorem 6 implies:

Theorem 12. The structure spaze of a biregular ring is locally compact and 0-dimensional. ∎

3. The theorem of Arens and Kaplansky

Let R be a biregular ring which is also a right algebra over the simple ring K with identity 1. Suppose moreover that for each primitive ideal P of R, the natural monomorphism (defined in Theorem XII.1.3)

$$\chi_P: k \to (e_P + P)\, k, \qquad k \in K \tag{1}$$

of K into R/P is an epimorphism. Then, by Theorem XII.1.4, to each element $a \in R$ there corresponds a function a' from the structure space $\mathscr{P}$ of R to the ring K where the value of a' at P is defined by the equation

$$a + P = (e_P + P)\, k \tag{2}$$

The mapping

$$a \to a' \qquad \text{for} \quad a \in R \tag{3}$$

is a monomorphism of R into the ring of functions on $\mathscr{P}$ with values in K.

The support of a function f on $\mathscr{P}$ with values in K was defined in XII.1.(7) to be the set

$$T(f) = \{P \mid P \in \mathscr{P}, f(P) \neq 0\}$$

By Theorem XII.1.5, $R' = \phi R$ contains, together with a function a', the characteristic function e_a' of its support $T(a')$. Since R is biregular, the principal ideal (a) of R is generated by a central idempotent e_a. This gives rise, by virtue of Theorem IX.1.6, to a representation

$$R = e_a R \oplus B_a, \qquad B_a = \{x - e_a x \mid x \in R\}$$

of R as a direct sum of two ideals with the support $T(a')$ consisting of those primitive ideals P containing B_a. By the very definition of the topology on $\mathscr{P}$, $T(a')$ is therefore closed. By Theorem XII.1.6, its complement $CT(a')$ consists of those P containing $e_a R$ and is therefore also closed. This shows that the *support of each function a' in $R' = \phi R$ is both open and closed. It is, moreover, compact.* In fact, the P in $T(a')$, (i.e. the P containing B_a), considered modulo B_a, form the structure space of the ring R/B_a with identity $e_a + B_a$. By Theorem XII.2.10, this is compact. Hence $T(a')$ is a compact subset of $\mathscr{P}$.

The complement $CT(a')$ of the support of a' is the set of points at which a' takes on the value 0. Therefore it is both open and closed. The same is also true of the set of points at which the function $a' \in R'$ takes on a fixed value $k \neq 0$. In fact, this is the set

$$\{P \mid a - e_a k \in P\} \cap T(a') = \{P \mid (a - e_a k) \subseteq P\} \cap T(a') \tag{4}$$

where, because of the biregularity of R, the principal ideal $(a - e_a k)$ is generated by a central idempotent. Therefore, again by Theorem XII.1.6, the set in curly brackets in (4) is both open and closed. As the intersection of two sets, each both open and closed, the set in (4) is also both open and closed.

This situation may be conveniently described in terms of the concept of continuity of a function provided we endow K with the discrete topology (cf. Definition III.2.3).

Definition 1. Let $\mathscr{P}$ and $\mathscr{P}'$ be topological spaces and f a function from $\mathscr{P}$ to $\mathscr{P}'$. The function f is said to be *continuous* if the preimage

$$f^{-1}(\mathscr{A}') = \{P \mid P \in \mathscr{P}, f(P) \in \mathscr{A}'\}$$

of each set $\mathscr{A}'$ open in $\mathscr{P}'$ is open in $\mathscr{P}$. Since the open sets of a topological space are the complements of closed sets we obtain an equivalent definition of continuity by replacing "open" by "closed" throughout.

Our discussion above leads to the following:

Theorem 1. The elements a' of the ring of functions $R' = \phi R$ (cf. Theorem XII.1.4) have compact supports and are continuous provided we endow K with the discrete topology. ▌

Conversely, the continuous functions on $\mathscr{P}$ with compact supports are contained in ϕR as we shall now prove. First, however, we need the following:

Theorem 2. The characteristic function of each open and compact subset $\mathscr{B}$ of $\mathscr{P}$ is contained in $R' = \phi R$. Recall that the characteristic function $g'_{\mathscr{B}}$ of $\mathscr{B}$ is defined by

$$g_{\mathscr{B}}'(P) = \begin{cases} 1 & \text{if} \quad P \in \mathscr{B} \\ 0 & \text{if} \quad P \notin \mathscr{B} \end{cases}$$

Proof. 1. If the subset is of the form

$$\mathscr{B}_e = \{P \mid P \in \mathscr{P}, e \in P\} = \{P \mid P \in \mathscr{P}, P \supseteq B_e\} \tag{5}$$

where e is a central idempotent and $B_e = \{x - ex \mid x \in R\}$, the assertion follows by Theorem XII.1.6 since then $g'_{\mathscr{B}_e} = e' = \phi e$ *is the characteristic function of* $\mathscr{B}_e$.

2. *Now let* $\mathscr{U}$ *be an open-and-closed subset of the set* $\mathscr{B}_e$ *in* (5). Then by Theorem XII.1.6

$$R = eR \oplus B_e \qquad \text{where} \quad B_e = \{x - ex \mid x \in R\} \tag{6}$$

The structure space of the factor ring R/B_e consists therefore of ideals of the form P/B_e where $P \in \mathscr{B}_e$. The set $\mathscr{B}_e$ of primitive ideals containing B_e may be divided into two subsets, namely, the subset consisting of those ideals contained in $\mathscr{U}$ and the subset of those ideals contained in $C\mathscr{U}$. Let

$$D_{\mathscr{U}} = \bigcap_{P \in \mathscr{U}} P \qquad \text{and} \qquad D_{\mathscr{U}}' = \bigcap_{P \in C\mathscr{U} \cap \mathscr{B}} P \tag{7}$$

We show that

$$eR = (D_{\mathscr{U}} \cap eR) \oplus (D_{\mathscr{U}}' \cap eR) \tag{8}$$

as follows: By virtue of (5) an element of $D_{\mathscr{U}} \cap D_{\mathscr{U}}'$ is contained in each ideal $P \in \mathscr{P}$ containing B_e and hence, by Theorem XII.1.6, is contained in B_e. The intersection of eR and B_e is (0) since e is a central idempotent. Denote the right-hand side of (8) by V. To prove (8) we must show that V fills the whole of eR. The rings eR and R/B_e are isomorphic by (6). By Theorem XII.1.2 eR (and hence also eR/V) is biregular and therefore semisimple. Suppose now that $V \neq eR$. Then there exists a primitive ideal of the ring eR/V. Corresponding to this ideal, there exists a primitive ideal of the ring

eR containing V. Finally, since $eR \approx R/B_e$, the image of this ideal under the isomorphism is a primitive ideal P of R containing V and B_e. Since P contains B_e it cannot contain eR. From $D_{\mathscr{U}} \cap eR \subseteq V \subseteq P$ it follows that $D_{\mathscr{U}} \cdot eR \subseteq P$, whence $D_{\mathscr{U}} \subseteq P$, since P is prime. Since $\mathscr{U}$ is closed, P belongs to the set $\mathscr{U}$. In a similar manner, using the fact that $C\mathscr{U} \cap \mathscr{B}_e$ is closed and since $D_{\mathscr{U}}' \cap eR \subseteq V \subseteq P$, we may show that $P \in C\mathscr{U}$, contradicting what we have just proved. Therefore $V = eR$. The ring eR contains the identity e. By (8) this is expressible as the sum of the identities e_1 of $D_{\mathscr{U}} \cap eR$ and f of $D' \cap eR$. *The corresponding function f' in $R' = \phi R$ is then the characteristic function of the set $\mathscr{U}$* as we may see as follows:

$$f'(P) = 0 \Leftrightarrow f \in P \Leftrightarrow D_{\mathscr{U}}' \cap eR \subseteq P \Leftrightarrow D_{\mathscr{U}}' \subseteq P \quad \text{or} \quad eR \subseteq P \tag{9}$$

since P is prime. By (5) and (7) $D_{\mathscr{U}}'$ is the intersection of all P in $C\mathscr{U} \cap \mathscr{B}_e$. Since this is a closed set in $\mathscr{P}$, $D_{\mathscr{U}}' \subseteq P$ is equivalent to $P \in C\mathscr{U} \cap \mathscr{B}_e = \mathscr{B}_e \backslash \mathscr{U}$. Finally $eR \subseteq P$ is equivalent to $P \notin \mathscr{B}_e$. It therefore follows from (9) that

$$f'(P) = 0 \Leftrightarrow P \in (\mathscr{B}_e \backslash \mathscr{U}) \cup C\mathscr{B}_e \Leftrightarrow P \in C\mathscr{U} \tag{10}$$

Since f is a central idempotent, the value of f' at each P in $\mathscr{U}$ is 1.

3. Now finally let $\mathscr{B}$ be an arbitrary open and compact (and hence by Theorem XII.2.5 also closed) subset of $\mathscr{P}$ and let P be a point in $\mathscr{B}$. Since R is biregular, there exists a central idempotent e in $R \backslash P$. The subset $\mathscr{B}_e$ of (5) corresponding to e is an open and closed neighborhood of P. Its intersection with $\mathscr{B}$ is also an open and closed neighborhood $\mathscr{U}(P) = \mathscr{B} \cap \mathscr{B}_e$ of P. Since $\mathscr{B}$ is compact by hypothesis, there exist a finite number of these $\mathscr{U}(P)$ which cover $\mathscr{B}$, say

$$\mathscr{B} = \mathscr{U}(P_1) \cup \cdots \cup \mathscr{U}(P_s)$$

Each set $\mathscr{U}(P_i)$ is contained in a set $\mathscr{B}_{e_i}$, $i = 1, \ldots, s$. Intersecting the $\mathscr{U}(P_i)$, $i = 1, \ldots, s$ in pairs we obtain a disjoint decomposition of $\mathscr{B}$ into open and closed sets:

$$\mathscr{B} = \mathscr{U}_1 \cup \mathscr{U}_2 \cup \cdots \mathscr{U}_t \qquad \text{with} \quad \mathscr{U}_j \cap \mathscr{U}_k = \varnothing \tag{11}$$

Each of the sets $\mathscr{U}_j$ is also contained in one of the sets $\mathscr{B}_{e_i}$. By what we showed in Part 2 of this proof, *the ring $R' = \phi R$ of functions* contains the characteristic function f_j' of $\mathscr{U}_j$ for each $j = 1, \ldots, t$. Therefore R' contains the sum $g_{\mathscr{B}} = f_1' + \cdots + f_t'$, i.e. by virtue of (11), R' contains the characteristic function of $\mathscr{B}$. ∎

It is now quite easy to prove that the ring $R' = \phi R$ consists of all continuous functions with compact supports since such functions may be constructed from the characteristic functions $g_{\mathscr{B}}'$ of the open-and-closed subsets $\mathscr{B}$ of $\mathscr{P}$ in the following fashion:

Let f be a continuous function on $\mathscr{P}$ with values in K and let the support $T(f) = \{P \mid P \in \mathscr{P}, f(P) \neq 0\}$ of f be compact. Since K is endowed with the discrete topology, the set $\mathscr{U}(k) = \{P \mid P \in \mathscr{P}, f(P) = k\}$ is open and closed for each $k \in K$. Therefore $T(f)$ is also open and hence, by Theorem XII.2.5, closed. Finitely many $\mathscr{U}(k)$ therefore over $T(f)$; in other words, f takes on only finitely many values different from 0, say $k_1, ..., k_s \in K$. Let $g_1, ..., g_s$ be the characteristic functions of the sets $\mathscr{U}(k_1), ..., \mathscr{U}(k_s)$. Then $g_1k_1 + \cdots + g_sk_s = f$ and is contained in R since by Theorem 2 $g_1, ..., g_s$ are in the right K-algebra $R' = \phi R$. This shows that the functions fill the ring $R' = \phi R$ completely and proves that the three conditions in the conclusion of the following theorem are sufficient.

Theorem 3 (Arens and Kaplansky). Hypothesis: Let K be a simple ring with identity 1 and let K be endowed with the discrete topology. Let R be a ring. Conclusion: R is isomorphic to the ring of continuous functions with compact supports defined on a locally compact totally disconnected space $\mathscr{P}$ with values in K if and only if the following three conditions hold:

1. R is biregular.
2. R is a right algebra over K.
3. For each primitive ideal P of R, the natural monomorphism of K into R/P is an epimorphism.

Proof of the necessity of conditions 1, 2, 3. 1. Since R is not the ring of all functions from $\mathscr{P}$ to K, we do not know whether R is biregular. Let $a \in R$ and $T(a)$ the support of this function. Since K is endowed with the discrete topology, $T(a)$ is open and closed. The characteristic function e_a of $T(a)$ is therefore continuous and its support is the compact set $T(a)$. Hence $e_a \in R$ and since $a = a \cdot e_a$, $a \in (e_a)$, where the function e_a is a central idempotent of R. To show that $e_a \in (a)$ let $\mathscr{U}(k) = \{P \in \mathscr{P} \mid a(P) = k\}$. By the compactness of $T(a)$, a takes on only finitely many values $k_1, ..., k_s$ different from 0. Therefore $\mathscr{P}$ may be decomposed as a disjoint union:

$$\mathscr{P} = \mathscr{U}(k_1) \cup \cdots \cup \mathscr{U}(k_s) \cup CT(a)$$

Since $\mathscr{U}(k_j)$ is open and closed (and hence compact by Theorem XII.2.5), the characteristic function $g_j = g_{\mathscr{U}(k_j)}$ belongs to R for $j = 1, ..., s$. Since K is simple the ideal generated by k_j is the whole of K. Therefore there exist elements u_{jp} and v_{jp} in K such that

$$\sum_p u_{jp} k_j v_{jp} = 1$$

This implies that the function

$$\sum_p g_j u_{jp} a g_j v_{jp} = g_j$$

i.e. the characteristic function g_j of (k_j), is also contained in (a) (cf. XII.1(13)). Hence $e_a = g_1 + \cdots + g_s$ lies also in (a).

2. To prove that R is a right K-algebra we merely observe that if $a \in R$, ak is a continuous function for all $k \in K$ and furthermore the support of ak is the union of those $\mathscr{U}(k_j)$ above for which $k_j k \neq 0$; hence the support of ak, as a closed set, is also compact.

3. The primitive ideals $\mathscr{P}$ of the biregular ring R of functions are precisely the maximal ideals of R by Theorem XII.1.2. At first sight it appears that these ideals have nothing to do with the given topological space $\mathscr{P}$. However, we may set up a one-one correspondence between the points P of $\mathscr{P}$ and the maximal ideals $\mathfrak{P}$ of R as follows: *To each maximal ideal $\mathfrak{P}_0$ of R there is exactly one point $P_0 \in \mathscr{P}$ such that*

$$\mathfrak{P}_0 = \{a \mid a \in R, a(P_0) = 0\}$$

Proof: Assume that, given an arbitrary point $P \in \mathscr{P}$, there exists $a \in \mathfrak{P}_0$ with $a(P) \neq 0$. Then, as in Part 1, we may prove that the characteristic function e_a of the support set $T(a)$ of a is also contained in $\mathfrak{P}_0$. Therefore we may assume without loss of generality that given an arbitrary point $P \in \mathscr{P}$ there is a characteristic function e_P contained in $\mathfrak{P}_0$ such that $e_P(P) = 1$. Denote the support of e_P by T_P. Now let b be an arbitrary function contained in R and let $T(b)$ be its support. The set $T(b)$ is covered by the family

$$\mathscr{F} = \{T_P \mid P \in T(b)\}$$

of open sets T_P. Since $T(b)$ is compact, a finite subfamily of $\mathscr{F}$ covers $T(b)$, say

$$T(b) = T_{P_1} \cup \cdots \cup T_{P_s} \tag{12}$$

The characteristic function e_{P_1} and e_{P_2} of T_{P_1} and T_{P_2}, respectively, are contained in $\mathfrak{P}_0$ and hence so is the characteristic function

$$e_{P_1} + e_{P_2} - e_{P_1} e_{P_2} \qquad \text{of} \quad T_{P_1} \cup T_{P_2}$$

By induction we may show that the characteristic function e of the set occurring on the right-hand side of (12) is contained in $\mathfrak{P}_0$. But then eb is contained in $\mathfrak{P}_0$ and by virtue of (12) coincides with b on the whole of $\mathscr{P}$. It follows that $R = \mathfrak{P}_0$, a contradiction. Hence there exists at least one point $P_0 \in \mathscr{P}$ for which $a(P_0) = 0$ for all $a \in \mathfrak{P}_0$. Thus $\mathfrak{P}_0$ is contained in

$$\mathscr{T}(P_0) = \{a \mid a \in R,\ a(P_0) = 0\} \tag{13}$$

This set of functions is an ideal of R containing $\mathfrak{P}_0$ but not R as may be seen as follows: Since $\mathscr{P}$ is locally compact, there exists a closed neighborhood $\mathscr{V}$ of P_0 which is compact. By Theorem XII.2.6 there is an open and closed neighborhood $\mathscr{D}'$ of P_0 contained in $\mathscr{V}$. By Theorem XII.2.5, $\mathscr{D}'$ is also compact. Its characteristic function f is therefore a continuous function defined on $\mathscr{P}$ with compact support $T(f)$ and $f(P_0) = 1$. This proves that, since f is contained in R, the ideal $\mathscr{T}(P_0)$ is not the whole of R. Since $\mathfrak{P}_0$ is a maximal ideal of R contained in $\mathscr{T}(P_0)$, it follows that

$$\mathfrak{P}_0 = \mathscr{T}(P_0)$$

On the other hand we cannot have two distinct points P_0 and P_1 with $\mathfrak{P}_0 = \mathscr{T}(P_0) = \mathscr{T}(P_1)$ since the neighborhood $\mathscr{D}' \subseteq \mathscr{V}$ may be so chosen as not to contain P_1.

The equality $\mathfrak{P}_0 = \mathscr{T}(P_0)$ shows that, with the help of the characteristic function f constructed above, we may express each element $a \in R$ in the form

$$a = fk_0 + p_0 \tag{14}$$

where $p_0 \in \mathfrak{P}_0$ and k_0 is the value of a at P_0. Since f is a central idempotent of R not contained in $\mathfrak{P}_0$, it follows from (14) that the natural monomorphism

$$k \to fk + \mathfrak{P}_0 \qquad \text{for} \quad k \in K$$

of K into $R/\mathfrak{P}_0$ is an epimorphism. Thus the ring of functions R also satisfies Condition 3 of the Theorem. ▮

4. Boolean rings

The rings we discussed in Section VI.2, namely, rings R with the property that for each $a \in R$ there exists an exponent $n(a) > 1$ such that $a^{n(a)} = a$, are examples of biregular rings. Indeed from $a^{n(a)-1}a = a$ it follows that $a^{n(a)-1}a^{n(a)-1} = a^{n(a)-1}$ and the idempotent $e = a^{n(a)-1}$ generates the principal ideal (a) since $e \in (a)$ and $a = ea \in (e)$. Since such a ring is commutative by Theorem VI.2.1, it is its own center and hence each principal ideal is generated by a central idempotent.

In the case that the ring R satisfies $a^2 = a$ for all $a \in R$ we may prove that R is commutative without appealing to the results of Chapter VI.

Definition 1. A ring R with identity 1 is a *Boolean ring* if $a^2 = a$ for all $a \in R$.

Theorem 1. A Boolean ring R is commutative. Moreover

$$2a = 0 \qquad \text{for all} \quad a \in R \tag{1}$$

Proof. We do not make use of the existence of an identity. Since $(a + b)^2 = (a + b)$,

$$a + b = (a + b)^2 = a^2 + ab + ba + b^2 = a + ab + ba + b$$

Hence

$$ab + ba = 0 \tag{2}$$

Putting $a = b$ in (2) we get $2a^2 = 0$ and hence $2a = 0$. This is equivalent to $a = -a$, and this applied to (2) gives $ab - ba = 0$ proving the commutativity of multiplication in R. ▮

A Boolean ring R is a right algebra (in fact even an algebra) over the field $C_2 = \mathbb{Z}/(2)$ consisting of the two elements $\bar{0}$ and $\bar{1}$ provided we define the operation of $K = C_2$ on R by

$$a\bar{0} = 0, \qquad a\bar{1} = a \qquad \text{for} \quad a \in R \tag{3}$$

This definition makes sense since

$$a + a = 0 \quad \text{in} \quad R \quad \text{and hence} \quad a(\bar{1} + \bar{1}) = a\bar{0} = 0 = a + a = a\bar{1} + a\bar{1}$$

Since R is commutative, the primitive factor ring R/P, where P is a primitive and hence maximal ideal, is a field which is again a Boolean ring. In such a field it follows from $\bar{a} \cdot (\bar{a} - \bar{1}) = 0$ that $\bar{a} = \bar{1}$ if $\bar{a} \neq \bar{0}$. Therefore R/P is isomorphic to C_2. Since R contains an identity element 1, the natural monomorphism of $K = C_2$ into R/P may be defined by

$$\chi_P : k \to 1k + P \qquad \text{for} \quad k \in K = C_2 \tag{4}$$

Since R/P is isomorphic to C_2, it follows that χ_P is an epimorphism for all primitive ideals P of a Boolean ring R. Therefore all the hypotheses of Theorem XII.3.3 of Arens and Kaplansky are satisfied. The structure space $\mathscr{P}$ of R is even compact by Theorem XII.2.10 since R contains an identity. The support of each continuous function from $\mathscr{P}$ to C_2 is closed since it is the inverse image of the identity 1 of C_2. Thus by Theorem XII.2.5 it is compact and this proves the necessity of the following:

Theorem 2 (Stone). A ring R is a Boolean ring if and only if it is isomorphic to the ring of continuous functions from a compact totally disconnected space to the field $C_2 = \mathbb{Z}(2)$ endowed with the discrete topology. *A compact totally disconnected space is called a Boolean space.* ▮

That a ring of continuous functions as described in the theorem is indeed a Boolean ring follows immediately since each function which takes on only the values $\bar{0}$ and $\bar{1}$ is idempotent and the identity of this ring is the constant function $\bar{1}$ which has the whole compact space as support.

An example of a compact totally disconnected space is the Cantor Discontinuum (cf. Hermes, Section 23, or Franz [1960], Section 9).

There is a very close relationship between Boolean lattices, i.e. complemented, distributive lattices—which we have already met in our discussion of completely reducible modules in Theorem IV.4.1—and Boolean rings. In fact if R is a Boolean ring we may define two operations $\cup$ and $\cap$ in R by

$$a \cup b = a + b - ab \qquad \text{and} \qquad a \cap b = ab \tag{5}$$

It is easily verified that the set R under $\cup$ and $\cap$ is a lattice with 0 as the null element and 1 as the all element. Moreover R is complemented with $1 - a$ as the complement of a and the distributivity follows from

$$\begin{aligned}(a \cap c) \cup (b \cap c) &= ac + bc - (ac)(bc) \\ &= ac + bc - abc = (a + b - ab)\, c \\ &= (a \cup b) \cap c\end{aligned}$$

This result will be better understood if we observe that the functions in Theorem 2 are the characteristic functions of subsets of the Boolean space and that $a + b - ab$ and ab are the characteristic functions of the union and intersection, respectively, of the two sets determined by a and b.

Conversely, starting with a Boolean lattice V with all element 1 *we may define two operations $+$ and $\cdot$ in V by*

$$A + B = (A \cap CB) \cup (CA \cap B) \qquad \text{and} \qquad A \cdot B = A \cap B \qquad \text{for } A, B \in V$$

where CA denotes the complement of A. V together with $+$ and $\cdot$ is a ring which, because of $A \cdot A = A \cap A$, is a Boolean ring with 1 *as the identity.* Definition (5) applied to $+$ and $\cdot$ yields the lattice V again and, in the other direction, (6) applied to $\cup$ and $\cap$ gives rise to the original ring R. Because of this, the theory of Boolean lattices, Boolean rings and Boolean spaces may be developed along parallel lines.

Guide to the Literature

The commutative ring $\mathbb{Z}$ of rational integers and the noncommutative ring $\mathrm{Hom}_K(M, M)$ are representative of the two principal directions which the development of ring theory has taken.

The theory of commutative rings is of importance in algebraic number theory and in algebraic geometry and has received considerable impetus from both these areas. Zariski and Samuel's "Commutative Algebra" gives an account of this branch of ring theory. In the main, however, this book deals with noncommutative rings and concepts from the theory of commutative rings only come within our scope of study when they may be generalized to the noncommutative case, as for example in Chapter X where we present noetherian ideal theory, or in Chapter XI where we embed a semiprime Goldie ring as an order in a semisimple artinian ring. Chapter XII gives an introduction to rings of continuous functions.

In the earliest investigations of rings of linear mappings it was assumed that these rings formed a vector space over a field; more precisely that they formed an algebra of finite dimension over a field or a hypercomplex system, to use the earlier terminology (M. L. Wedd [1908], Dickson [1914]). An account of the development until 1934, may be found in Deuring's book [1935]. Artin then weakened this condition and assumed the rings to have the minimal condition on left ideals. An exposition of the theory of these so-called artinian rings was published in 1955 by Artin, Nesbitt, and Thrall in a book entitled "Rings with Minimum Condition."

Finally in 1945 Jacobson was able to eliminate all finiteness conditions by introducing the concept of Jacobson semisimplicity. His results appear in his book "Structure of Rings" [1956]. However, a semisimple ring is only a subdirect sum (and in general not a direct sum) of primitive rings (Compare Theorem III.2.1 with Theorem III.3.1 due to Wedderburn–Artin). Therefore Bourbaki [1958] called a ring R semi-simple if ${}_RR$ is completely reducible as a left R-module. However, for this concept of semisimplicity no radical is known.

Artin, Nesbitt, and Thrall began the theory of rings which are not necessarily artinian semisimple in their book [1955]. Later it was recognized that a large portion of the theory could be developed under the weaker hypothesis that the rings R with identity be semiperfect, i.e. that the factor ring by its radical be artinian and that idempotents in R/W be liftable. This concept was introduced by Bass [1960] and subsequently applied in Lambek's book [1966]. In this book the full strength of this hypothesis is not needed. In fact it is sufficient for the identity elements of the simple direct summands of R/W to be liftable, as for example in the treatment of Behrens' distributively representable rings in Chapter IX.

With the exception of Chapter X, "ring" means a ring with associative multiplication. The reason for dealing at all with nonassociative rings in Chapter X is that Lesieur and Croisot's [1963] generalization of noetherian ideal theory to noncommutative rings can be further generalized (Kurata [1965]) to the nonassociative case without difficulty. From an associative ring R we may construct a nonassociative (Lie) ring by defining $a \circ b = ab - ba$ as the new multiplication or also by defining $a \circ b = ab + ba$ (Jordan ring). These rings play a part in the theory of continuous groups and in the foundations of geometry. These topics are dealt with in Jacobson's "Lie Algebras" [1967] and Braun and Koecher's "Jordan Algebras" [1966], respectively.

We now make a few comments on the individual chapters:

Chapter I. Since each ring is an Abelian group under addition it is natural to ask for a characterization of those Abelian groups on which a particular ring structure can be imposed. The reader will find a number of results in this direction in books by Fuchs [1958] and Kertesz [1968].

Chapter II. A further development of topological rings may be found in van der Waerden's book "Algebra II" [1959] in the chapter on topological algebra and also in Pontrjagin's book [1957/58]. Birkhoff's book [1948] on lattice theory gives the most comprehensive treatment of the subject. The introduction by Hermes [1955] makes easy reading. Not every right primitive ring is also left primitive; Bergmann's counterexample may be found in Herstein's book [1965]. Theorem II.3.5 on the antiisomorphism between the lattice of left ideals of a simple artinian ring and the lattice of subspaces of a finite-dimensional vector space gives a hint to the relationships developed by Baer in his book "Linear Algebra and Projective Geometry" [1952].

Chapter III. An axiomatic investigation of the concept of radical may be found in Divinsky's book [1965].

Chapters IV and V. Completely reducible modules are of great importance in representation theory. A detailed account may be found in Curtis and Reiner's book [1962] which is often referred to in here. Polynomials are introduced in the context of semigroup rings since this latter concept plays an important role in Chapter IX. We derive only those results of field theory which are absolutely essential for our purposes. A more complete account may be found in van der Waerden [1960]. Tensor products are freely used in the latter portion of this book. They form, together with $\text{Hom}_R(M, M)$, a starting point for homological algebra, a branch of modern mathematics about which more will be said below (Chapter VII).

Chapter VI. The concept of the Brauer group is based on that of a central simple algebra as follows: Two finite dimensional central simple algebras A and B over the ground field Φ are equivalent if each is a complete matrix ring over the same division ring D. These equivalence classes of algebras form an abelian group under multiplication induced by the tensor product $A \otimes_{\Phi} B$. This group is called the Brauer group. The inverse of the class represented by the division ring D is the class represented by D^*, the division ring opposite to D. The subgroup consisting of those classes of algebras which have as splitting field a given overfield $\sum$ of Φ may be obtained as a "restricted product" of $\sum$ with its Galois group over Φ (See for example van der Waerden [1959]). From a different point of view, these ideas are related to cohomology theory which is the cornerstone of Hochschild's proof of Wedderburn's Theorem in Section VI.4. More about division rings will be found in Jacobson's book [1956]. There and in Herstein's book [1961] one may find a number of results about rings satisfying a polynomial identity, for example $X^n = X$, where n is independent of X.

Chapter VII. It has already been mentioned above that a large portion of the known results of the theory of (not necessarily semisimple) artinian rings may be generalized to semiperfect rings. Maximal orders R in algebraic number fields Q are examples of Asano orders. (cf. for example van der Waerden [1959]). The author does not know of a short direct proof of the fact that in this case all R-ideals in Q are projective modules. To obtain this result one usually proves much more, namely, that each R-ideal is invertible, and then applies Theorem VII.2.3. Projective and injective modules form the basis of homological algebra as one can appreciate from the following: By Theorem VII.2.4 each R-module M is the image $\text{im} f_1$ of a projective module P_1 under an epimorphism f_1. The kernel $\ker f_1$ is a submodule of P_1. It too is the image of a projective R-module P_2 under an epimorphism f_2, and so on. In this way we obtain a sequence

$$\cdots \to P_{n+1} \xrightarrow{f_{n+1}} P_n \xrightarrow{f_n} P_{n-1} \to \cdots \to P_2 \xrightarrow{f_2} P_1 \xrightarrow{f_1} M \to 0 \qquad (1)$$

of projective R-modules P_i and homomorphisms f_i. This sequence is "exact" in the sense that $f_{i+1} = \ker f_i$, $i = 1, 2, \ldots$. The sequence (1) is called a projective resolution of the R-module M. A projective resolution of an R-module M is of finite length n if $P_n \neq 0$ but $P_{n+1} = 0$. The minimum (if this exists) of the lengths of all the projective resolutions of M is called the projective dimension of M. This gives a measure of how far M is from being a projective module. Dual to the concept of a projective resolution of a module M is the concept of an injective resolution

$$0 \to M \xrightarrow{g_1} I_1 \xrightarrow{g_2} I_2 \to \cdots \tag{2}$$

Here each I_j is injective (this is possible since each module M may be embedded in an injective module by Theorem VII.3.5) and once again the sequence (2) is exact. In the same way as above we obtain the concept of injective dimension of M. This may be used in the treatment of self-injective—and in particular Quasi-Frobenius rings (cf. Theorem VII.3.13). On the other hand projective resolutions play a part in the investigation of the question raised in Section V.3, namely, under what conditions does a monomorphism $f_1 : M_1 \to M$ of Φ-modules (Φ a commutative ring) induce a monomorphism $f_1{}^* : M_1 \otimes_\Phi N \to M \otimes_\Phi N$. To answer this we form from the sequence (1) for M the sequence

$$\cdots \to P_n \otimes N \xrightarrow{f_n{}^*} P_{n-1} \otimes N \to \cdots \to P_1 \otimes N \xrightarrow{f_1{}^*} M \otimes N \to 0 \tag{3}$$

where $f_i{}^*$ is the homomorphism induced by $f_i{}^*$. (3) is in general no longer exact. The factor groups $\ker f_n{}^*/\text{im} f_{n+1}^*$ depend on M and N only and not on the particular projective resolution (1) of M. $\text{Ker} f_n{}^*/\text{im} f_{n+1}^*$ is denoted by $\text{Tor}_n{}^\Phi(M, N)$. A good readable introduction to homological algebra may be found in Hu's book [1968]. The connection with quasi-Frobenius rings is explored in Jans [1964]. Many more results appear in the standard text by Cartan and Eilenberg [1956] and in Maclane's book [1963].

Chapter VIII. Frobenius algebras and their relationship to representation theory are dealt with in detail in Curtis and Reiner's book [1962].

Chapter IX. The concept of a distributively representable artinian ring was first introduced in Behrens [1960] and the connection with the theory of uniserial rings was established in [1961]. In the meantime it was found that it is possible to replace these artinian rings with rings having artinian factor rings by their radicals and having the further property that the identities of the simple direct summands of these factor rings are liftable modulo their radicals (Behrens [1970]). This allowed one to consider prime arithmetic rings in which the zero ideal is the intersection of powers of the radical without, however, the radical being necessarily nilpotent. This leads one to the relationship of these ideas with the theory of quasiuniserial semigroups without zero, which is presented in this book (Behrens [6], [8], [10], to appear). In the nonprime case we have to make use of the theory of topological rings (Behrens [1964]). These investigations may be extended to those D^*-arithmetic rings in which the factor ring of R by the radical is the direct sum of an arbitrary number of simple artinian rings. In this situation the concept of the direct limit of rings is used (cf. Behrens [1970]). Clifford and Preston's book gives a good introduction to the algebraic theory of semigroups [1961/1967].

Chapter X. This generalized noetherian ideal theory of Lesieur and Croisot is also valid for modules. The reader will find a treatment of the topic in the author's monograph [1963] or in Herstein's book [1965].

Chapter XI. Goldie's results on orders in semisimple artinian rings also have their origin in Lesieur and Croisot's work. A treatment may also be found in Divinsky's book [1965]. Lambek gives another approach and also deals with Utumi's complete quotient rings.

Chapter XII. The idea of considering the primitive ideals of a ring R as forming a topological space (the structure space of R) was introduced by Jacobson [1956]. The generalization of Stone's Theorem XII.4.2 to Theorem XII.3.3 of Arens and Kaplansky naturally raises the question of the structure of rings R of continuous functions on a topological space X taking on possibly infinitely many values in an algebra K. In Chapter XII the structure of such rings was determined provided the functions take on only finitely many values in K. Furthermore, we may also ask for the relationship of these rings to analysis in the case the ground field Φ of the algebra K is the field of real or complex numbers. These questions are dealt with in the book by Gilman and Jerison [1960].

Bibliography

Albert, A. A., Non-associative algebras, *Ann. Math.* **43**, 685–707 (1942).

Albert, A. A., "Structure of Algebras," Second Edition, American Mathematical Society, Providence, Rhode Island, 1964.

Artin, E., Nesbitt, C. J., and Thrall, R. M., "Rings with Minimum Condition," University of Michigan Press, Ann Arbor, Michigan, 1955.

Baer, R., Kriterien für die Existenz eines Einselements in Ringen, *Math. Zschr.* **56**, 1–17 (1952).

Baer, R., "Linear Algebra and Projective Geometry," Academic Press, New York, 1952.

Baer, R., Abelian groups which are direct summands of every containing group, *Proc. Amer. Math. Soc.* **46**, 800–806 (1940).

Bass, H., Finitistic homological dimensions and a homological generalization of semi-primary rings, *Trans. Amer. Math. Soc.* **3**, 466–488 (1960).

Behrens, E. A., [1] Zur additiven Idealtheorie in Nichtassoziativen Ringen, *Math. Zschr.* **64**, 169–182 (1956).

Behrens, E. A., [2] Distributiv darstellbare Ringe-I, *Math. Zschr.* **73**, 409–432 (1960); -II *Math. Zschr.* **76**, 367–384 (1961a).

Behrens, E. A., [3] Einreihige Ringe, *Math. Zschr.* **77**, 207–218 (1961b).

Behrens, E. A., [4] Die Halbgruppe der Ideale in Algebren mit distributivem Idealverband, *Archiv. Math.* **13**, 251–266 (1962).

Behrens, E. A., [5] Unendliche quasi-einreihige Halbgruppen, *Math. Ann.* **155**, 343–368 (1964).

Behrens, E. A., [4] Unendliche quasi-einreihige Halbgruppen, *Math. Ann.* **155**, 343–368 (1964).

Behrens, E. A., [5] "Algebren," Mannheim, 1965.

Behrens, E. A., [6] Partially ordered completely simple semigroups, *J. Algebra*, 1970.

Behrens, E. A., [7] D^*-arithmetical prime rings, Mathematics Reports McMaster University, 1970.

Behrens, E. A., [8] The quasiuniserial semigroup without zero, their arithmetics and their ϕ^* algebras, Semigroup Forum, 1970.

Behrens, E. A., [9] Distributiv darstellbare Ringe, *Math. Zschr.* **73**, 409–432 (1960).

Behrens, E. A., [10] The arithmetics of the quasiuniserial semigroups without zero.

Bergmann, G. M., A ring primitive on the right and not on the left, *Proc. Amer. Math. Soc.* **15**, 473–475 (1964).

Birkhoff, G., "Lattice Theory," American Mathematical Society, Providence, Rhode Island, 1948.

Bourbaki, N., "Algebra," Chapter 2, Algebra lineaire, Hermann, Paris, 1947.

Bourbaki, N., "Algebra," Chapter 8, Modules et anneaux semi-simples, Hermann, Paris, 1958.
Brauer, R. and Weiss, E., "Non-commutative rings," Part I, Harvard Univ. Press, Cambridge, Massachussetts.
Braun, H. and Koecher, M., "Jordan Algebren," Berlin, 1966.
Cartan, H. and Eilenberg, S., "Homological Algebra," Princeton Univ. Press, Princeton, New Jersey, 1956.
Clifford, A. H. and Preston, G. B., "The Algebraic Theory of Semigroups," Vol. I, American Mathematical Society, Providence, Rhode Island, 1961.
Curtis, C. W. and Reiner, I., "Representation Theory of Finite Groups and Associative Algebras," Wiley (Interscience), New York, 1962.
Deuring, M., "Algebren," Springer, Berlin, 1935.
Divinsky, N. J., "Rings and Ideals," Univ. Toronto Press, Toronto, 1965.
Eckmann, B. and Schöpf, A., Über injektiv Moduln, *Arch. Math.* **4**, 75–78 (1955).
Franz, W., "Topologie I," Springer, Berlin, 1965.
Fuchs, L., "Abelian groups," Budapest, 1958.
Fuchs, L., Über die Ideale aritmetischer Ringe, *Commentarii Math. Helv.* **23**, 334–341 (1949).
Gericke, H., "Theorie der Verbände," Mannheim, 1963.
Gillman, L. and Jerison, M., "Rings of Continuous Functions," Princeton Univ. Press, Princeton, New Jersey, 1960.
Goldie, A. W., Non-communtative principal ideal rings, *Archiv. Math.* **13**, 213–231 (1962).
Goldie, A. W., Semi-prime rings with maximum condition, *Proc. London Math. Soc.* (3) **10**, 201–220 (1960).
Hermes, H., "Einführung in die Verbandstheorie," Springer, Berlin, 1955.
Herstein, I. N., Theory of rings, Univ. of Chicago, Mathematical lecture notes, 1961.
Hu, S. Z., "Introduction to Homological Algebra," Holden-Day, San Francisco, California, 1966.
Jacobson, N., "The radical and semisimplicity for arbitrary rings, *Amer. J. Math.* **67**, 300–320 (1945).
Jacobson, N., "Lie Algebras," Wiley (Interscience), New York, 1962.
Jacobson, N., "Structure of Rings," *Amer. Math. Soc.*, Providence, Rhode Island, 1956.
Jans, J. P., "Rings and Homology," Holt, New York, 1964.
Johnson, R. E., Distinguished rings of linear transformations, *Trans. Amer. Math. Soc.* **111**, 400–412 (1964).
Kasch, F., Grundlagen einer Theorie der Frobenius Erweiterungen. *Math. Ann.* **127**, 453–474 (1959).
Kertesz, A., Vorlesungen über artinsche Ringe. To appear.
Koethe, G., "Topologische lineare Räume," Berlin, 1966.
Kurata, Y., On an additive ideal theory in a non-associative ring. *Math. Zeitschr.* **88**, 129–135 (1965).
Kurosh, A. G., "Lectures on General Algebra," Pergamon Press, New York, 1963.
Lambek, J., "Lectures on Rings and Modules," Blaisdell, Waltham, Massachusetts, 1966.
Lesieur, L. and Croisot, R., "Algèbre noethérienne non-commutative," Paris, 1963.
MacLane, S., "Homology," Springer, Berlin, 1963.
Müller, B., Linear compactness and Morita duality, *J. Algebra.*
Naimark, M. A., "Normed Rings," Groningen, 1959.
Nakayama, T., On Frobenius algebras, I and II, *Ann. Math.* **40**, 611–633 (1939); **42**, 1–21 (1941).
Redei, L., "Algebra," Part I, Leipzig, 1959.

Ribenboim, P., "Rings and Modules," Wiley, New York, 1969.
Robson, J. C., Non-commutative Dedekind rings, *J. Algebra* **9**, 249–265 (1968).
Schafer, R. D., "An Introduction to Nonassociative Algebras," Academic Press, New York, 1966.
Szasz, G., "Einführung in die Verbandstheorie," Leipzig, 1962.
van der Waerden, B. L., "Algebra I," Unger, Berlin, 1960.
van der Waerden, B. L., "Algebra II," Unger, Berlin, 1959.
Zariski, O. and Samuel, P., "Commutative Algebra," Vols. I, II. Van Norstrand, Princeton, New Jersey, 1958, 1960.

Index

Pure and Applied Mathematics

A Series of Monographs and Textbooks

Editors

Paul A. Smith and Samuel Eilenberg

Columbia University, New York

1: Arnold Sommerfeld. Partial Differential Equations in Physics. 1949 (Lectures on Theoretical Physics, Volume VI)
2: Reinhold Baer. Linear Algebra and Projective Geometry. 1952
3: Herbert Busemann and Paul Kelly. Projective Geometry and Projective Metrics. 1953
4: Stefan Bergman and M. Schiffer. Kernel Functions and Elliptic Differential Equations in Mathematical Physics. 1953
5: Ralph Philip Boas, Jr. Entire Functions. 1954
6: Herbert Busemann. The Geometry of Geodesics. 1955
7: Claude Chevalley. Fundamental Concepts of Algebra. 1956
8: Sze-Tsen Hu. Homotopy Theory. 1959
9: A. M. Ostrowski. Solution of Equations and Systems of Equations. Second Edition. 1966
10: J. Dieudonné. Treatise on Analysis. Volume I, Foundations of Modern Analysis, enlarged and corrected printing, 1969. Volume II, 1970.
11: S. I. Goldberg. Curvature and Homology. 1962.
12: Sigurdur Helgason. Differential Geometry and Symmetric Spaces. 1962
13: T. H. Hildebrandt. Introduction to the Theory of Integration. 1963.
14: Shreeram Abhyankar. Local Analytic Geometry. 1964
15: Richard L. Bishop and Richard J. Crittenden. Geometry of Manifolds. 1964
16: Steven A. Gaal. Point Set Topology. 1964
17: Barry Mitchell. Theory of Categories. 1965
18: Anthony P. Morse. A Theory of Sets. 1965

Pure and Applied Mathematics

A Series of Monographs and Textbooks

19: Gustave Choquet. Topology. 1966

20: Z. I. Borevich and I. R. Shafarevich. Number Theory. 1966

21: José Luis Massera and Juan Jorge Schaffer. Linear Differential Equations and Function Spaces. 1966

22: Richard D. Schafer. An Introduction to Nonassociative Algebras. 1966

23: Martin Eichler. Introduction to the Theory of Algebraic Numbers and Functions. 1966

24: Shreeram Abhyankar. Resolution of Singularities of Embedded Algebraic Surfaces. 1966

25: François Treves. Topological Vector Spaces, Distributions, and Kernels. 1967

26: Peter D. Lax and Ralph S. Phillips. Scattering Theory. 1967.

27: Oystein Ore. The Four Color Problem. 1967

28: Maurice Heins. Complex Function Theory. 1968

29: R. M. Blumenthal and R. K. Getoor. Markov Processes and Potential Theory. 1968

30: L. J. Mordell. Diophantine Equations. 1969

31: J. Barkley Rosser. Simplified Independence Proofs: Boolean Valued Models of Set Theory. 1969

32: William F. Donoghue, Jr. Distributions and Fourier Transforms. 1969

33: Marston Morse and Stewart S. Cairns. Critical Point Theory in Global Analysis and Differential Topology. 1969

34: Edwin Weiss. Cohomology of Groups. 1969

35: Hans Freudenthal and H. De Vries. Linear Lie Groups. 1969

36: Laszlo Fuchs. Infinite Abelian Groups: Volume I. 1970

37: Keio Nagami. Dimension Theory. 1970

38: Peter L. Duren. Theory of H^p Spaces. 1970

39: Bodo Pareigis. Categories and Functors. 1970

40: Paul L. Butzer and Rolf J. Nessel. Fourier Analysis and Approximation: Volume 1, One-Dimensional Theory. 1971

41: Eduard Prugovečki. Quantum Mechanics in Hilbert Space. 1971

42: D. V. Widder: An Introduction to Transform Theory. 1971

43: Max D. Larsen and Paul J. McCarthy. Multiplicative Theory of Ideals. 1971

44: Ernst-August Behrens. Ring Theory. 1972

In preparation

Werner Greub, Stephen Halperin, and James Vanstone. De Rham Cohomology of Fibre Bundles: Volume 1; Manifolds, Sphere Bundles, Vector Bundles.

Morris Newman. Integral Matrices

Glen E. Bredon. Introduction to Compact Transformation Groups